THE STUFF OF SOLDIERS

A volume in the series

Battlegrounds: Cornell Studies in Military History
General Editor: David J. Silbey

A list of titles in this series is available at cornellpress.cornell.edu.

THE STUFF OF SOLDIERS

A HISTORY OF THE RED ARMY IN WORLD WAR II THROUGH OBJECTS

BRANDON M. SCHECHTER

CORNELL UNIVERSITY PRESS
Ithaca and London

First published 2019 by Cornell University Press
Printed in the United States of America

Library of Congress Cataloging-in-Publication Data

Names: Schechter, Brandon M., author.
Title: The stuff of soldiers : a history of the Red Army in World War II through objects / Brandon M. Schechter.
Description: Ithaca [New York] : Cornell University Press, 2019. | Series: Battlegrounds : Cornell studies in military history | Includes bibliographical references and index.
Identifiers: LCCN 2018053748 (print) | LCCN 2018054969 (ebook) | ISBN 9781501739804 (pdf) | ISBN 9781501739811 (epub/mobi) | ISBN 9781501739798 | ISBN 9781501739798 (cloth: alk. paper)
Subjects: LCSH: Soviet Union. Raboche-Krestanska'a Krasnai'a Armi'a—Military life. | Soviet Union. Raboche-Kresti'anska'a Krasna'a Armi'a—Equipment. | Soviet Union. Raboche-Kresti'anska'a Krasna'a Armi'a—History—World War, 1939–1945. | World War, 1939–1945—Soviet Union—Equipment and supplies. | Soldiers—Soviet Union—Social conditions. | Military paraphernalia—Soviet Union. | Material culture—Soviet Union. | Personal belongings—Soviet Union.
Classification: LCC UA772 (ebook) | LCC UA772 .S276 2019 (print) | DDC 940.54/1247—dc23
LC record available at https://lccn.loc.gov/2018053748

Миляуше и Булату, без вас—никуда

Contents

Prelude: Outgunned and Outmanned

The Soviet Union literally had a script for World War II—a popular film showed how the next global conflict would unfold. The film, *If War Comes Tomorrow* (Esli zavtra voina, Mosfilm, 1938, dir. Efim Dzigan), seen by millions, shows an unclear enemy—their uniform combined French, British, and German elements, their symbol resembled a triangular swastika, and they spoke German—launching a sneak attack on the Soviet Union. This dastardly assault is crushed in short order by the overwhelming might of the Soviet military and Revolution in Europe. The enemy seems too silly to be truly threatening, often represented by graying, bespectacled men well past their prime. The film opens and closes with images of virile Red Army soldiers and Red Navy sailors standing in line and features long shots of innumerable Soviet planes, tanks, artillery pieces, and well-equipped soldiers, making sure to highlight the variety of weapons and equipment available to Soviet troops. Using footage of real Red Army maneuvers, the film reassured Soviet citizens that the government provided everything necessary for the upcoming war, which it promised would be swift, with few losses and on enemy territory. The next war would bring the promised worldwide Socialist Revolution within reach.

Millions of Red Army soldiers confronted a different reality when the war actually came. It resembled *If War Comes Tomorrow* in that it started with a sneak attack on the Soviet Union and one side enjoying seemingly endless resources of planes and tanks. Unfortunately, it was the enemy, now clearly German, who had vast armadas of planes and tanks and marched with such confidence deep into Soviet territory. They were not silly but terrifying and would visit brutality far beyond the imagination of Communist Party propagandists or the Soviet peoples, looting, destroying, raping, and murdering at will. It would take years for the Red Army to bring the conflict to enemy soil, by which time most of the equipment and many of the men shown in the 1938 film had been destroyed. The war, which was supposed to finish the work of the Revolution with a brief, glorious battle, had come to resemble an apocalypse in which Fascism would emerge victorious.

Many soldiers in 1941 and 1942 felt themselves hopelessly outgunned. Harassed from the air without their own planes in sight, facing enemy tanks with rifles and bottles of incendiary liquid that could burst into flames in their hands, and suffering from hunger if not starvation, Red Army soldiers confronted the terrors of modern warfare at a distinct disadvantage. Their foe was better armed, fed, and trained for killing. In these critical years there was often scarce time to train soldiers; most learned on the job at a terrible cost in human lives.

One of these soldiers in the making was Daniil Granin, a young factory worker of Jewish descent from Leningrad, a political officer and sometimes translator in a hastily thrown together People's Guard unit (*narodnoe opolchenie*). He recalled his first encounter with a German prisoner of war in late July 1941. The young Red Army soldiers expected to find much in common with their captive, who was the same age as Granin and of working-class background. Instead, despite being a prisoner, it was the German that felt and looked in control:

> He was young, tall, and blond. From underneath his tunic a shirt white as snow peeked out, his boots shined, and the mouth of leather gloves stuck out from under his belt. We were the same age. He stood before me, taking a wide stance, swaying a bit, looking somewhere above our heads at the tops of trees. I started to question him, and he squinted, slowly lowering his eyes to me, grinned, and looked me up and down in such a way that I became ashamed. I felt my puttees, beat up, used boots with rope laces, my dirty tunic, and most importantly, my bottles with burning liquid, old rifle and RG grenades—crappy grenades that were totally useless—my cotton ammunition pouch—really just a shapeless bag with bullets. I still remember how his gaze made the bag feel heavy and my clumsiness, how I was not at all a soldier and he was a soldier. Yes, the difference was clear to anyone, he was a soldier, and we, the People's Guard with our incompetently rolled bedrolls, we looked like some sort of rabble.

Granin expected the German to come to his senses and realize that his real enemy was Hitler and the bourgeois world. Instead, Granin was told that he and his comrades would be destroyed by the order of the Führer. The prisoner refused to acknowledge his captors, smoking in Granin's face and laughing as he coughed. These soldiers couldn't even hit their prisoner, so overcome were they by shock and offense at the difference between what was and what should have been.[1] Granin and his comrades would eventually retreat all the way back to Leningrad, which would endure a nearly nine-hundred-day blockade. Elsewhere, whole cities and armies were surrounded and submitted.

Kharis Iakupov, a young Tatar artist who had been serving in the army since before the war, described the common experience of falling into a "bag" (*meshok*)—a term for being encircled by the enemy—in 1941: "The 'bag'—that was something awful. The bag is weapons that have become useless—no shells; it is machines abandoned—neither roads nor fuel for them; it is hunger spasms in your stomach and the last crusts in your pack—there's nowhere to get rations."[2] Weapons became useless without munitions; soldiers became feckless without food. They eventually came to rely wholly on German arms and food they could scrounge. He and his comrades could become prisoners at any moment, a fate that befell millions in 1941–1942, most of whom would die of starvation in crudely improvised barbed wire enclosures. Iakupov was lucky; he and some of his friends managed to escape after weeks of wandering through Byelorussian forests. By that time they had ceased to be a military unit, reduced to small bands trying to survive.

In September 1941, Oleg Reutov, a chemistry student from Donetsk oblast', arrived at the southern flank of the massive front and began keeping a diary. As he retreated through Ukraine, he despaired at the future of his country. Time and again he recorded how a few German tanks could sow panic among green troops. He lamented: "it's clear that the soldiers leave a lot to be desired. 'The village,' of course, is the most cowardly. They are particularly afraid of mortars and tanks." Nowhere were there enough planes or armor. Panic before tanks and the fear caused by artillery were running themes in his diary.[3] Reutov retreated past the Soviet Union's breadbasket, through its industrial heartland, and even past his hometown, sickened and enraged. By the end of November he had decided, "I personally have little chance of survival."[4] German plans for the Soviet Union centered on colonization, with the locals at best having a place only as slaves, but more likely facing extermination.

Around the same time, the artillerist Nikolai Inozemtsev, whose shattered unit was receiving reinforcements of men and material, recorded his mix of shame and pride as the disastrous summer of 1941 came to a close:

> The bitter taste of retreat, the sorrow of shame for our country (we remember the films *We Are Undefeatable*, *If War Comes Tomorrow*, etc., etc.), the awareness of many shortcomings of our system and way of life, resentment that our lives are being cut short just as they are beginning—it seems that all this should overcome your soul and morally knock you to the ground—but nonetheless I want to straighten up

> and shout out: "But life is devilishly good! Precisely because we love life so much, we have a calm attitude toward death. We want to die, remaining ourselves to the end, not poisoning our consciousness for a day or an hour with cowardice or pettiness. If we have to die—then with pride![5]

Proud men like Inozemtsev offering up their lives would not be enough to win the war. If soldiers like Reutov, Iakupov, Granin, and indeed the Soviet Union itself were to survive, if men and women serving in the Red Army were to see their loved ones again, millions of people from all walks of life (including the "cowardly village"—peasants who made up the majority of the population and had been viciously exploited to achieve Soviet modernization) and all corners of the country had to learn to use the tools of war with alacrity.

In 1941 the army barely held things together. Millions of soldiers surrendered to the enemy in despair. Reutov recorded with shame stories of long columns of Soviet prisoners lead by a few Germans and that men from his hometown had deserted to the enemy.[6] The professional cadres that began the war were decimated, and the Red Army was completely rebuilt. In 1942, after having stopped and even pushed back the Germans near Moscow, the army again retreated even deeper into Soviet territory. Desperate to arm and motivate soldiers, the Soviet government simultaneously simplified the production of its equipment, made weapons as quickly as possible and minted new medals to motivate soldiers to feats of bravery.

The Red Army managed to halt the enemy in Stalingrad, a feat celebrated by a massive makeover at the end of a counteroffensive that marked the turning point of the war. Red Army soldiers would advance westward in a uniform adapted from that of the old regime, better armed, better fed, and with hard won experience. In 1943 the army held its own for the first time during a major summer offensive by the Wehrmacht, the last on the Eastern Front. Iakupov would compare the German dead to the locusts, so thick did they lay and rapacious they had been while alive.[7] From that point on the Red Army would only advance, first to the 1941 borders, and then beyond, into the strange world of capitalism.

Inozemtsev could write in his diary in June 1944 that the situation at the front "seems more like a well-organized military parade than war." He reflected that few had survived long enough to see the "colossal changes" and could truly understand "How much easier it has become to wage war! . . . Yes, now we are strong, now we are a great, undefeatable army, worthy of our Motherland, inspiring universal awe in all the honest world!"[8]

This book tells the story of that dramatic change—from a desperate, retreating band to a victorious army—as experienced by soldiers. The years 1941–1945 replayed in real life a universal tale that had become a major trope of Bolshevism: "the standard exodus and construction stories about the transformation of a motley crowd into a holy army."[9] Indeed, one of the war's anthems was "Holy War." Yet with the sacred was mixed the profane—human weakness, dirt, blood—and a variety of mundane objects that made the war possible—spoons, shovels, rifles, and so much more. This is a story that can be properly told only through the objects that had to be mastered to make the victory possible. For people like Granin, Iakupov, Inozemtsev, and Reutov to become soldiers, they needed to learn to feel comfortable and formidable in their humble uniforms, to dig trenches deep into the earth with small spades, and to kill using a variety of weapons.

Acknowledgments

I owe a tremendous debt to a small army of people in the realization of this project and am happy to list many of them here.

New York University has been an ideal environment to finish this book. Without Elihu Rose, who generously sponsors my position here, you would not be reading these lines. Jane Burbank, Herrick Chapman, Stefanos Geroulanos, Stephen Gross, David Ludden, Molly Nolan, and Leslie Pierce have gone out of their way to make me feel at home in the History Department. Conversations with Yanni Kotsonis about the state, Anne Lounsbery about *meshchanstvo*, and Anne O'Donnell about property allowed me to clarify key ideas as I finished this work. Tatiana Linkhoeva, an old co-conspirator, continues to be a good friend, confidant, and fellow student of empire.

Before arriving in New York, I spent a year at the Davis Center for Russian and Eurasian Studies, where Xenia Cherkaev, Alex Diener, Anna Ivanova, Jackie Kerr, Terry Martin, Kelly O'Neill, Serhyi Plokhyi, and many others helped me develop this book. Special thanks to Kelly and Anna, who were particularly insightful.

I had the great fortune to study at the University of California Berkeley's History Department. Yuri Slezkine initially drew me to Berkeley with his ability to tell a story with humor, which I hope rubbed off on me a bit. He has been an exceptional mentor who has challenged me to think about structure and narrative in ways that have made me understand history with greater depth and become a more interesting human. Victoria Frede-Montemayor challenged me to seriously consider the words I use and is one of the warmest people I have known. John Connelly reminded me to see Stalinism through non-Soviet eyes. Alexei Yurchak was a kind mentor and along with Melanie Feakins, a generous friend.

While at Berkeley I enjoyed the benefit of a superb Russian History Working Group ("The Kruzhok"), which provided fellowship and intellectual sustenance. I would like to thank all the members and single out Rhiannon Dowling, Nicole Eaton, Eric Johnson, Jason Morton, Alexis Peri, Charles

Shaw, Yanna Skorobogatova, Mirjam Voerkoelius, and Katherine Zubovich in particular. Charles provided invaluable feedback on the penultimate draft and has been a great friend since I landed in California. I am very happy that our paths have had so many happy parallels. Alice Goff has served as an outside perspective on my work and source of much needed levity on innumerable occasions. James Skee and Gene Zubovich were true comrades at Berkeley and beyond. Bathsheba Demuth has forced me to consider things I usually ignore, especially ice and calories. Most chapters in this book were born in research seminars under the tutelage of Yuri Slezkine, Tom Laqueur, Kerwin Klein and Stefan-Ludwig Hoffmann. I thank them and my colleagues in those seminars.

This project received generous support from a Fulbright Institute of International Education Grant, The Dean's Dissertation Research Grant, several Foreign Language and Area Studies (FLAS) Fellowships and innumerable small grants and fellowships distributed by Berkeley's Institute of Slavic, East European, and Eurasian Studies (ISEEES) and The Berkeley Program in Eurasian and East European Studies (BPS). I thank Ned Walker for support as the head of BPS, his critique and input from the ground floor of this project and most importantly his generosity as "Godfather." I would also like to thank Zach Kelly and Jeff Pennington for all of their advocacy.

Before I arrived in California, a number of people shaped my interests and continue to encourage them. Nikolai Firtich, Alexei Klimov, Mikhi Pohl, and Dan Ungurianu introduced me to Russian culture, history, and language at Vassar College. Bair Irincheev was my guide to the battlefields of the Leningrad and Karelian Fronts and is a wonderfully supportive friend. Alexander Semyonov introduced me to the conundrum of empires and convinced me to become a historian. I thank Boris Kolonitskii, the late Oleg Ken, Mikhail Krom, Vladimir Lapin, and Ekaterina Pravilova for their mentorship while I studied at European University at St. Petersburg and Sonia Chuikina and Elena Zdravomyslova for introducing me to issues of gender. A special thanks to Geoff Hass and Nikita Lomagin for convincing me to stick with this and their continued mentorship. Rufina Adilzianova, Foma Campbell, Sasha Beliaev, Ilya Chuikin, Danis Garaev, Ilya Orlov, Sasha Sotov, and Anna Zhelnina have all informally contributed to this project. Il'mira Gainutdinova provided shelter and friendship in Moscow and Kazan. Since before I knew a word of Russian, Ethan Bien, Matt Galvin, Jon Li, Drew Rodgers, and Ben Sozanski have provided lasting friendship and outside perspectives on work and life.

While doing research, I have acquired debts to a number of people. Oleg Vital'evich Budnitskii of the International Center for the History and

Sociology of World War II and Its Consequences sponsored me in Moscow and introduced me to many sources. He also gave me several opportunities to share my research with a Russian audience. The International Social Science School in Ukraine was a unique opportunity to run my ideas by a wide range of international scholars, and I thank in particular Anna Colin-Lebedeva for organizing it and François-Xavier Nérard and Amandine Regamey for their comments and continued engagement with my work. I am grateful to Tarik Cyril Amar, Dominique Arel, Leora Auslander, Katja Bruisch, Nikolai Mikhailov, Harriet Murav, François-Xavier Nérard, Jan Plamper, John Randolph, Gábor Rittersporn, Mark Steinberg, Richard Wortman, and Tara Zahra for inviting me to present papers in Champaign-Urbana, Chicago, Moscow, New York, Paris, St. Petersburg, and Zhitomir. Conversations with Rachel Applebaum, Betty Banks, David Brandenberger, Jonathan Brundstedt, Masha Cerovic, Donald Filtzer, Ilya Gerasimov, Sergei Glebov, Wendy Goldman, Jochen Hellbeck, Lisa Kirschenbaum, Natalia Laas, Steve Maddox, Erina Megowan, Harry Merritt, Alexandra Oberländer, Anatoly Pinsky, Jan Plamper, Ethan Pollack, Giulio Salvati, and Vanessa Voison have helped shape this text, and I am grateful to each of them for their generosity of spirit. I also thank the participants of the Harvard and Columbia kruzhki as well as the NYU graduate student seminar, for their feedback on chapters 1, 6, and 7.

I would like to thank the archivists at the State Archive of the Russian Federation (GARF), Central State Archive of Historical-Political Documentation of the Republic of Tatarstan (TsGA IPD RT), National Archive of the Republic of Tatarstan (NART), Russian Academy of Science's Institute of Russian History (NA IRI RAN), Russian State Archive of Social and Political History (RGASPI), and Moscow Central Archive-Museum of Personal Collections (TsMAMLS). Galina Koroleva at the Russian State Archive of Film and Cinema provided excellent photographs in record time. (Anna Ivanova made sure that I got the photos.) Galina Tokareva at RGASPI was exceptionally kind and helpful. Konstantin Drozdov at NA IRI RAN was a researcher's dream—an archivist who worked with lightning speed and was invested in my project—without him, this work would be much weaker. I thank Jochen Hellbeck for introducing me to this archive.

On the production front, I wish to thank Emily Andrew for being an ideal editor who saw the potential of this project and has been exceptionally kind and supportive through the entire process. Bethany Wasik has been patient with my innumerable questions. Carolyn Pouncy provided superb copy editing. Angelina Lucento, David Shneer, and Erika Wolff all guided my navigation of image permissions. A special thanks to David for his above and beyond help! I would like to thank the NYU Humanities Center for their generous Grant-in-Aid, which provided funds for images. I also thank Indiana University Press for granting permission to publish an updated version of chapter 3, which originally appeared in Wendy Goldman and Donald Filtzer, eds., *Hunger and War: Food Provisioning in the Soviet Union during World War II* (Bloomington: Indiana University Press, 2015), 98–157.

Mark Edele read this manuscript twice, offering insightful criticism and convincing me that it was ready, for this and his friendship I am most grateful. I thank Mark von Hagen, formerly an anonymous reviewer, whose work I build on and whose counsel was key to realizing this project. I also wish to thank my as yet unmasked anonymous reviewer, your insights were most appreciated, and I hope that you are satisfied with the result.

Finally, I would like to mention my family. My parents, Cynthia and Lee, have supported my interests, wherever they should take me and whatever sacrifices they would entail, since before I could talk. My late father's work ethic and mother's skepticism and curiosity are in the DNA of this text. My in-laws, Al'fiya and Marsel, provided support while in Russia. Most importantly, without the love, support, and critique of my wife, Milyausha Zakirova, this project would not have been possible. I cannot count the number of times her knowledge has helped shape this project or her encouragement has kept me going in moments of doubt. Her final proofreading prevented me from great embarrassment, though any mistakes left are my own fault. In the course of writing this, our ranks have expanded to include a son, Bulat. It is to Milya and Bu that I dedicate this text.

Archival Sources and Their Abbreviations

*Arkhiv Prezidenta Rossiiskoi Federatsii (AP RF): Presidential Archive of the Russian Federation

*Galuzevii derzhavnii arkhiïv Sluzhba bezpeki Ukraïni (GDA SB Ukraïni): Main State Archive of the Security Services of Ukraine

Gosudarstvennyi arkhiv Rossiiskoi Federatsii (GARF): State Archive of the Russian Federation

Fond R-5446 Sovet ministrov SSSR

Natsional'nyi arkhiv Respubliki Tatarstan (NART): National Archive of the Republic of Tatarstan

Fond R-3610 Prezidium Verkhovnogo soveta Respubliki Tatarstan

Nauchnyi arkhiv Instituta rossiiskoi istorii Akademii nauk Rossiiskoi Federatsii (NA IRI RAN): Scientific Archive of the Institute of Russian History of the Academy of Sciences of the Russian Federation

Fond 2 Komissiia po istorii Velikoi Otechestvennoi voiny Akademii nauk SSSR

Razdel I, Istoriia voinskikh chastei i podrazdelenii

Razdel III, Oborona gorodov

Razdel X, Raznye materialy

Razdel XIII, Gazety, vyrezki, listovki, biulleteni po istorii Velikoi Otechestvennoi voiny

Rossiiskii gosudarstvennyi arkhiv kinofotodokumentov (RGAKFD): Russian State Archive of Film and Photo Documents

Rossiiskii gosudarstvennyi arkhiv sotsial'no-politicheskoi istorii (RGASPI): Russian State Archive of Social and Political History

Fond 17 Tsentral'nyi komitet KPSS (TSK KPSS) (1898, 1903–1991), op. 125, Upravlenie propagandy i agitatsii TSK VKP(b)

Fond 74, op. 2, Voroshilov Kliment Efremovich (1881–1969), lichnyi fond

Fond 84, op. 1, Mikoian Anastas Ivanovich (1895–1978), lichnyi fond

Fond 88, op. 1, Shcherbakov Aleksandr Sergeevich (1901–1945), lichnyi fond
Fond M-7, op. 2, Dos'e komsomol'tsev—geroev Sovetskogo soiuza
Fond M-33, op. 1, Pis'ma s fronta i na front

*Rossiiskii gosudarstvennyi voennyi arkhiv (RGVA): Russian State Military Archive
*Tsentral'nyi arkhiv Federal'noi sluzhby bezopasnosti Rossiiskoi Federatsii (TsA FSB RF): Central Archive of Federal Security Services of the Russian Federation
*Tsentral'nyi arkhiv Ministerstva oborony Rossiiskoi Federatsii (TsAMO RF): Central Archive of the Ministry of Defense of the Russian Federation
*Tsentral'nyi arkhiv obshchestvennykh dvizhenii Moskvy (TsAODM):
Central Archive of Social Movements of Moscow
Tsentral'nyi gosudarstvennyi arkhiv istoriko-politicheskoi dokumentatsii Respubliki Tatarstan (TsGA IPD RT): Central State Archive of Historical-Political Documentation of the Republic of Tatarstan

f. 8250, Dokumenty geroev Sovetskogo soiuza i veteranov Velikoi Otechestvennoi voiny

Tsentral'nyi moskovskii arkhiv-muzei lichnykh sobranii (TsMAMLS): Central Moscow Archive-Museum of Personal Collections

* Consulted in published volumes.

Terms and Abbreviations

agitator: low-level political cadre tasked with inspiring troops and disseminating propaganda

fel'dsher: paramedic, physician's assistant, medical assistant; in the army also a field nurse

kolkhoz: collective farm (*kollektivnoe khoziaistvo*)

kolkhoznik: collective farmer or collective farm member (plural *kolkhozniki*)

Komsomol: Communist Youth League, formally The All-Union Leninist Communist Union of Youth (Vsesoiuznyi Leninskii kommunisticheskii soiuz molodezhi—VLKSM)

kulak: "fist"—an elastic term used to denote peasants who participated in "exploitative" or "anti-Soviet" practices, such as hiring labor

listovki, listki: short-form, flyer-like print propaganda

Main Political Directorate of The Workers and Peasants Red Army (Glavnoe politicheskoe upravlenie Raboche-krest'ianskoi Krasnoi armii—GlavPURKKA): central political apparatus of the Red Army, headed by Aleksandr Shcherbakov

Medsanbat: Medical Sanitary Battalion, unit that provided medical services at the divisional level

meshchanstvo: roughly translates to bourgeois philistinism, adjectival form *meshchanskii*, noun form *meshchanin* (bourgeois philistine). (See chapter 7)

NKVD: People's Commissariat of Internal Affairs (Narodnyi komissariat vnutrennikh del), in the army present as the *osobyi otdel*—the special section, and from April 1943 SMERSh—Smert' shpionam! (Death to Spies).

non-Russians: term for non-Slavic national minorities in the Soviet Union

paëk: ration

pogony: shoulder boards distinctive to the tsarist army, and from January 1943 the Red Army

portianki: footwraps used in place of socks

POW: prisoner of war

PPZh: Mobile Field Wife (*Pokhodno-polevaia zhena*); derisive term applied to women in the Red Army who had a wartime relationship with their commanding officer

samostrel: a self-inflicted wound to get out of combat

schët, boevoi schët: tally of enemy soldiers a Red Army soldier had killed

SSSR/USSR: Union of Soviet Socialist Republics (Soiuz sovetskikh sotsialisticheskikh respublik)

starshina: highest level of noncommissioned officer in the Red Army, responsible for supply of basic units (company, section, etc.)

Explanatory Notes

Images

The images in this volume have been placed in the chapter in which they are most relevant. That being said, they almost all show objects that are relevant to multiple chapters (e.g., figure 1.3 shows the most common type of belts and the padded jackets soldiers wore). Where the photographer has given the photograph a name, I cite it as such. Otherwise, I give the maximum information that the archive provided, giving the author's name in parentheses after the caption.

Historiography

Throughout the text, most of my discussion of historiography appears in the endnotes.

Transliteration and Translation

I have followed Library of Congress formatting for transliteration. All translations from Russian to English are my own unless otherwise noted. On occasion I have brought English-language sources' transliterations into Library of Congress format (e.g., Gabriel Temkin).

Names

All people in this volume are referred to by the names they identified by, rather than their names at birth. For example, Iosif Stalin is referred to by his assumed name, rather than Dzhugashvili, and David Samoilov by his nom de plume, rather than birth name (Kaufman). I have maintained Russian spelling of names throughout the text (e.g., Aleksandr Tvardovskii instead of Alexander Tvardovsky).

Scope

Strictly speaking, the Soviet Union entered World War II on September 17, 1939, when it invaded Poland, engaged in further operations with the invasion of Finland on November 30, 1939, and ended its participation on September 2, 1945, with the Japanese surrender to Allied Forces, although (as of January 2019) neither the USSR nor Russian Federation has formally signed a peace treaty with Japan. The USSR invaded Manchuria on August 9, 1945, decisively defeating Japanese forces there, but disputes over the Kuril Islands have hampered attempts to formally end the war. However, for most Russians and Russian narratives of World War II, the Great Patriotic War from June 22, 1941, through May 9, 1945, is the central event. This book focuses on that period, with minimal discussion of events of World War II before or after. Furthermore, this book focuses only on Red Army ground forces (*sukhoputnye voiska*), excluding the Red Navy (Raboche-krest'ianskii Krasnyi Flot, RKKF) and Soviet Air Forces (Voenno-vozdushnye sili RKKA), as the scope of the book would have become unmanageable. Sailors and pilots interacted with a very different material culture and had lifestyles quite distinct from ground troops. I leave it to other scholars to tell their stories.

THE STUFF OF SOLDIERS

Introduction

Government Issue

All of our things are government issue [*kazennye*], we put our own stuff into storage.

—Nikolai Chekhovich

So wrote Nikolai Chekhovich, a young Muscovite, to his mother on June 27, 1941. The purpose of his letter was to warn her that she would soon receive his civilian clothes and not to consider him dead when she got them.[1] Over the next four years he would write his mother, and later his fiancée, many letters, describing his hopes, aspirations, and day-to-day life. He sometimes wrote about his clothing, what he ate, and the trenches he and his comrades inhabited. Chekhovich was fatally wounded liberating an anonymous village ("P") from the enemy early in 1944. His letters were collected in 1945 and published, the forward describing Chekhovich as an "authentic hero of our times."[2] His story was indeed typical of his age. Chekhovich was one of over thirty-four million men and women to serve in the Red Army between 1941 and 1945, putting their civilian things in storage and their previous lives on hold. They would trade their disparate civilian worlds for one in which all their things were government issue.

Objects were the tools that turned yesterday's civilians into soldiers, as they first learned to wear their uniforms, care for and use their weapons, pack their knapsacks, and dig shelter. At the beginning of their service, these things—the weapons they carried, the uniforms they wore, the dugouts they lived in, and the army rations they ate—were all that distinguished soldiers like Chekhovich from civilians. For "non-Russian" soldiers (a catch-all used

FIGURE I.1 A Guards senior lieutenant issues a rifle to a newly arrived soldier, July 1943. Rossiiskii gosudarstvennyi arkhiv kinofotodokumentov [RGAKFD] 0-359076.

to describe non-Slavic cadres in the Red Army), the first words of Russian they learned could well be "rifle," "grenade," and "spade."[3] Objects were so essential to the acculturation of soldiers that by regulations even the password at a military position was drawn from the list of army goods.[4] These objects were the quotidian material that made the epic events of the war possible.

The Red Army was the largest army the world has yet known. The front on which it fought stretched from the Barents Sea to the Black Sea, the Caucasus Mountains to Bøkfjorden in Norway, from the Volga to the Oder. The scale of the war was difficult to imagine. An article for agitators (low-level propagandists) from June 1942 complained that "[a] single soldier occupies some one three-millionth part of this huge front," and that despite his or her tiny stature, the fate of the Motherland on all fronts depended on what *he* or *she* did.[5] The massive scale of events was belied by the mixture of mundane and epic. One soldier reflected in his diary: "A person in a lambskin coat with a pistol on his belt, who sleeps on louse-ridden straw in a clay hut—is it really true that I too am making history?"[6] While Soviet citizens, particularly during the war, were encouraged to see themselves as part of a historic process, the Marxist and Stalinist narratives they were supposed to conform to had little room for such messy details as lice, dirt, and threadbare clothing. And yet soldiers' letters home, diaries, wartime interviews, and later autobiographical fiction and memoirs are filled with references to the everyday, to the ordinary things they used. Millions of soldiers at war were accompanied by even more objects. Each soldier developed bonds with the standard issue items they were allotted—they could distinguish their spoon, rifle, tunic, and spade from scores of others.

This book tells the story of the central event in Soviet history, the Great Patriotic War, through objects. It explores how soldiers used and the government provided a plethora of things from underwear to tanks, tracing major changes in Soviet society via government-issue objects.[7] The standardized world of uniforms, rations, weapons, and trenches render explicit many aspects of the meaning of the war for the Soviet government and its citizens. The material culture of the Red Army embodied the obligations shared by state and soldier. It was the milieu in which members of an often largely segregated society became acquainted with each other.[8] Finally, soldiers' stuff was the medium in which the Soviet leadership and soldiers worked out a variety of pragmatic and symbolic problems—from how to survive bombardment to which aspects of the past were usable. This work provides both an ethnographic sketch of life in the Red Army and a narrative of how the war changed the meaning of the Soviet project and the content of Soviet citizenship.[9]

Objects constrain and make possible human action on both an individual and a societal scale.[10] From Prometheus to the present, the ability to craft and use complicated tools has set Homo sapiens apart from other beasts and allowed humans to survive. The objects that people create are also imbued with meaning. They can connect a network of humans and carry associations both consciously crafted, such as the faces on coins or logos on T-shirts, or as an accident of history, for example the rise of denim from work clothes to couture.[11] This book inventories the material culture of the Red Army, treating items both as tools and as bearers of meaning. Another goal is more prosaic. The Red Army is often portrayed as either a faceless barbaric horde or a flatly heroic collection of square-jawed heroes that could only be cast in bronze. This work, by focusing on the quotidian, highlights the fact that these historical actors were very human; they ate and slept and defecated. They were neither saints nor monsters but products of their society and humanity as a whole, with all of the accompanying flaws and glories.

Red Army soldiers were modern people, citizens of arguably the most consciously modernizing state of the twentieth century. The phenomenon of modernity is notoriously slippery, having swum through seas of ink spilled to define it.[12] Modernity is always aspirational and in the eye of the beholder. Its practitioners invoke Enlightenment traditions that posit humanity is capable of improvement, seeking to mold people who are "civilized" through state intervention in the quest to create a rational society. A "civilized person" is self-disciplined, capable of functioning in a rationalized society, and dedicated to that society's values (whatever the rationale and values of that society may be). In the Soviet case that meant becoming socialist, which was itself a slippery term when applied to everyday life.[13] Stalinism eventually arrived at its own set of practices that defined its own version of modernity and civilizing processes. Of particular interest to this book, *things* tend to elicit a visceral response as being modern or backward. Roads, toilets, food, clothes, vehicles, computers, and firearms all elicit knee-jerk reactions that are often accompanied by condescension and moral judgments if they are deemed backward. These reactions show how "modernity" is often a shorthand for the various ideas of how people should adapt to face changing conditions brought about by technological, economic, and political change. From the very beginning of the Soviet project there was a keen interest in the capacity of everyday items (*predmety byta*) and practices to transform Soviet people.[14] Soviet modernity was a struggle against "backwardness" (embodied in the village) to overcome capitalism (i.e., the enemy states that encircled the Soviet Union).

The Soviet Union defined itself in contradistinction to the property relations existing in the rest of the world. Conceptions of property were distinct from those in capitalist societies. All major forms of production were controlled by the government, and personal property was a weak institution: most living space and many things people used were state-owned. All forms of property could be confiscated if one ran afoul of socialist justice, which sometimes could be as simple as being born into the wrong class or ethnicity. Shortages were common due to the massive destruction of infrastructure in the course of the Great War and Civil War as well as the vicissitudes of a planned economy staffed largely with people learning on the job. A society that was in a constant state of mobilization placed emphasis on sacrifices today for a brighter future tomorrow. Access to goods was closely tied to status within a framework of categories based on class, party membership, and state service, in which connections were key. In this system of shortages and strong government intervention, the acquisition of commodities worked very differently from the way it did in capitalist economies, relying much more on special access and connections than cash.[15]

In Soviet society in general, and in the Red Army in particular, objects were the embodiment of the relationship between the government and citizens, and they could serve as both markers of status and instruments capable of reshaping habits and attitudes.[16] Certain items, such as tsarist officers' *pogony* (shoulder boards), the bowler hat and spats of bourgeois fat cats, the vest or calico tunic of kulaks (rich peasants who exploited others), and priest's vestments all became markers of enemies slated for the dustbin of history.[17] The state unleashed citizens on various categories of people (the bourgeoisie in 1918, "kulaks" in 1929), allowing them to participate in the expropriation of the enemy's wealth while (in theory) punishing those who attempted to enrich themselves in the process. Other items, such as tractors, soap, radios, books, and urban forms of dress served as indicators of the Soviet enlightenment. Before the war, everything from new apartment buildings and the Moscow Metro to sausages were lauded as signs of the new regime's ability to provide for the people and the rising *kul'turnost'* (culturedness) of Soviet citizens. *Kul'turnost'* was something that could be accumulated if one acquired the right things and knowledge.[18] Access to most of the fruits of Bolshevik plenty was available only to those who lived in the cities, in which the right to reside was controlled by a passport regime. And even within the relatively privileged cities, party members enjoyed special access. Distribution of goods was a key part of how the Soviet government ruled, and it was not bashful about prioritizing key demographics while leaving others to fend for themselves.[19] Then the war came, and the Soviet Union found itself functioning under conditions that were much less forgiving.

During the war, the army, as the body defending the state and the people, was given top priority in the allocation of increasingly scarce resources, as territory fell to the Germans and was later devastated as the Wehrmacht retreated. Even given the dramatic shortages that the war forced on the Soviet Union, the investment of symbolism into objects could be so important that it outweighed immediate pragmatic concerns. For example, the Red Army fundamentally refashioned its uniforms at a moment when the war looked lost. The government needed to invest every object with meaning and convince soldiers that they should defend the state and could win the war.

Soldiers did not formally own the wide variety of objects they were issued, yet they could use them with owner-like autonomy. A soldier could be executed for loss of a weapon or be sent to prison or a penal unit for destroying or stealing government property. However, soldiers were also armed individuals who could make choices about how to dispose of what they had been given, trading rations or articles of clothing or discarding things that they felt were useless, such as gas masks, helmets, and bayonets. Soldiers were increasingly encouraged to become frugal masters of what the government provided, learning how to repair and maintain their equipment and other supplies and to improvise in order to improve their conditions. They cobbled together tasty dishes at field kitchens, organized barbershops in the trenches, and recycled ration cans or artillery shells to make lamps.

This is the story of how a host of people from what was still a largely peasant society, led by a government that was deeply hostile toward the peasantry, came together around a set of objects, mastered their use, and defeated the Third Reich.[20] The peasant origins of those filling the ranks seeped through into the objects they carried. The uniforms they wore were militarized versions of peasant dress, from their tunics down to the archaic foot wraps they used instead of socks. The food they ate was drawn from the peasant menus of the peoples of the Soviet Union and often eaten with a peasant's wooden spoon, which had been a powerful symbol of backwardness and potential corruption.[21] The knapsacks they wore were traditional peasant bindles. When exposed to foreign goods, their mix of disdain and fascination could easily be described as peasant or socialist in origin. This is also the story of how that process changed the way that soldiers thought about their government and each other and how the government positioned itself in regards to the past and its citizens. Arguably the most popular fictional character to emerge from the war, the everyman *Vasilii Tërkin*, was clearly a peasant.[22] Soviet leadership was keen to use traditional symbols such as peasant huts and churches, particularly in ruins, to motivate soldiers in defense of something sacred and ancient. Indeed, religious institutions key to peasant life before the Revolution enjoyed a wartime renaissance. But improving the image of the peasantry was not the only shift taking place. Noblemen of the old regime were enrolled into the pantheon of Soviet heroes and a keen interest in prerevolutionary history and Russian imperial symbolism became central to how the war and the essence of the Soviet project were understood. Key symbols of the old regime would find themselves reborn in medals and insignia worn by soldiers. What had been seen as backward was drafted into modernity.

In order to understand the importance of objects to this story, we have to grasp the centrality of the war in the Soviet experience. Soviet citizens—whether they had been raised in the Soviet Union, Russian Empire, or republics formed in the wake of the Great War—embarked on one of the most epic, tragic, and paradoxical wars in human history. The Great Patriotic War was one of the most anticipated wars of all time, but it began with a surprise attack. The Bolsheviks had been waiting for the contradictions of capitalism to lead to a world war and worldwide revolution since they had seized power in the fall of 1917. However, they were shocked when the Third Reich, a country that had entered an alliance with the Soviet Union in order to start World War II, invaded Soviet territory. The war almost destroyed the Soviet Union but became the basis of its legitimacy. A party of Marxist internationalists dedicated to global revolution waged a Great Patriotic War, consciously echoing the epic Patriotic War that lead to Napoleon's defeat, the key legitimizing victory of the old regime. Victory in 1945 secured and expanded Soviet rule, while bringing heroes of the Russian imperial past into the pantheon of progressive humanity. It muddied the difference between "Russian" and "Soviet," while integrating millions of "non-Russians" into the Soviet project as never before. Millions of people who had been outcasts in 1941 found themselves accepted into the fold in 1945 and vice versa. Increasingly, in an institution that dubbed itself the Workers' and Peasants' Red Army (Raboche-krest'ianskaia Krasnaia armiia—RKKA for short), nationality trumped class background as a key factor in determining reliability. This army of amateurs, composed mostly of peasants, defeated the most professional, modern, and terrifying army in the field, one that had already conquered Europe.

The war was also paradoxical in that it both highlighted what made the Soviet Union different and rendered Bolshevik practices "normal." As they had during the Great War, liberal states dramatically expanded the power and reach of the government during World War II. Mobilization for war led to rationing and the reorientation of business, media, and education to meet the state's needs in the United States and Britain. Extensive propaganda, censorship, and the evocation of self-sacrifice were utilized by all belligerents. The war also led to the redrawing of boundaries of citizenship, as both the United States and the Soviet Union deported people based on national categories but also integrated people who had previously been peripheral. Victory in 1945 fundamentally shifted the place of each of the Allies in the world, as well as how these states saw their citizenry. The losses among the Allies were distributed unequally, becoming a bone of contention. In much of the space of the former Soviet Union, the war remains at the center of how people interpret the Soviet legacy and their own national histories.

All roads in Soviet history lead to or from the war. If one were to trace the life cycle of the Soviet Union, the war coincided with the Soviet regime's adulthood and was fundamental to how events before and after were perceived.[23] Before the war happened, millions of people had already been sacrificed for victory. Crash industrialization and collectivization, which led to millions of deaths by famine and the destruction of the peasant way of life, were rationalized by the need to modernize the country in order to avoid defeat in the upcoming war. The Great Terror of the late 1930s was seen as a necessary purge of potentially traitorous elements before the next war. Once the war began, propaganda shifted to draw parallels between the current war and past invasions of Russia, casting the Great Patriotic War as a repetition of an age-old battle between good and evil. After the victory, the war retroactively justified everything that had come before and stood as a testament to the power of the Soviet system. Victory Day would eventually become a major holiday, and Soviet official culture was imbued with a respect for veterans and a fear of future war up until the regime's unraveling. In modern Russia and pockets of the post-Soviet space, the Victory of 1945 is seen as the one indisputable accomplishment of the Soviet regime, celebrated and mobilized to various political ends to this day. How much of this victory is attributed to Stalin, communism, excellent weapons, the "Friendship of the Peoples," or a timeless Russian character varies on who you ask, but wartime propaganda supported all of these theses.

The war was the first true test of the Soviet people who had been subject to Marxist-Leninist-Stalinist policies of social engineering for nearly a quarter-century. From the very beginning, the Soviet leadership sought to refashion humankind to create a new world. Faced with the fact that the Revolution happened in an overwhelmingly rural society with a low level of education, they nonetheless strove to make a new kind of modern person. Who the new Soviet person should be was subject to change, but certain aspects remained constant: the Soviet person was urban as opposed to rural, sought out educational opportunities, and displayed political consciousness. Consciousness implied a certain familiarity with the texts of Stalin, Lenin, Marx, and Engels, accompanied by an obsessive work ethic, but more than anything else it required a fanatical loyalty to the Soviet Union and merciless hatred of its enemies. History was understood to have objective laws with inexorable results that Marxism made clear and which made the destruction of those who would obstruct the path of history a necessity. From 1924 on, almost all enemies were internal, loosely defined, hidden, and often simply fabricated, making Bolshevik attempts at mobilization against enemies risky and highly divisive. Soviet people were constantly preparing or mobilizing for campaigns of various scales, from the liquidation of enemy classes to higher levels of culture in retail. Despite this, the Soviet leadership was often disappointed in its ability to penetrate Soviet society and create a new type of person.

The war offered new opportunities to influence citizens and changed what the government saw as a Soviet person. In the army millions of men and women lived under state surveillance, with political commissars appointed to explain the world and party policies. The army did its best to organize a soldier's free time in a way that maximized the dissemination of Soviet propaganda. However, the idiom of this propaganda was increasingly patriotic in a traditionalist sense and easy to explain. Rather than a loosely defined internal enemy, Red Army soldiers fought a clearly foreign foe who was an obvious existential threat. What the state needed was something much more functional and universal than earlier iterations of the Soviet person—it needed soldiers. Whether they were fighting out of a personal desire for vengeance or a sense of national pride, because they thought Hitler was the Antichrist or to defend what the Bolsheviks had built, was not as important as that they killed the enemy. To inspire soldiers, the government showcased atrocities committed by the Nazis and mobilized the romantic national past of the Russian Empire. If the Soviet person could have spoken any number of languages in the 1920s, he or she clearly needed to speak Russian by 1943. Medals, tanks, and newspapers would display the names of noblemen who had distinguished themselves in battle for the old regime often against a backdrop of the destruction that German occupation had wrought. This would all merge with Bolshevik revolutionary traditions to create a new sense of what it meant to be Soviet.

Both the Soviet government and citizens were keen to note their reforging in the fires of Mars. Soviet modernity was located in the city, where key sites of refashioning—schools, universities, museums, and factories—were concentrated. War became the new means of legitimizing oneself by service to the state but also a new space of enlightenment.[24] Soldiers attempting to explain to the uninitiated the difference between a person who had been to the front and those who had not might do so in terms of the transformation from a rural to an urban being. Hero of the Soviet Union Guards Major Malik Gabdullin reflected on this change. He wrote to the head of the Literature Department at the Kazakhstan Filial of the Academy of Sciences of the Soviet Union in September 1942 about how "the front changes a person's character": "If you liken the ideas that we had back when we lived in the aul [village] to one mountain, the ideas that we developed after we moved to the city—after studying, when we learned to tell black from white, have our own opinions and views, when we joined the ranks of conscious people—could be likened to a second mountain, and we see a huge difference . . . the front, where life battles death, dramatically changes a person's character."[25] Service in the military was a testament to loyalty to the state and a way to become a "fully conscious" person. Wartime service was also a common experience that connected millions of people. All things Soviet came to be seen as the force of life and light against Germano-Fascist death and darkness. German goals—which included starving millions of people in the western parts of the Soviet Union to death and enslaving those who remained—and actions—which included looting on a colossal scale, mass rape, and indiscriminate mass murder—meant that risking your life at the front at least gave you a fighting chance, something soldiers became increasingly cognizant of.

However, what soldiers saw at war could also undermine the Soviet government's claims. The Nazis penetrated deeply into Soviet territory in a war that was supposed to be waged on enemy soil. Millions of soldiers surrendered in the initial months of the war. That the Red Army was incapable of defending the borders created an existential crisis for many. However, according to some, these challenges "gave the system what it had heretofore lacked, competition—something like a market, on which its products (regiments and divisions) collided with foreign ones," forcing the system to work more effectively.[26] After the Red Army proved a match for the Wehrmacht, exposure to foreign contagion became a major issue. Millions of people had lived under occupation, and the war ended with millions of soldiers leaving Soviet territory. Coming from a country that—for the vast majority of its citizens—had been nearly hermetically sealed from the outside world in the decade before the war, the campaign abroad was a revelation. In 1939–1940 and then again on a much grander scale in 1944–1945, Soviet citizens were exposed to the wealth of the bourgeois world, a world of undeniable material affluence as compared to the everyday life of Soviet citizens. This world would both attract and repulse, sensations that largely played out through objects.

This work builds on existing scholarship of the Soviet Union and imports some approaches from the historiography of the United States and Europe. Much of the most important work on Soviet history ends just before the war, in part because of archival access. The scholarship of the 1990s and early 2000s focused primarily on the rise of various aspects of Stalinism and their establishment prior to its great trial. In the wake of the "archival revolution"—the declassification of many materials in the 1990s—and the collapse of the Soviet Union, a variety of subfields emerged within Soviet history, many of which this book traces through the war. Key among these subfields is the study of nationalities policy and empire.[27] Another tracks the shifting contours of propaganda.[28] The study of everyday practices and how they highlight the relationship between the state and individuals is crucial to *The Stuff of Soldiers*.[29] Scholars have shed considerable light on the processes of social engineering, as well as the effects of these policies, such as collectivization or the Great Terror, on Soviet society.[30] Creating a new Soviet person was a key Bolshevik goal, and historians have begun to examine how this process felt, taking seriously the subjective experiences of those living through the Soviet experiment.[31] Others have explored the differences in experiences of a variety of groups—peasants, workers, women, ethnic minorities, and déclassé elements—in the prewar period, providing a multidimensional view of Soviet society.[32] Following these threads of examination into the period of

the Soviet Union's greatest peril and triumph, this book explores the fates of people who were shaped by Soviet policies and realities.

The Soviet Union was a country born of war in which the army played a central role. Historians have demonstrated the power of the military as a vehicle of social transformation, as either a nationalizing project or a way of inculcating a certain sense of modernity in what was still a largely peasant society.[33] They have also shown the transformative effect that the Great War and Civil War had on society and how war affected governance.[34] The historiography of the Great Patriotic War is nearly as vast as the conflict itself, so I dwell here only on those works with which I directly engage or attempt to build on. David M. Glantz, Roger Reese, and Walter Dunn have set an impressive foundation with their operational histories detailing how the army functioned.[35] A growing body of work examines the social history of the war. Elena Seniavskaia has developed a subfield she calls "military anthropology," heavily focused on folklore and the lived experience of soldiers, a tradition carried forward by Aleksei Larionov.[36] Amir Weiner has produced groundbreaking work on how war service became a new form of legitimization, one that could wipe away past sins or cast the formerly celebrated into infamy, taking a key border region as his case study.[37] Mark Edele has provided the sequel to the story this book tells and is himself writing cutting-edge work on the social history of the war.[38] Filip Slaveski and Robert Dale have written detailed accounts of the immediate aftermath of the war, focusing on occupation and demobilization, respectively.[39] Jochen Hellbeck has taken his study of subjectivity—a person's sense of self and the social, political, and economic factors that shape its formation—into the war years and introduced me to the archive that proved most vital in this study. Anna Krylova, Amandine Regamey, Roger Markwick, and Euridice Charon Cardonna have investigated the experience of women in combat, opening up discussion about how the war had an impact on gender, with Krylova also highlighting the special relationships between human and machine at war.[40] Oleg Budnitskii has written excellent and provocative work detailing reactions to the Third Reich and the complexities of responses to the war and positing a wartime sexual revolution on a par with that of the 1960s in North America.[41] Finally, there has been one fairly recent attempt to write an overarching social history of life in the Red Army, Catherine Merridale's *Ivan's War,* written as a sort of exposé of life in the Red Army with a popular audience in mind. While Merridale's work is an important foundation for my own, I take a very different approach. *The Stuff of Soldiers* does not offer an overall narrative of the war, its major battles and inflection points, but rather focuses more narrowly and intensely on soldiers' everyday lives. Readers of

both books will see that some tangential points in Merridale's work are fundamental to mine and vice versa and that I am much more interested in the diversity of experiences.[42]

Thus far, little attention has been paid to the Red Army's material culture and the everyday practices surrounding it. While collectors' guides abound, few have bothered to interpret the meaning of the objects themselves or explore what stories they might tell. Elsewhere, the potentially transformative nature of objects and the role of objects in social processes is well established in, for example, studies of the American and French revolutions, the New Deal, and the Cold War. Leora Auslander, T. H. Breen, and Lizabeth Cohen have utilized objects to highlight pivotal changes in US or European history—notably the role of conscious market choices in shaping "imagined communities," the growing penetration and nationalization of markets, the rise of consumerism, and the birth of the suburbs—all with their accompanying accoutrements, from tea to clothing to radios to shopping malls.[43] Historians of the Soviet Union, while not shying away from histories of consumption, have been more reticent to pick up objects themselves, particularly during Stalin's rule. There are some notable exceptions—including Lewis Siegelbaum, Vladimir Lapin, and Jukka Gronow—but *things*, particularly the more mundane stuff, have not gotten the attention they deserve.[44] In-depth study of the material culture of the Soviet Union has largely been the purview of anthropologists, art historians, and literature specialists, tending to focus more on the postwar and post-socialist periods and often concerned less with everyday stuff, with a few works such as Emma Widdis's *Socialist Senses* bucking this trend.[45]

This book is based on a variety of officially and personally produced sources. Official documents, such as the minutes of meetings from the Political Directorate of the Red Army, reports from the front, orders, speeches, newspapers, and army journals and manuals provide information about the production and use of objects as well as rhetoric surrounding them. Manuals in particular provide a wealth of information on the use, purposes of, and kinds of communities built around objects. It would have been impossible to write about weapons and trenches without referencing manuals. Internal documents captured failures and moments that would later be erased from the official narrative. Political showmanship was key to Stalin's way of rule, as he positioned himself as a force that could swoop in, make a pronouncement, and set things right, something we will see at several key points in the text. Stalin provided an authoritative voice that would be echoed in the press, which translated the evolving meaning of the war as seen by the Communist Party.[46] His profile would come to adorn the "Victory over Germany" medal, casting his authority in bronze.

Personal sources such as memoirs, fiction written by veterans, letters, diaries, and interviews, especially when used in combination with reports and corrective orders, provide invaluable information about how rhetoric played out in everyday situations and how soldiers understood and interpreted events. Interviews and diaries are particularly elucidating. Diaries, which were very dangerous and difficult to keep, are full of surprises, furnishing us with a wealth of personal reflections on the war and references to the material culture of soldiers. Diaries and letters are also of interest as objects created during the war, and their production will be discussed in chapter 6. The interviews that I draw on include some of the work of *Ia pomniu*, an oral history website centered in Russia and Israel, but are more heavily weighted in the works produced by the Commission on the History of the Great Patriotic War at the Russian Academy of Sciences Institute of Russian History, popularly known as the Mints Commission (as I refer to it throughout the text) after the academic who headed it. This wide-ranging interview project took place during and immediately after the war. Many of the interviews were recorded on the battlefield, just after or even between battles. There is a surprising lack of self-censorship in many of these texts. The interviewees often speak in a Stalinist idiom and are drawn from those most dedicated to the regime, but they describe events before they have been hewed into a foundational myth. As a result, they provide details that would later be written out of the narrative of the war (e.g., the killing of prisoners, desertion, fear, and the particular problems faced by female soldiers).[47] The largely autobiographical fiction written by veterans, many of whom were low-level commanders during the war, often referred to as "lieutenant's prose" (*leitenantskaia proza*) is another important source, rich in ethnographic detail. One reason for this wealth of sources is that the war was clearly exceptional, understood as a great historical event of unprecedented magnitude, inspiring millions of people who otherwise saw themselves as ordinary to chronicle what they had done and witnessed.[48]

All of these texts—whether official reports, diaries, or memoirs—are neither considered to be distillations of the truth or inherently false. Some of their authors could be lying or misremembering what they have seen. Whether true or false, the stories they chose to tell and their perceptions create the meaning of these events and are what we have to reconstruct them. For our purposes, there is no such thing as reality divorced from the person recording and mediating it. I am also less concerned with how representative one or another of these authors are, for as one veteran came to understand in the course of his service "the people are not the homogenous ground meat of history."[49] Despite being an overwhelmingly peasant army, those recording its deeds skewed urban and educated. Among diarists and later writers, Jews were overrepresented but also arguably culturally the most Soviet and Russian.[50] I have done my best to read as widely as possible and take my subjects' claims seriously. These texts have been read with attention to how soldiers and those in the government describe the army's material culture and everyday life and how soldiers used, made, or destroyed things together. Two categories—"non-Russians" and female soldiers—receive special attention due to their outsider status, making their experience a distillation of issues faced by all soldiers as they adjusted to the idiosyncrasies of army life. This work highlights the variability of experiences in the ranks; being a Red Army soldier in 1941 was a fundamentally different experience from serving in 1944, while the competence of one's commander and which front one ended up on could lead to vastly different fates.

In crafting this narrative, I have made it a point to give roughly equal billing to ethnography and argument. The details of everyday life receive serious attention throughout this text as something worth recording in and of themselves. It is in everyday life that politics is worked out. The Soviet leadership became increasingly aware of this point as the war continued and it became obvious that making trenches homey and keeping soldiers well fed, well armed, and more or less free of lice were necessary to achieve victory. The government would eventually make good on its promise to provide soldiers with the things they needed to defeat the enemy.

Violence is central to this story. Soldiers in any society are mobilized, trained, and equipped in order to kill the state's enemies. In the Red Army, a variety of incentives were used to motivate often undertrained soldiers to kill. These included better food for those at the front, decorations that carried social and monetary benefits, and the humiliation and corporal punishment of those who failed to successfully prosecute violence. Soldiers lived under the constant threat of death, which forced them to build a myriad of structures to survive. All of this happened in the presence of the dead. The Soviet state celebrated killing and exorcised any clinical idea of trauma from the official narrative. The vast majority of these soldiers had no language or concept of trauma to frame their experience. They often describe horrific events matter-of-factly or frame them using Soviet official language. Witnessing, but also causing, violent death was a normal experience for a massive swath of Soviet society. As far as Soviet leadership was concerned, the soothing presence of a woman and inspiring literature were supposed to assuage a soldier's "wounded soul" after demobilization.[51]

This work is organized thematically in roughly the order soldiers would proceed through their service and encounter various material objects. Each chapter is organized around a group of related objects (e.g., weapons, uniforms), which form the lens through which the experience of the war is examined. Some chapters tell a story of change over time that is directly linked to the overall arc of the war, others concentrate more on transformations that could be experienced by soldiers whenever they entered the army.

Part I focuses on biological needs and their political dimensions, engaging the body itself, the soldier's uniform, and what soldiers ate. Chapter 1 does three things. First, it uses soldiers' bodies to discuss the diversity of cadres entering the Red Army. Second, it shows how and to what extent those bodies became state property. Third, it provides a brief "life cycle" of soldiers in service, from induction through training to the front and eventual wounding into the system of hospitals and back again. This chapter leaves the soldier naked and shivering before the state. Chapter 2 dresses the soldier. This chapter provides an ethnography of soldiers' clothing and explores the meaning of the uniform's iconography. It shows how soldiers' biographies and the state's self-presentation changed via medals and uniforms, which themselves became readable texts. Chapter 3 focuses on provisioning in the Red Army, tracing how soldiers ate, the ways the government positioned itself as provider, and the logic of who deserved more or less food under conditions of extreme shortage.

Part II explores the issue of violence. Put bluntly, chapter 4 is about not getting killed, while chapter 5 is about killing. Both focus on how evidence of what the Germans had done was used to incite hatred and help remove the taboo of killing. Chapter 4 foregrounds the vulnerability of soldiers on the modern battlefield, which led to constant labor in order to survive. It takes us through the destroyed landscape of war and the cities of earth that soldiers built to survive, crafting a semiurban space using standard issue spades and attempting to recreate aspects of civilian life. Finally, this chapter discusses how the living were forced into close association in the trenches, and how the dead became part of the landscape. Chapter 5 tells the story of how Red Army soldiers learned to kill. It explores the arsenal of the Red Army, treating weapons as tools for specific tasks and highlighting the soldiers' symbiotic relationship with weaponry. The chapter concludes with a discussion of social differentiation among different branches of service and a brief examination of changing tactics as the army became a professional fighting force.

Part III focuses on the miscellany of the soldier's kit. Chapter 6 takes us into the soldier's knapsack, exploring the few personal items that soldiers carried, from knickknacks to print propaganda to personal letters. Given the nomadic nature of soldiers' existence and the lack of free space in their packs, these items had to be either useful or precious. This chapter is ultimately about the creation of meaning under nomadic conditions, in which mail would come to play a central role. Chapter 7 focuses on all manner of trophies, from German prisoners of war to objects looted from houses in the Third Reich. It shows how a Soviet understanding of jurisprudence and a particular perception of the bourgeois world combined with a desire for vengeance to both justify looting and frame Soviet understandings of the Third Reich.

The Stuff of Soldiers is the story of individual and political transformation through objects. These objects were both standard and personal, as well as carriers of meaning that was both prescribed by the government and inscribed by soldiers. One such object was the simple spoon carried by one soldier during the Battle of Moscow. While recovering from wounds in a hospital near the capital in the fall of 1941, *Starshina* (Quartermaster Sergeant) Nikolai Pavlovich Donia had a rare opportunity to see his wife. He gave her a spoon into which he had carved "In memory of the days: 1.1.38; 22.5.39; 10.7.41; 17.8.41; 19.10.41; 5.11.41." Each day was a milestone in their life together—their wedding day, the birthdays of their son and daughter, the day he was drafted, the day he was wounded, and the day of their meeting in the hospital. A month later Donia fell in battle, one of over 8.5 million Red Army soldiers to perish in the war.[52] That he gave his wife a spoon—one of the very few objects that Red Army soldiers owned and one of the most intimate items a soldier had—is telling. It was during moments of eating that soldiers took stock of their situation and realized that they were alive. The spoon would remain as a reliquary in his family's house, a visceral connection with a man that his children barely knew. This spoon was one of millions of objects that did not merely bear silent witness to the war but made the waging of war possible.

Part One

Mortal Envelopes

Figure P.1 Red Army soldiers eating in the field, Third Ukrainian Front, August 1944. RGAKFD 0-167316 (L. Ivanov).

CHAPTER 1

The Soldier's Body

A Little Cog in a Giant War Machine

I am a little cog in the massive, creaky, and ungreased machine called the Army.

—Vladimir Stezhenskii

In the course of the war the Red Army had to transform millions of Soviet citizens into usable components of its war machine. The immense scale of the war would bring people into the army who would otherwise never have served. These soldiers were faceless to those at the top, who mobilized them as numbers on paper, but a good commander knew those under his command intimately and treated them as individuals. In 1943 Lieutenant Mikhail Loginov walked the ranks of his platoon of thirty-seven men and reflected: "They are all dressed in the same uniform. To the outside observer they all look the same, but these are different people. Even from afar you can recognize each one by his gait. They wear their helmets, carry their pack and rifle each in their own way. Each has his own personality. Every soldier is a particular, separate world, and each deserves respect."[1] He was a school teacher, a Russian from Kazan. His sergeant was a middle-aged Russian veterinary assistant, his corporal an Uzbek textile worker, his machine gunner a giant man who had been a mid-level manager on a collective farm. His other soldiers included a worker from Moscow, a boy who graduated from high school on the eve of the war, and two shepherds—a young Ukrainian and an aged Uzbek. This last soldier, the Uzbek shepherd Dzhuma, caught Loginov's attention. He was his worst soldier (a sectarian pacifist) and had traveled the longest distance, both literally

FIGURE 1.1 Red Army soldiers on the march, 1944. RGAKFD 0-92459 (Velikzhanin).

and figuratively, to be in the army. Loginov imagined the naked, shaken Dzhuma standing before a medical commission, shamed by the presence of a woman, and then stepping onto a train (probably for the first time) that would take him and other recruits to Central Russia. Dzhuma's body and fate had ceased to belong to him. The shepherd's experience could describe that of a peasant from any remote region, as he left the comfort of familiar territory and lifestyle for an unknown fate.[2] Dzhuma and his comrades—men and women, urban workers, students, peasants, and shepherds—all became cogs in a giant military machine. Their bodies were subject to a new disciplinary regime and way of life as numbered, largely interchangeable components in the Red Army.

The government laid claim to the bodies of these men and women, handing them over to the commanders who were deputized to use these human resources to wage war. Commanders were tasked with training, tracking, and properly exploiting their soldiers and given almost total control over their subordinates' bodies. They often saw their units as fiefdoms over which they had total dominion.[3] With this power came great responsibility: a good commander was supposed to be able to turn anyone into a soldier.

Both the government and its deputies were forced to reckon not only with the physical bodies of soldiers but also with the souls that animated them. This was made all the more difficult by the demographic diversity of those serving, men aged seventeen to fifty-five, women, former criminals and almost all of the ethnic groups of the heterogeneous USSR.[4] The army drafted a document, the Red Army booklet, which turned these diverse citizens into legible mechanisms of a military machine. Past experiences could be negated or key to the fate of a soldier. Whole ethnic groups were deported from the ranks while formerly déclassé elements entered them. Professional skills such as being a cook, tailor, or poet could define how a soldier served or be ignored completely. The war reshaped the contours of belonging and exclusion in Soviet society in unpredictable ways.

The diversity observed by Loginov was in part by design, but much more the result of massive losses in the first months of the war. The casualties suffered by the army in 1941 and 1942 were catastrophic—from June 22 through December 31, 1941, 3,137,673 were killed, missing, or captured.[5] On top of this, over 5.6 million draft-age males were left behind enemy lines, and the last prewar draft had only managed to reach 28 percent of those eligible for service.[6] Between January 1, 1942, and January 1, 1943, 11,245,740 men and women were sent to the front, over 4 million of whom had recovered from wounds and returned to the ranks. By January 1, 1943, the army had suffered 5,639,782 permanent losses (killed, missing, captured, sick, etc.), and 7,543,004 recoverable casualties. At the same time, 2 million men went undrafted on enemy territory, and there were 10,000,942 people in the ranks of the army. In that year alone the average rifle division had gone through 234 percent of its combatants (*boevoi sostav*). It was estimated that there were only 3.7 million men left to be drafted into the army.[7] By war's end, 11,273,026 were permanently lost, and 34,476,700 had been drafted (on average there were about 11,000,000 persons in uniform every year). In the active army at the front, there was an average of 5,778,500 people in ranks during any one month. The army at the front had gone through 488 percent of its average monthly strength from 1941 to 1945.[8] In other words, it had been rebuilt five times.

The process of rebuilding the army required the government to take the raw material of a variety of civilians and turn them into the refined product of the soldier. Stalin's famous 1945 toast in which he praised "simple Soviet people" as "the little cogs of history" betrayed key truths about soldiering, particularly in the Red Army.[9] Millions of men and women would become soldiers, but before they became "little cogs" in the vast war machine of the Red Army, they had been civilians of every stripe, from shock workers to Gulag inmates.

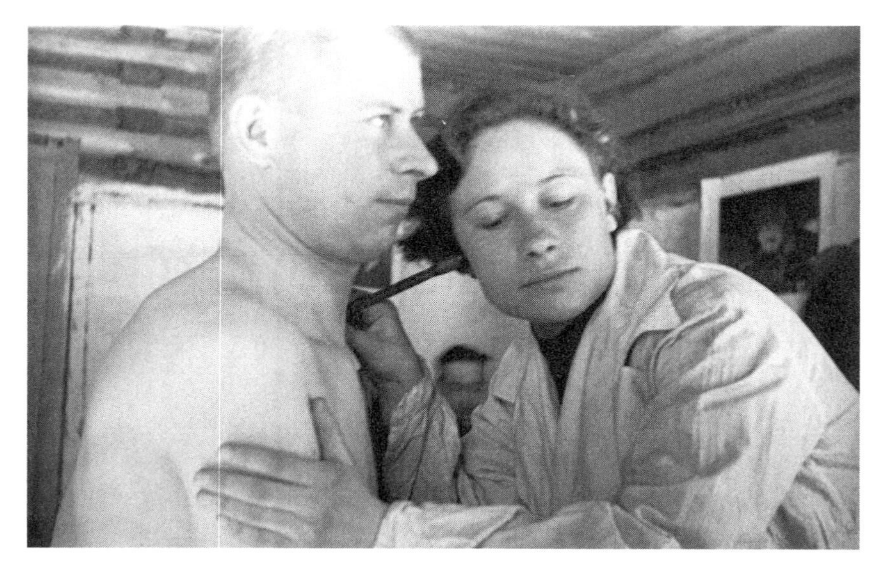

FIGURE 1.2 Junior Sergeant D. A. Iablokov and Dr. E. I. Kolpakova, Frontline Rest Home, Kalinin Front, 1942. RGAKFD 0-60235.

The Draftee's Body

After begging to be allowed to join the army several times, Grigorii Baklanov (who would go on to become an artillery officer and famous novelist) finally found an officer who saw his potential, declaring, "A person is such material that you can sculpt anything from him."[10] The bodies of new recruits were raw material for the army. The army wanted either to negate or to utilize prior identities and could not ignore the prewar experience that was often imprinted on soldiers' bodies and recorded in their personal documents.

The Red Army drew from a diverse population. According to the 1939 census, roughly 170,000,000 people lived in the USSR. Peasants outnumbered urbanites by about two to one (56,125,139 urban vs. 114,431,954 rural). Russians were the largest single ethnic group, constituting a majority with almost 60 percent of the population (99,500,000 people), followed by Ukrainians (almost 16.5 percent and about 28,000,000 people) and Byelorussians (a little over 5,000,000 and just over 3 percent). Other major ethnic groups included Uzbeks (4,800,000, almost 3 percent), Tatars (4,300,000, a little over 2.5 percent), Kazakhs (3,100,000, 1.8 percent), Jews (3,000,000, 1.78 percent), Azeris (2,300,000, 1.3 percent), Georgians (2,200,000, 1.3 percent), and Armenians (2,100,000, 1.3 percent). More than fifty officially recognized languages were spoken in the Soviet Union. There were thirty-seven million men aged twenty to thirty-nine (the usual age of mobilized soldiers). Ninety percent of men were literate (this could vary dramatically by ethnic group and age), although only around 1,092,221 people had a higher education and 13,272,968 had received a full secondary education. These figures shifted slightly with the annexation of Bessarabia, Eastern Poland, and the Baltic Republics on the eve of the war, but the Slavic, rural core continued to predominate.[11]

The ideal soldier of the Red Army was young; educated; Russian, Ukrainian, or Byelorussian; and a member of the Communist Party or Komsomol. By 1942 the number of these cadres fell far short of demand.[12] The decimation of the regular army, together with the grueling realities of mechanized warfare, led to the Soviet government significantly widening the scope of potential cadres. The prewar categorization of soldiers' bodies included three basic categories: suitable for combat positions, suitable for noncombat tasks, and unsuitable, as well as a note on whether these men had received training or not. Exemptions existed for specialists, and men could receive a deferment for study, illness, or work, although many of those who were eligible for exemption enlisted.[13] Accounting for the war saw a variety of new categories added. Sex, age, criminal convictions, and nationality all became classifications.

Wide swaths of the population that had been excluded from the honor of military service before the war became included as the war dragged on. "Former people" or déclassé elements left over from the old regime found their way into the army, including those who had served in the White Army.[14] Criminals had their sentences commuted at enlistment. Bandits and political criminals were not eligible for this redemption.[15] In 1943 there was a further reexamination of cadres, with more petty criminals and even the children of political criminals allowed to serve.[16] The need to fill the ranks gave many people a chance to escape their pasts.[17]

Prisoners of war (POWs), despite officially being considered traitors, factored into the army's plans from the winter of 1941, when significant numbers were liberated. Between 1942 and 1945, 939,700 soldiers who had been POWs or missing were called back into the ranks.[18] All soldiers escaping encirclement (*okruzhentsy*) or liberated from POW camps were subjected to filtration in special camps.[19] This could be a harrowing experience, as those who had just undergone life in POW camps, where many starved, were submitted to humiliating interrogations similar to those of enemy prisoners.[20] Food was often poor, conditions crowded, and lice rampant. However, it seems that these operations were concerned primarily with supplying the army with as many soldiers as possible. This is implied by the estimates of numbers of soldiers to come out of filtration in the reports of Red Army officers in charge of staffing and reports by soldiers who went through the process.[21]

POWs and petty criminals were not the only group invited back into the fold. Peasants had been viewed with distrust and forced onto collective farms a decade before the war. "Kulaks," a category that crystallized the Soviet hostility toward the peasantry as a whole, were not allowed to serve in the army in 1941. Defined by their supposed wealth, exploitation of others' labor, and backwardness, those described by this catchall term could include anyone who local officials perceived as dangerous or resisting collectivization. Initially heavily taxed, this category of people was eventually repressed, with some being executed, some fleeing for the cities, and many being sent to resettlement in remote regions of the USSR. All kulaks lost their property as a form of punishment. However, in May 1942 dekulakized peasants became subject to the draft. In the ranks they were often treated with condescension, sometimes described as errant children, but their family members were to be freed from the restrictions that had forced many to live in special settlements in some of the least desirable areas of the Soviet Union.[22] This great reversal signaled that loyalty was increasingly mapped onto nationality rather than class.

The identification of Soviet power and Russian patriotic traditions was one of the most significant shifts of the prewar period and only intensified during the war. The very name of the war harkened back to the struggle against Napoleon. As the war continued, the government took major efforts to identify itself with traditional Russian ways in an attempt to secure the loyalty of soldiers. Restrictions against religious practices were significantly loosened. Churches and mosques were reopened during the war, the governing bodies of religious institutions reconstituted, and religious leaders recruited to help propagandize the war.[23] However, the reification of the Motherland and Fatherland, connecting soil and people, could lead to exclusion. At the beginning of the war, Germans, Finns, and other peoples belonging to groups whose countries of ethnic origin were at war with the Soviet Union (estimated at 250,000 men in 1942) were exempted from service in the active army and sent to the labor army, a much worse-supplied organization that performed menial labor. As the Wehrmacht occupied large portions of the Soviet Union, the categories of those who would not be accepted into the ranks expanded, while several whole nationalities were labeled "traitor nations."[24] Entire nations that were perceived as collaborators (e.g., Crimean Tatars, Chechens, and Kalmyks) were later deported from their homelands to internal exile in Central Asia. Soldiers from these nationalities were expelled from the ranks to join their co-ethnics in exile. But even before the emergence of "traitor nations," nationality was a complicated subject in the army.

Unprecedented numbers of "non-Russians" served in the army during the Great Patriotic War. The Russian language became a mandatory subject in schools throughout the Soviet Union only in 1938, and millions of soldiers had little to no knowledge of the language of command. Before the war men drafted from "non-Russian" regions would have months to integrate into the army or serve in special territorial units. However, territorial units were officially abandoned on the eve of the war, only to be revived again in 1941–1942, largely as a stopgap measure. The lack of experience, training, and supplies led many of these units to disaster at the front. In 1942 some staffers initially refused to use cadres from the Caucasus or Central Asia. Then an array of extraordinary events forced the army to recognize these cadres as needing special attention, and "non-Russians" and more specifically "potential inductees from Central Asia and the Caucasus" appeared as a new category in internal memos circulating among staffing officers.[25] Some commanders were so frustrated with these alien cadres that they eagerly awaited heavy casualties that they hoped would lead to a new batch of Slavic soldiers.[26]

The army eventually created a support network for "non-Russians" that included special propagandists and translators, political officers, and a print network dedicated to providing "non-Russians" with information in their native languages. Some commanders and soldiers believed that it was better to keep co-ethnics together so that they could help each other, while others were certain that they fought better when separated from one another.[27] Either way, everyone was supposed to melt into an organization where nationality was often muted, and soldiers of all nationalities were affectionately referred to as *slaviane*.[28] Some "non-Russians" even adopted Russian or Russified names at the front: for example, Rafgat Akhtiamov wrote his parents that his Russian comrades called him Raphael (Rafail′), while Talgat Genatulin became Anatolii, a name that stuck through his whole life.[29] This could be seen as an act of conscious identification with the culture of the army and later its victory, rather than a negation of ethnic particularism.[30] Experience showed that when properly trained, attended to, and given time to learn Russian, these troops performed as well as others. By August 1943 internal memos of the army's political department could boast that "non-Russians" had become true soldiers.[31] However, after the "non-Russians" had been integrated, a different category of draftees presented a new set of problems.

"Non-Russians," even if they had not been fully integrated into Soviet society, had been part of the Soviet Union since its formation. Soldiers drawn from regions that had become Soviet only on the eve of the war (Western Ukraine, Western Byelorussia, Moldova, Latvia, Lithuania, and Estonia) presented a different set of problems. They had not had time to become integrated into the Soviet system, and many were hostile to it, as they had fresh memories of sovereignty. Soviet power was associated with violence, deportations, and the nationalization of property. The Baltic Republics all fielded their own national military units, each of which had a special corporate identity. These Baltic formations, all of which had been given adequate time to train, fought well. However, draftees from Bessarabia and Western Ukraine from 1944 on often presented a serious challenge to political officers and commanders as differences in culture and language were dramatic.[32] Aleksandr Shcherbakov, head of the Political Directorate of the Red Army, warned high-level political officers of the special case that these recruits presented. They were overwhelmingly illiterate and deeply religious, and many had never seen such trappings of modernity as films, farm machinery, or a radio. Political officers perceived these men as coming from a different time, not having progressed through the stages of history through which the October Revolution and Stalin had ushered the Soviet Union. This perception led to specific propaganda aimed at integrating them into the Soviet fold and motivating them to fight.[33] These people were socialized into Soviet society in the ranks of the army, under unusually difficult conditions. Much like Dzhuma, our Uzbek draftee, they were in for a shock even greater than most of their comrades and were likely to receive duties in the most expendable positions.[34]

At the opposite end of the spectrum were the college students, factory workers, and other "politically conscious" people who volunteered to serve in the *narodnoe opolchenie*—People's Guard units that formed as the Wehrmacht closed in on major cities at the beginning of the war. This practice traded lives for time, as these poorly trained and armed formations were often devastated in short order, decimating the ranks of those who would have gone on to become scholars, writers and artists, although some People's Guard units went on to become elite formations.[35] The People's Guard led many college-educated people into the army for the first time. Commanders and veteran soldiers had a variety of responses to the sudden presence of the intelligentsia in the ranks. Grigorii Baklanov recalled the antipathy of his drill sergeant toward educated soldiers, while a commander interviewed by the Mints Commission said that despite initial misgivings that "an educated person won't last long," *intelligenty* under his command had become "the real deal."[36] Often worker and scholar, young and old, served together in the People's Guard, offering a preview of what became general trends.

The age of soldiers in the ranks expanded dramatically, ranging from seventeen to fifty-six.[37] Older soldiers were often celebrated: The World War I veteran who had returned to the ranks and sometimes slipped into old regime language was a stock figure of wartime propaganda.[38] Young men who were of the more traditional draft age were sometimes viewed with derision as they acted "just like children"—playing with shrapnel and constantly complaining.[39] General Erëmenko, commanding the Stalingrad Front, told the war correspondent Vasilii Grossman that "The youth have no life experience; they are like kids; wherever you send them, that's where they die. The smartest soldier is twenty-five to thirty years old. And the older soldier—he is not quite healthy; his family is on his mind."[40] These older soldiers were often assigned to noncombat roles such as cooks in order to free young men for combat, but older men were not the only ones fulfilling household duties.

Starting in 1941, women were introduced into the army in uneven ways and into specific specializations on a mostly voluntary basis.[41] One of their main tasks was to free men for combat duties. Women also served in front-line positions—most notably as pilots, snipers, anti-aircraft gunners, communications specialists, medics, and traffic controllers.[42] Women's performance at the front was often highly rated. Moreover, it was believed that women made the front a more civilized place: "it is simply pleasant to meet a neat, clean girl among the dirty soldiers at the front." Some feared that service would disrupt women's ability to fulfill their role as mothers and provide another generation of Soviet warriors.[43] Women's bodies presented a particular problem to the army. Their figures were often too small for military clothing, and the army was indecisive on whether to order clothing be made unisex or to make concessions to women. All military documents used the grammatically male form. Yet traditional gender roles would affect women's service, and women were often faced with what we would now call sexual harassment, a theme that I discuss in more detail later.

Soldiers' bodies could speak to their histories, which in turn could have an impact on their present. Despite the expansion of cadres, specific branches of service still had precise physical standards. Deranged and mentally underdeveloped men were rejected from the service. NKVD (Narodnyi komissariat vnutrennikh del—People's Commissariat of Internal Affairs, i.e., the secret police) men couldn't suffer from hemorrhoids, commanders couldn't have speech impediments, paratroopers couldn't weigh more than 80 kilograms, and tankers had to have powerful hearts and lungs.[44] It was common to find former criminals in reconnaissance units, as the skills necessary in both endeavors overlapped.[45] Peasants ended up overwhelmingly in the infantry, while urban cadres, who were generally more educated, often found themselves in technical branches of service with longer life spans. Prewar profession could determine wartime specialization, because drivers, cooks, writers, artists, thespians, and others could ply their trades in the army. Differences

FIGURE 1.3 "Combat girlfriends" anti-aircraft gunners M. Gerasimova and E. Bukhanova, May 1943. A 76mm m.1938 anti-aircraft gun is visible in the background. RGAKFD 0-154797 (Gribovskii).

among soldiers were enhanced by the various forms of body modification that could be observed among soldiers, including tattoos among criminals and circumcision among Jews and Muslims. Soldiers from different regions might vary by height, particularly those who were less severely or wholly unaffected by collectivization and wartime hunger. A commander in the Latvian Rifle Division remarked that "you looked like a kid next" to Latvians, with size 45–46 [12–13 US sizes] boots being the norm and one soldier having size 49 [US 16] boots. Later the unit received reinforcements from Leningrad and Tashkent, starving men who needed to be brought back to strength before serving.[46]

The army was interested principally in bodies, but it could not ignore souls. Commanders needed to inspire men and women from a wide variety of backgrounds to risk their lives saving a state that some of them despised. There were numerous ways in which the government attempted to appeal to those in the ranks. One side of the coin was positive motivation. Drawing on Russian and "non-Russian" national traditions, as well as revolutionary traditions, showcasing German atrocities and by attempting to provide for their families, the state provided a variety of reasons to fight. The other side of this coin was coercion and discipline. While discipline is key to any army's success, the Red Army had a particular problem of trying to cobble together a fighting force from inexperienced civilians to fight a professional enemy who was rapidly advancing deep into Soviet territory. These prerogatives, combined with the Bolshevik readiness to sacrifice lives now for a brighter future, led to a draconian disciplinary regime. Despite this, the army kept remarkably poor track of its soldiers at the beginning of the war.

Becoming State Property

The living and the dead were often undocumented in the war's early days. Although Red Army regulations laid out a clear system of documenting the dead, these proved inadequate, and there was no consistent system in place to document living soldiers.[47] Every soldier was issued a "personal medallion" or "death medallion," a Bakelite cylinder with two long thin pieces of newsprint recording all of a soldier's personal information. Worn in a special pocket on the trousers, this was the Red Army's equivalent of a dog tag, one copy staying in the capsule with the body, the other being collected by the commander as a record. The capsule was exclusively meant to document the dead and not the living, being shut tight at all times to prevent moisture from destroying the flimsy paper inside. Many soldiers refused to fill out the information, believing that if they used them as directed, they were sealing their own fate. The documents were popularly referred to as "death passports." Soldiers often ritualistically threw them away for good luck, and the capsules proved so unpopular that they were discontinued in November 1942, because all the same information could be found in a new document introduced in 1941.[48]

The Red Army began the war without a single, army-wide document that could be used to identify soldiers at the front. A Red Army book that contained detailed information about the soldier and his service was cancelled in 1940, and in practice was to be collected from the soldier before going to the front anyway. By the fall of 1941 it had become apparent that this situation was a serious danger to the army. As Stalin noted, the lack of documents meant that "our division, which ought to be a closed fortress, impenetrable to outsiders, has become in practice a public thoroughfare." Enemy agents could easily penetrate the army and already had. Without proper documentation, it was impossible to see if legitimate soldiers were receiving and taking care of their allotted uniforms, equipment, and weapons. In the chaos of encirclements, retreats, and rushed mobilizations of 1941, it was often impossible to control and document the movement of soldiers, let alone find deserters or spies.[49]

On October 7, 1941, a new Red Army booklet (*Krasnoarmeiskaia knizhka*) was instituted that was to be immediately issued to all soldiers in the ranks. Signed by the soldier and his immediate superior, showing (in theory) the bearer's photograph and stamped with the unit's seal, these flimsy sheets of newsprint became the only proof of identity. Any soldier caught without a Red Army booklet was to be arrested as a suspected spy. The Red Army booklet tied soldiers to their unit to every piece of equipment they were issued. It was a document that stated identity and responsibility very clearly. In the rear these documents were to be checked every day during morning inspection, and at the front at least every three days.[50] A propaganda campaign surrounded its introduction with the two-pronged messages of vigilance against impostors and responsibility for all state property one had been issued.[51] While its introduction highlighted the chaos and carelessness that could be found in the army, the categories that composed it and the way in which it was used reveal how the state had come to view its "little cogs."

A soldier's Red Army booklet contained all of the information the army deemed necessary about its cadres. Name, education, nationality, birthdate, prewar occupation, next of kin and their address, and blood type were all that was written of soldiers' prewar life. Their wartime life was recorded via date and place of induction, specialization, unit, ID number, notes on where they had served, when they had taken their oath of service, wounds they had received, which medals they had been awarded and weapons issued along with their serial numbers, uniforms and equipment issued, and on the final page, their sizes. On the pages of the Red Army booklet, the biography of a soldier was reduced to the army's parameters and tied to the objects issued to a soldier. The Red Army booklet also became the primary document used to record soldiers' deaths, collected from the dead on the battlefield and in hospitals.

In some ways this document was simply a continuation of the prewar internal passport, documenting people and fixing them to a specific territory and profession—in this case, their military unit. However, it was issued much more widely than passports, and the Red Army booklet was the first passport-like document that the vast majority of peasants received. Often filled out via dictation, under circumstances where confirmation of details was impossible, many soldiers, including wanted criminals, had a chance to reinvent their biographies using the Red Army booklet.[52] Aside from citizens' editorial choices, the government was seriously reevaluating which points of biography mattered. Class had no place on the pages of this document, despite its predominance in the lives of Soviet citizens and on the pages of the Soviet passport and application forms. While class would still be recorded in the files of the NKVD, both the Red Army booklet and official propaganda encouraged soldiers and commanders to ignore the class origins of its fighters, focusing instead on what they had done at the front.[53]

The exclusion of class and inclusion of such categories as profession, ethnicity, and education reveal what the state needed from its soldiers. While soldiers were interchangeable components in a military machine, not every soldier was the same sort of part. An uneducated soldier could not be used as a translator, while a linguist could not be used as an engineer. Both formally and informally, on the level of the army and individual unit, the ranks were periodically swept for specialists—writers, translators, cooks, tailors, cobblers, carpenters, and others.[54] Every commander of a small unit wanted a cobbler and a woodworker. However, in dire situations, everyone could become cannon fodder, as the army gambled specialists' lives for time. Nationality remained important, as an educated "non-Russian" could act as a translator for rural co-ethnics. Finally, as we shall see, the soldiers' next of kin were used as a means to pressure them into compliance and sometimes as a determinant of reliability. Soldiers whose families were on nearby occupied territory might be highly motivated to liberate their home town or to desert to the enemy, depending on the situation. In May 1942 NKVD units in the Red Army were ordered to place special surveillance on soldiers with family on occupied territory, former special settlers, former convicts, prisoners of war, and those who escaped from encirclement.[55] This Red Army booklet carried with it a set of responsibilities that were simply draconian. Essentially, the soldier's body became property of the army and the Red Army booklet could be seen as a user's manual for the convenience of the commander who took control over this or that soldier. From the moment they entered the ranks, soldiers were made to understand that they had entered a different world.

The Civilian Body Transformed

Soldiers received notices to go to the draft board or came themselves to volunteer. They arrived carrying a few belongings according to instructions and were often accompanied by their families. Soldiers underwent a physical assessment and an interview about their past and family that some compared to a "difficult exam," particularly if there was a skeleton in the closet such as a repressed family member.[56] What followed was both a somber event and a celebration. In some cases the traditions of dancing and drinking with family members were revived, and generally soldiers were transported miles away from their point of induction to camps, physically separating them from their homes and families, often forever.[57]

Literature and propaganda from the period drew a clear line between one's prewar and wartime biography. On induction into the army, soldiers literally shed their previous identities. Passports were surrendered. They would eventually give up their civilian clothing, whether they wore suits, summer dresses, prison uniforms, or silken Central Asian robes.[58] Some soldiers, such as Old Believers or Orthodox Jews, had their identities challenged in fundamental ways as their beards were shorn to meet army regulations.[59] Some sent their clothes home to younger siblings or instructed their families to sell their clothing.[60] Other soldiers eagerly awaited the issue of military clothing so that they could sell or trade their civilian clothes for food.[61] Guards Major Baurdzhan Momysh-uly, the son of a nomadic Kazakh herdsman who found his calling in the army, explained to inductees that they had entered a different world: "Yesterday you were people of different professions, different means. Yesterday there were among you rank-and-file *kolkhozniki* and directors. From today on you are fighters and junior commanders of the Workers' and Peasants' Red Army."[62] Induction into the army was meant to partially erase one's prewar identity. One diarist records consciously avoiding any thoughts of home and his loved ones in order to be as present with his comrades as possible.[63] An article for political officers reminded them that the loss of status could be crushing for those who had been pillars of their community suddenly thrust into a situation where they were no one, particularly for those who didn't speak Russian.[64]

Men's heads would be shaven so that differences in age disappeared, as the poet Aleksandr Tvardovskii wrote: "Look—they are just like kids! / So who is, in truth, a greenhorn / Single or married [you can't tell] / [Among] this shorn people."[65] The privilege of having longer hair belonged to officers and those who had already served their standard term in the army (*sverkhsrochniki*), but with lice infestations this benefit could vanish.[66] Women had varied fortunes, from having their heads shaved or cropped to being allowed to wear long braids.

Soldiers learned to stand, move, and speak according to a set of regulations that were universal to the army and alien to the civilian world. They moved in ranks, regularly counted off to check that all were present, and became subject to constant surveillance by their superiors.[67] Manuals demonstrated everything from the proper way to brush one's teeth to how to form a marching column. Soldiers learned to move as massive biological machines, to hold their fists along the seams of their pants, with their feet forming a perfect "v."[68] They were to sing on the march, and automatization of their actions was a sign of their perfection.[69] Every day they were to shave; a moustache was seen as a claim to special status.[70] Soldiers had lost autonomy over their bodies on the level of posture and hygiene, which were small outward indicators of a more fundamental shift in their status as resources to be used.

Discipline

Many soldiers failed to grasp the full extent of what the army demanded of them. Reports from early in the war show that soldiers who did not speak Russian were often unaware of the consequences of their actions, and in general discipline could be remarkably lax.[71] Many soldiers and commanders were simply not used to military discipline.[72] Anatolii Genatulin, a seventeen-year-old orphan who spoke almost no Russian when he entered the army, reflected on his coming of age in the barracks:

> We had to have the boyish carelessness, laziness, childish sleepiness, and all we had learned about motherly pity toward physical and spiritual weakness beaten out of us and have our unripe bodies hammered into the yoke of military ranks, to habituate our consciousness to the harsh ways of the army. In this crusty masculine society—where instead of the affectionate call of mother and gentle nagging you hear commands that resonate like metal, yells, and swearing—we were brutalized and ripened for the front, for killing and death.[73]

The pouring of the soldier's body into the "yoke of military ranks" carried with it the implicit understanding and explicit indicators that soldiers were no longer the masters of their own bodies. Momysh-uly warned recruits: "perhaps you were a good person before, perhaps people loved and praised you, but whoever and whatever you were earlier, if you commit a crime, an act of cowardice or treason, you will be executed."[74] However much the prewar Stalinist state used militaristic rhetoric and imposed harsh discipline on its citizens, the army, in which commanders regularly ordered soldiers to risk their lives and could take their lives for refusing commands, was orders of magnitude more stringent.

While the fact that soldiers were expected to risk their lives seems obvious, other aspects of how their bodies became the preserve of the state were less so. Soldiers were expected to eat only what the state provided them. While on guard duty, they were forbidden "to leave their post, sleep, sit, lean against anything, talk, eat, drink, smoke, sing, answer the call of nature, accept any object from any person, become distracted from continuous observation of their position."[75] Soldiers were forced to exert themselves on the march and in combat in ways that pushed them to the limits, as one commander wrote home to his wife: "it would be enough to tell you that in three days I walked ninety-two kilometers and slept for an hour and a half. This fundamental test of nerves has ended, and I didn't break."[76] There was no option but to keep moving during forced marches, and soldiers learned to sleep while walking.[77] According to Vasil' Bykov, this could lead to a state of "total indifference" and near stupor punctuated by moments of fear and commands mixed with profanity.[78] However, none of this was unusual for armies at the time.

What made the Red Army particular was a readiness to use force against its own soldiers and to treat their families as hostages. The authors of the 1940 Code of Discipline reasoned that discipline should be "higher, stronger, and more severe" in a society without class conflict, which demanded absolute loyalty. "The order of a commander or leader is law for his subordinate." Commanders were also reminded that they were responsible for any breakdowns in discipline and that they were allowed to use any methods, including fists and bullets, against those who "failed to follow orders, openly resisted, or maliciously failed to observe discipline."[79] While most infractions led to less severe punishment, the threat of violence was always present. Most soldiers witnessed a show execution during their service, in which soldiers were assembled to watch and sometimes required to pull the trigger themselves. These executions often served as punishment for self-mutilation or attempted desertion.[80] These were highly ritualized actions in which the army made clear its right to take the lives of soldiers in its service. These displays often left witnesses shocked and disgusted, even when they agreed with the sentence.[81]

Despite an already harsh regime, in both 1941 and 1942 the army issued orders to make the disciplinary regime more severe and threat of violence more explicit. Even so, the state remained ambiguous as to how important violence was to its strategy. For example, in September 1941 an order was issued that gave the commanders of divisions and their commissars the right to sign death sentences, but within a month another corrective order decried the use of repression as opposed to "educational work."[82] These orders could be seen as either a series of improvisations by a regime that was on the verge of collapse or as purposefully leaving room for maneuver for commanders. The two most famous orders expanding the use of violence, no. 0270 and no. 227, deserve closer attention, as these were the two key rulings on the use of violence as a motivator.[83]

Order 0270, issued on August 16, 1941, gave subordinates the right to execute their superiors if they should panic.[84] Order 227, issued on July 28, 1942, the (in)famous "Not one step backward" order, which was read to all troops and became a motto for the army, established a new, harsher disciplinary regime. This order stated that panicky soldiers were contributing to a "retreating mood" in the army. It called for the destruction of panic mongers and cowards "on the spot" and labeled anyone retreating without orders a traitor. The order established blocking detachments, well-armed formations placed in the rear of "unstable divisions" to keep soldiers from retreating, by opening fire if necessary. Declaring that they were adopting German tactics, it established penal units for commanders and soldiers who had disgraced themselves; such units were to take on the most dangerous tasks in order to "expiate their sins with blood." Penal units provided an alternative to executions for serious crimes and thus a more practical way to use the lives of soldiers and commanders who had failed to live up to the Red Army's superhuman standards.[85] Both orders gave an immense amount of discretionary power to commanders, who had an expansive right to kill their subordinates.

Damnation of cowards went beyond death. The families of soldiers who surrendered to the enemy were also to be punished, a fact announced to soldiers, occasionally at special meetings.[86] Propaganda stressed repeatedly that there was no future for the traitor and that even his loved ones would forsake him.[87] Soldier's responsibilities, and the legal consequences of failure, were common themes of agitation. Copies of the *Military Oath*, in Russian and a variety of other languages, were common pieces of print propaganda. To the very end of the war, this document remained a key part of agitation and propaganda.[88] The last line of the *Military Oath* reads, "If I maliciously break this solemn oath, then I will suffer severe punishment by Soviet law, the total hatred and scorn of the working people."[89] Executions carried with them the confiscation of all property, leaving beneficiaries with nothing.

In moments of panic or under extreme duress soldiers sometimes shot themselves in the hand or foot in order to be evacuated to the rear. In the Red Army, this act—called *samostrel*, "self-shooting," or *chlenovreditel'stvo*, "self-mutilation"—was considered treason, punishable by execution or being sent to a penal unit. On August 2, 1941, the army's Special Section (Osobyi otdel, i.e., secret police) was given permission to arrest and if deemed necessary execute practitioners of self-mutilation.[90] Those attempting to injure themselves to escape the front found increasingly creative means. Usually these soldiers, dubbed *samostrely* or *levoruchniki*, shot themselves in the left hand (their right was still necessary to work after demobilization), through a piece of wood, bread, bucket, or cloth, so as not to leave a telltale powder burn on their hand. Surgeons, officers, and political organs learned to see through these methods, sometimes assigning specialists to look out for *samostrely*.[91] Some soldiers "voted," sticking their left hand above the trench at dawn until enemy soldiers shot it. Eventually military units statistically tracked left-hand injuries and all soldiers wounded in the left hand became suspect; those who "voted" could be caught and punished.[92]

Many soldiers, particularly commanders, endorsed harsh punishments for *samostrely* and those who showed cowardice. The artillery officer Vasilii Chekalov reported that when confronted with a soldier who had shot himself in the hand because he feared for his wife and child, another soldier, almost in tears, screamed: "And you think that I don't have a wife and child. All of us have someone that we need to fight for." Chekalov himself called for the execution of this man, which was carried out immediately.[93] Momysh-uly imagined the thought process that led to acts of cowardice: "He loves life; he wants to enjoy air, land, and sky. And he decides that you can die and he will live. That's how parasites live, at someone else's cost." However, he also discussed how difficult it was to kill one of his soldiers: it was like "cutting a piece of flesh from your own body"—but one that had become poisonous and dangerous to the rest.[94] Violence against those who failed to meet Bolshevik standards reified the community by purging it of the unworthy, highlighting the righteousness of those who had fulfilled what was expected of them, while providing a graphic demonstration of the price of failure. Commanders could use lethal force to ensure that their orders were fulfilled, and this threat of violence gave them a level of control over soldiers' lives that bordered on ownership. Not everyone thought that the use of force was reasonable, and occasionally an understanding voice would defend the actions of a soldier who panicked in the first battle or whose nerves were worn out.[95] Furthermore, soldiers often perceived a commander who was too willing to wave his pistol at his subordinates as hysterical or cowardly.[96]

The threat of violence was very real for soldiers in the Red Army, and 994,300 were prosecuted by military tribunals and ultimately tallied as a permanent loss for the army, although 400,000 of those prosecuted had their sentence commuted to service in a penal unit.[97] An estimated 135,000 Red Army soldiers were executed after trials.[98] This was orders of magnitude higher than the Wehrmacht, which executed an estimated 15,000–20,000 soldiers, and the US Army, which executed only 142, almost all for murder or rape, with only one soldier, Eddie Slovik, executed for desertion.[99] According to the military lawyer Iakov Aizenshtat there were few options open to members of a tribunal other than execution, penal units, or full acquittal.[100] No official statistics concerning battlefield executions exist, but references to them are found in interviews, official reports, diaries, and memoirs. This appeal to violence was largely an act of desperation and sign of weakness, as the government used terror as a means to force poorly trained and supplied troops to do the impossible.[101]

The Subversion of Discipline

Despite the intensity of the disciplinary regime, there was a variety of ways in which soldiers subverted it. Among them were extreme acts such as "voting" and *samostrel*, though most forms of subversion were less dramatic. The fact that the army consisted largely of yesterday's civilians meant that basic forms of discipline could be undermined by more informal structures, such as when soldiers on guard duty failed to stop people they knew or chatted with local civilians.[102]

The disciplinary regime was functional only if the army could control soldiers, which in 1941 and 1942 was often impossible. In 1941 the single largest category of permanent losses were missing soldiers and POWs (2,335,482 or 52.2 percent).[103] Millions of soldiers refused the state's claims over their persons, surrendering to the enemy in encirclements or crossing no man's land to surrender. Their motivations for doing so varied from fear and a desire to help loved ones on enemy territory to principled opposition to the Soviet state. Often those whose families had suffered during the Revolution or collectivization crossed over to the enemy. In 1942 "non-Russians" were considered to be particularly susceptible to this temptation.[104] The Germans, for their part, put considerable resources into luring soldiers to desert to them, including leaflets and radio broadcasts.

Acts of desertion required explanation. State organs understood that terrible conditions would encourage desertions and, as subsequent chapters show, put significant effort into improving soldiers' material conditions. Often desertion was seen as the work of internal enemies preying on the ignorant. One soldier interviewed by the Mints Commission was certain that the "non-Russian" deserters in his unit were villagers "manipulated by the *basmachi* [anti-government bandits]" as opposed to those who had attended Soviet school and were reliable.[105] Acts of desertion and *samostrel* evoked a sense of failure in political officers akin to a priest who has failed to save a soul.[106]

If soldiers were near their homes, the temptation to desert could be too much. Political Officer Reutov grudgingly noted in his diary in October 1942 that whole groups of soldiers from occupied regions, including political officers, were deserting to the enemy. He grudgingly concluded: "The Germans have a smart policy: they let them go home."[107] At times when the army was in massive disarray, this could occur with virtually no consequences for the soldiers involved. The special war correspondent A. Gutman complained that a large number of soldiers near Voronezh returned to their villages as the Red Army retreated past them but were "forgiven everything" and drafted into the army with no questions asked once the army returned in 1943. He was disturbed that few were properly filtered and that "in a number of cases being home during the period when one's village was occupied by the Germans was counted almost as a service: it was everywhere believed that every soldier drafted from formerly occupied territory would hate the Germans and earnestly beat them."[108] While this laxity was surely based on the desperate need for manpower, it was not lost on these soldiers that if Soviet power *did* return to their village, it would be impractical to punish everyone. In addition, this sort of desertion had a logic more complex than individual survival, because it gave soldiers a chance to assist their families under occupation. Furthermore, during 1941 and 1942, when desertion was at its peak, the future existence of the Soviet Union looked unlikely.

Over 376,000 soldiers were prosecuted for desertion (and those were the ones the army could get its hands on).[109] Desertion by a few soldiers could lead to investigation and serious trouble; on an army-wide level it threatened collapse. As a result, commanders came up with creative methods to keep soldiers in place. Many tried to put Communists and Komsomol members on guard, because they were considered more reliable. An officer in the Latvian Rifle Division would make sure at least one Jew stood watch, knowing the Nazis would not take a Jew prisoner.[110] On the eve of the Battle of Kursk, the army conducted a special operation code-named "Treason" (*Izmena Rodine*), in which specially trained soldiers pretended to be surrendering, then opened fire on the Germans who tried to take them prisoner. This had the expressed purpose of making surrender to the enemy more difficult.[111] But desertion remained a significant problem. Desertion and self-mutilation were the most severe affronts to the government's claims on soldiers' bodies, but other, less severe forms of subversion came from interpersonal relationships.

Men, Women, and Discipline

Mixing men and women in the ranks of the army led to particular disciplinary problems. First, there was the issue of physical difference and traditional ideas of gender. While demanding that women fulfill all regulations, "commanders and political officers should always remember that they are dealing with a warrior-girl, and not a man. They should take into account the peculiarities of the physical condition, character, and needs of a girl."[112] Second, the issues of sex and romance were unclear. On one hand, the Komsomol organization, which had mobilized more than half of the women into the army, made it clear through print propaganda and its activists that sex was forbidden. Virginity, or at least abstinence, was considered by many to be a part of military discipline for women, and women in the ranks were invariably referred to as "girls," rather than women, with "woman" sometimes used as a term of derision that implied promiscuity.[113] On the other hand, as early as 1942 members of the Army's Main Political Directorate had made it clear that commanders were not to be punished for pursuing relations with their subordinates. As the head of the Main Political Department of the Army, Aleksandr Shcherbakov, stated in 1942 at a meeting of high-ranking political officers: "If people come together—a commander and a woman—then what is the big deal?"[114] Shcherbakov made this statement after a series of complaints were lodged against commanders by the party organizations of various Red Army units. Commanders were clearly given the go ahead for consensual relations with their subordinates. A commander's word was law, and as a result it was not obvious to some female soldiers and their commanders whether women in the ranks had the right to refuse advances by their superiors. Female soldiers who resisted unwanted advances could have their service made much more unpleasant—in one way or another, their commanders controlled their bodies. One sniper described a situation in which her commander literally ordered her to have sex with him and later threatened to shoot her if she did not acquiesce to his advances. When she resisted and had him punished, he retaliated by sending her to the most dangerous parts of the front.[115] A scorned superior could get his subordinate killed or make her life miserable, and many commanders believed that they had a right to sex. In fact, the expectation of sex with subordinates was so common that one soldier recorded in his diary a particular system of organization in which commanders staked their rights to the favors of the medical personnel of their units: "A regimental doctor—if, of course, it's a woman—lives with the regimental commander, a battalion doctor with the battalion commander . . . and so on . . . regulations breed habits of such strength in

the army: that is, to always give preference to one's seniors."[116] As we see from this quotation, sex and love could be regulated by military discipline in unexpected, informal ways. Women were forced to negotiate a very difficult situation, and a new category of soldier developed as a negative expression of their plight.

"PPZh" (*Pokhodno-polevaia zhena*—Mobile Field Wife) emerged during the war as a term of derision.[117] Such women were considered by some in the ranks and at home to have prostituted themselves to get easier assignments in the army, and some veterans recalled: "Soldiers looked at these women with cheerful spite, saying 'for some war is a stepmother, for others their tender mother,' and most often with envy for those whom such a beauty kept warm."[118] Sex could be a way out of the army; pregnant women were discharged from service until September 1944, at which point maternity leave was established.[119]

The experiences of women, perhaps more than any other group, bring home the gravity of the claims that the army made on soldiers' bodies. The state claimed and then transferred to its deputies total control over a soldier's body. In the case of women, there were specific issues that men were unlikely to face. Total control was claimed, yet always negotiated among individuals, even if recourse to violence could ultimately turn this negotiation into a dictation of the commander's will. The commander was given such discretion because he would be forcing people to risk and sometimes sacrifice their lives in order to defend the state. Commanders were charged with assembling and then operating a functioning military machine from the various components the government provided.

Military Machines and Their Use

Staffing officers occasionally referred to units as "military organisms," but Stalin's famous quip about soldiers as "the little cogs of history" is more accurate.[120] An organism that loses its heart dies. Most living things can never fully recover from the loss of a limb, while a machine can have its motor, tracks, or turret replaced, often without noticeable change. A military unit, when provided with either well-trained troops or given time to train them, could recover from massive losses in much the same way as a machine. However, unlike the components of a machine, soldiers are living, breathing organisms, all of whom had longer and more complicated histories and connections than a lathed or stamped piece of metal. These details could have an effect on when the state drafted and how it used what were supposed to become interchangeable parts in the machinery of war.

Most military units came off the battlefield as a shadow of their former selves after suffering heavy losses in personnel and equipment.[121] As a rule, units were chronically understrength; it was not uncommon for a platoon, which would normally have thirty to forty soldiers, to have ten to twelve soldiers or even five, and not unheard of for a battalion, which should number over seven hundred, to have a only a few dozen soldiers in its ranks.[122] With such dramatic turnover rates, the Red Army was concerned primarily with providing a stable foundation for a unit via a *kostiak*, or backbone of unchanging commanders and rear-area personnel who kept a unit's traditions alive. As Mikhail Kalinin, chairman of the Presidium of the Supreme Soviet of the Soviet Union, told agitators, "a regiment or division can reconstitute itself after any battle as long as its backbone, embodying in itself the highly developed battle traditions of the unit, has survived."[123] Some of those making up the "battle traditions" were ghosts, deceased men and women who were permanently added to the rolls of a unit for an extreme act of bravery, with their names called at reveille and inspection.[124] The commander and surviving veterans were to teach new, often undertrained soldiers how to fight, while the feats of the current and previous members of a unit were to serve as inspiration and create a sense of continuity and responsibility.[125] That many of the heroes were mentioned posthumously could have a disquieting effect on soldiers. From the standpoint of some frontline soldiers, the process of forming a *kostiak* was a sort of "natural selection" of rear-area personnel—people who found relatively safe positions who were sending frontline soldiers to their death.[126]

Heavy losses among combat arms led to Red Army units becoming very dynamic social bodies. The constant motion of cadres in the Red Army created a different corporate culture from that in other armies. Mechanized combat ensured that people directly at the front would be killed or wounded in a short period of time. Aleksandr Lesin recalled in his diary how one officer criticized a soldier: "You really think that you are a good soldier? A good soldier doesn't spend a long time in a company: he either gets wounded, or—he softened—or he dies heroically."[127] While short lifespans at the front were not specific to the Red Army, the experience of wounded Red Army soldiers could differ from those of other armies.

In the US Army or the Wehrmacht, soldiers would generally be sent back to the units where they had previously served, but Red Army soldiers were usually sent first to training units, then to marching companies to be assigned to new units.[128] This could be devastating, as Valentina Chudakova remembered: "You'll go where they send you. Does it really matter where they send you?' No, it's not all the same, not at all the same! . . . Frontoviks in the hospital say "We want to go home!'"[129] Lightly wounded soldiers would go to the Medical Sanitary Battalion within their own division, but a more serious wound would send one to an evacuation hospital further to the rear, and only those with specializations such as translators or elite Guardsmen and cadets would stand a good chance of returning.[130] Some soldiers forged documents to return "home." The deputy commander of the Thirty-sixth Guards Corps, General Maslov, described how his soldiers would even desert to be reunited with comrades, creating "scandals": "He arrives, announces: 'I was in such-and-such unit, I have come to serve.' You write the commander of that unit: don't worry, and don't look for him, don't count him as a deserter, he is serving with us."[131]

This circulation of cadres meant that soldiers were constantly in flux, much like labor at Soviet construction sites. Soldiers would be forced to establish relationships every time they found themselves in a new unit. Letters of commendation and identity papers could be lost as they disappeared into the army's bureaucratic apparatus.[132] Some took advantage of this situation to reinvent themselves, falsely claiming rank and awards.[133] Soldiers' military units were also their addresses, so wounded soldiers would often lose correspondences from loved ones. Going to the hospital was a loss of identity, often accompanied by the theft of personal property.[134] Even as the army was attempting to foster unit pride via the "strengthening of combat traditions," it was clear to soldiers that they were interchangeable parts to the army.[135]

The soldier's unit was supposed to serve as a surrogate family, with commanders as father figures and comrades as siblings.[136] In practice this varied dramatically. One junior commander lamented that his men were faceless strangers: "You become a different person—a platoon commander. It is bad when you stop seeing a living person and see only an unbuttoned collar. But under these conditions, with such a massive turnover of people, any other way is almost impossible."[137] Commanders often failed to learn the names of their subordinates.[138] Conversely, soldiers in elite units truly felt a sense of community, some even stating that after the war they spent time primarily with their frontline friends.[139] As the war progressed, more and more units gained Guards status and more soldiers found a stable home.

The flow of cadres could lead to a certain amount of horse-trading, as soldiers who had recovered from wounds were mixed with new recruits either to be sent as reinforcements to a unit or to collection points (*sbornye punkty*), where commanders gathered replacements and specialists. Mansur Abdulin, a veteran of Stalingrad and Kursk, recalled an informal arrangement in which men from the same region were allowed to serve in the same crew or squad. Abdulin, who was "a Tatar, Siberian, Urals local, Central Asian, and Muslim—all in one!" could claim soldiers with any of those connections. Old sergeants listened carefully when the *Starshina* filled out the soldier's Red Army booklet in order to hear who they could claim. Abdulin described the joy of these meetings leading to "such happiness! Noise! the *Starshina* gets mad: 'Shut down your market fair! [*Prekratit' iarmarku!*].'"[140] While some common past could help soldiers adapt to their new milieu, the military was unconcerned with and at times even hostile to such arrangements.[141] Informal structures could soften the blow of entering a massive, impersonal institution, as people sought out any sort of common ground with their comrades in arms. It was, after all, with these strangers that one would face the dangers of combat.

Being Used: Soldier's Bodies as Currency

In any army, generals gamble with their soldiers' lives and accept losses as inevitable. Soldiers offer their bodies to harm in battle, becoming the currency of warfare.[142] In training, Red Army soldiers were to "be prepared for self-sacrifice." Propaganda frequently stipulated that a soldier's duty was "To kill the enemy and stay alive yourself, and if you die, then to sell your life at a high price."[143] The soldier's body became a form of currency with which commanders tried to buy or maintain territory, destroy or protect machinery and other resources.[144] Some soldiers expressed bitterness about this, decrying that "Soldiers have always been manure."[145] Others accepted high casualties as an inevitability and a sacrifice made for future generations and loved ones at home, stating fatalistically, "There is not and cannot be any other solution."[146]

Soldiers were just another type of resource for military planners. Losses were tallied in such a way that mortars, tanks, machine guns, and people were simply categories of resources that the enemy had lost or taken.[147] Grigorii Baklanov reflected in prose on the tension between the army's commodification of the soldier's body and the family ties of the individual inhabiting the uniform:

> before an operation has begun, it is known—approximately of course, not down to the exact number—how many will be killed, how many sent to the hospital, and how many of those will return to the ranks. And I am part of this, like any other unit, but me and no one else. Lt. Motovilov, a graduate of some year of the Second Leningrad Artillery School, can be replaced by another graduate of that school, and that won't be a problem. But to you, Mother, I am irreplaceable.[148]

People had volition, personalities and families that would mourn them, but to the army they were still ultimately resources to be used.

The modern machinery of war could irreversibly impact the human body. The Red Army instituted a disciplinary regime and claimed dominion over soldiers' bodies in order to compel them to face the possibility of death and maiming to defend the state. Propaganda celebrated soldiers who struggled on after being fatally wounded or consciously sacrificed themselves in order for the army to advance. Heroes such as Tulegen Tokhtarov—who, according to his comrades, after being disemboweled "picked up his insides and shoved them into the wound, holding it with his left hand, while shooting with his right"—or Aleksandr Matrosov, who used his body to block a German machine gun bunker, were held up as examples for all to follow.[149]

Soldiers faced the possibility of death, which could be abstract, but they saw very concretely what shrapnel, bullets, and flames could do to the human body. Some reflected on how they or their friends had been maimed in combat. One commander asked rhetorically how a jocular friend "could smile with a shattered jaw."[150] A sniper described how an explosion ripped through his body: "The whole right side of my body torn open—my face, head, hands, and feet . . . even my bones were visible—the meat was torn off them. But now none of that hinders my ability to hold a sniper rifle."[151] This soldier, echoing official tropes, blithely described terrible wounds from which he had recovered. Some reflected on how pieces of shrapnel became a part of them, sometimes being kept as souvenirs or being removed years after the war.[152] In the macho culture of the army, scars were a way to prove one had done their part, exemplified by phrases like "my résumé is on my hide" and a "a scar decorates a man."[153] Soldiers feared being crippled or disfigured and the military censored both complaints by invalids and images that showed their injuries.[154] Nonetheless, wounds carried a certain ambiguity, as Political Officer Nagim Khafizov, examining the piece of shrapnel that hit him in the spring of 1945, reflected: "I can't be angry with it—you see it could have been worse—it could have killed or permanently disfigured me. So it turns out that this fragment 'saved' me. However, I am upset with it, because it took me from the ranks in the most interesting days of the war."[155] Wounds could be terrifying but were not necessarily unwelcome. Khafizov was not alone in thinking that a piece of shrapnel had "saved" him.

Loginov's men all eagerly anticipated an offensive in part because they might "receive a light wound and finally get enough sleep in the hospital. Nobody thinks about the fact that he could be killed."[156] The Red Army had no regular system of furloughs or leave, so being wounded was the only exit from a system of service in which units fought until they suffered heavy losses and were then rebuilt.[157] Hospitals were often referred to as "heaven"—a place where soldiers could rest.[158] Being wounded meant that the army could make fewer claims on the body and was a visceral demonstration that you were fulfilling your duties.

The army's strict disciplinary regime touched the wounded as well. Soldiers were provided with a bandage to patch themselves up, and regulations gave specific instructions to soldiers who were wounded: "If wounded, bandage yourself and continue to fight. **Leave the battlefield with the permission of your commander**. Take your personal weapon and one pack (magazine) of cartridges; if it is necessary to move, crawl with your weapon to cover and wait for a medic. **It is forbidden to leave the battlefield to accompany the wounded**."[159] Given that prisoners and the wounded were often disposed of on the battlefield, this policy had a practical side. This order highlights the gray zone that wounded soldiers inhabited as government property—they were no longer active defenders and thus no longer top priority, and Red Army medical treatment often left men and women more or less on their own.

As we see from combat regulations, soldiers were instructed to actively wait for or effect their own evacuation. Both postwar and wartime medical literature points to the fact that soldiers were often expected to walk to medical units on their own.[160] Medics provided basic medical treatment (i.e., stopped bleeding, made splints, etc.) and helped nonambulatory patients evacuate. But medics suffered high casualties and were often too few to provide effective services.[161] Once soldiers arrived at either a medical point, the first relatively safe place where the wounded could rest, or a Medsanbat (Medical Sanitary Battalion), they received further treatment. At the Medsanbat, the decision was made to treat soldiers within their own unit (light shrapnel and bullet wounds, burns, and minor frostbite) or to evacuate them away from the unit and further to the rear. Medics also sorted soldiers by the character of their injuries, and a variety of specialized wards and hospitals was set up to deal with particular types of wounds at the front level.[162]

Regulations provided for an adequate system, but reality created a much messier situation. One senior medical officer pointed to a 250-bed hospital on the Kalinin Front that in March 1942 received 13,335 wounded and another that functioned at quadruple capacity for a month.[163] Services at such overcrowded hospitals were sparse, to say the least. Boris Komskii, wounded at the Battle of Kursk, complained to his diary that medical services were "disgracefully disorganized." His hospital was "an empty hut with broken windows," where soldiers slept "on the floor, four men on two mattresses," which they, all suffering from arm wounds, had to stuff themselves. There was nothing to do, nothing to read, and meals were poorly organized, making his "soul ache."[164]

Experienced medical personnel were often in short supply, as were all resources. Creative doctors and nurses learned to adapt to these conditions, but exhausted and inexperienced personnel were pushed to the limits during heavy fighting.[165] Under these conditions, hospital personnel often failed to keep proper records on both the living and the dead, neither notifying the next of kin nor providing a proper burial.[166] Even worse, in the massive retreats of 1941 and 1942 the wounded were often abandoned by retreating comrades, as Aleksei Shtin reflected while convalescing: "In general the situation in the first days was such that if you were wounded and couldn't walk, that meant you were dead."[167] Conversely, when the army advanced rapidly, it could outrun its hospitals, leaving wounded to fend for themselves for extended periods.[168]

Wounded soldiers were at the mercy of the situation at the front, the level of organization in their unit, and the skills of those providing treatment. The Medsanbat of the 322nd Rifle Division could boast of ideal organization in 1944, with soldiers immediately being washed and shaved, their hair cut, their clothing disinfected, and the nature of their wounds recorded. Men and women were even separated to prevent the spread of venereal disease.[169] However, not all units were this well organized. Some hospitals evacuated soldiers too severely injured to safely transport—for example, with a shattered jaw, crushed trachea, or gut shot—worsening their agony. Others needing only minor treatment were sent to rear-area hospitals, depriving the severely afflicted of their place in overcrowded wards.[170] Last but not least, wounded soldiers were often left to move from one hospital to another without any transport or food, sometimes while still under enemy fire.[171] Steppe Front war correspondent Rostkov wrote to party officials in the fall of 1943 complaining that passing trucks refused to take wounded soldiers and that commanders and medical personnel frequently had no idea where to direct groups of wounded soldiers. This lack of attention to the needs of the wounded prompted soldiers to complain: "We are so needed in battle, but as soon as we are wounded we aren't needed by anyone."[172] Salvageable bodies were much more valuable than those that the war had used up, and the Red Army medical system was geared to get soldiers back into combat as quickly as possible.

At the end of the war it was claimed that 77.5 percent of the wounded were returned to the ranks, and during the war wounded soldiers were reminded that "the vast majority of soldiers return to the front, so it is not suitable to get accustomed to the gentle atmosphere [*teplichnaia obstanovka*]

FIGURE 1.4 Senior Lieutenant A. I. Chadaev and Sergeant S. Mamedov, wounded while crossing the Dnepr, 1943. They have fashioned slings from their belts. RGAKFD 0-65230.

of the hospital."[173] The lightly wounded were left in the Medsanbat so that they could return to their unit as quickly as possible. This practice led some soldiers who were badly wounded to stay in the Medsanbat rather than evacuate, so that they could remain with their comrades and maintain their status.[174] Some units reported that the Medsanbat was their primary source for replacement manpower.[175]

While recuperating in hospitals, returning to their old unit, or making a place for themselves in a new unit, soldiers learned of the death of their comrades and had time to realize the dual cycles that governed the use of their bodies.[176] The first was the attrition of military units, a constant cycle of filling the ranks, training, and losses at the front until a unit was taken off the line to rebuild. The second was the soldier's individual path through this system, from training to the front to being wounded and filtered through various hospitals until returning to the front via either a marching company or a recovery team, special units for soldiers returning from hospitals.[177] This second cycle led to soldiers constantly establishing links with new comrades: in trains, wards, and on the march, and soldiers were encouraged to share their experiences and learn from each other while traveling and recovering.[178] This cycle would continue until a soldier was killed, captured, or crippled or the war ended.

The Fragile Little Cog

The goals of the government and soldier both overlapped and contradicted each other. Most soldiers wanted to defeat the Nazis, but they also wanted to survive. The state was prepared to pay a massive cost in the lives of its citizens in order to endure. This fostered the fatalism common among soldiers that David Samoilov ascribed to the peasant background of most of those in ranks, epitomized by the folk saying: "Don't volunteer for anything, and don't refuse anything."[179] Both violent coercion and peer pressure to do one's fair share weighed on soldiers as their bodies were offered in exchange for territory, enemy machinery, and the lives of enemy soldiers. To accomplish all of this, the government invested commanders with a power over soldiers that verged on ownership, declaring commanders to be the face of the state, even if this power could be highly conditional.

In return for everything that the government demanded from soldiers, it would feed and clothe them better than average citizens, arm them, show genuine concern for the conditions in which they lived, and invest the soldier's every action with meaning. In the following chapters, we will see how Dzhuma, whose shivering, naked body opened this chapter, and millions of his comrades were clothed, fed, armed, sheltered, entertained, and later given license to take enemy property. Before these men and women would preserve the Soviet Union and defeat the Third Reich, the government took possession of their bodies. The state and soldier would negotiate their new relationship via objects in the course of the war. These soldiers would also be forced to come to terms with each other on an intimate, everyday level in which their lives were in constant danger. The common ground of such diverse people was the sparse assortment of army issue objects and practices surrounding their use. Their experiences would significantly alter what it meant to be "Soviet."

Chapter 2

A Personal Banner

Life in Red Army Uniform

> Fellow, you have the whole history of your life at the front on your chest.
>
> —Vasilii Subbotin

In the summer and fall of 1941, hundreds of thousands of Red Army soldiers and commanders surrendered to the enemy or abandoned their uniforms while fleeing the battlefield. On August 16, 1941, Stalin issued Order No. 0270, in which he posed the question: "Can the Red Army tolerate cowards who desert to the enemy and surrendering or cowardly leaders, who rip off their rank and desert to the rear at the first sign of danger? No, we can't! If we let these cowards and deserters have their way, they will very quickly demoralize the army and destroy our Motherland."[1] The army was on the verge of disintegrating, but many soldiers continued to believe in their duty to fight on. Kharis Iakupov, escaping encirclement in 1941, recalled: "There was only one thing that was considered reprehensible and punished: when someone among the soldiers stole through to their own having changed into civilian clothes. For us, the uniform was like a banner, and to abandon it was deemed to be an act as shameful as any display of cowardice."[2] Iakupov's memoirs echoed Order 0270 and the image of uniform as a banner that had been developed in the Soviet military press during the war. This rhetoric was closely associated with a major change in style in 1943, when the army abandoned the uniform that so many soldiers had themselves cast-off or worn into German captivity, introducing a "new-old" uniform that reproduced late tsarist styles.[3]

Pogony (shoulder boards), a key old regime symbol, were recast as a point of pride: "Having put on pogony, Soviet soldiers and officers will carry them

FIGURE 2.1 Soldiers M. V. Kantariia and M. A. Egorov with the banner of the 150th Order of Kutuzov Rifle Division, which was lifted over the Reichstag in Berlin, 1945. RGAKFD 0-291381 (V. P. Grebnev).

through the fire of battles as their own small, personal banners, which the Motherland solemnly gave them."[4] Along with pogony, a variety of decorations, many of which referenced old regime heroes, formed a rich text that was readable by both soldiers and civilians. As one correspondent noted in a profile of a heroic soldier: "you have the whole history of your life at the front on your chest."[5] In contrast to the uniforms worn by other belligerents, the Red Army did not wear distinctive insignia for individual units. The Red Army uniform truly was a personal banner, showcasing the wearer's accomplishments.[6]

What follows is an ethnographic sketch of life in uniform in the Red Army. The reader will progress layer by layer from underwear to overcoat. In so doing, we will have a chance to explore often overlooked details of life in the Red Army, including the changing symbolism employed by the Soviet regime and soldiers' reactions to it.

The clothing, heraldry, arms, and armor of ancient Roman soldiers and gladiators, medieval knights, Sioux warriors, and modern gang members are all readable texts that speak to the practical concerns of the roles their wearers fulfill and showcase the identity ascribed to and claimed by their wearer.[7] Uniforms act as a "certificate of legitimacy," guaranteeing that the bearer will submit to the rules and hierarchy of the institution that issued it.[8] The proper wearing of uniforms, keeping them clean and in repair, is often more important than the articles themselves and includes the mastery of a particular etiquette.[9] While uniforms are supposed to subsume identity, they are also separate from it. In saluting a commander, regulations demanded that soldiers acknowledge rank but not the name of the commander, as the soldier was saluting the commander's uniform, and by extension the polity that the uniform represented.[10]

There were no "officers" in the Red Army in 1941. Formed as a revolutionary army, drawn from a society that had been purged of "exploiting" classes, authority centered on the role one fulfilled (e.g., *kombrig*—brigade commander), rather than rank as a status in and of itself.[11] Titles such as lieutenant and major returned to the military's lexicon on the eve of the war. In 1941, one could still see traces of the Civil War uniform echoed in soldiers' headgear and insignia. As the war progressed, uniforms became at first more practical, then displayed an entirely different symbolism.

Proper attire could be hard to come by, as shortages haunted the Red Army throughout the conflict. Warehouses that had been concentrated in the western Soviet Union fell into German hands at the beginning of the war just as millions of new cadres had to be outfitted. The army was forced to make items last, replacing them less frequently. This led to criminal responsibility for the mistreatment of equipment, an eventual fine of 250 percent of the value of anything a soldier ruined or lost, and major efforts to keep items in service as long as possible.[12] The most salient example of this was the recycling of uniforms and equipment, taking from the rear to give to the front and from the dead to give to the living. Various orders circulated demanding that scarce items such as helmets and overcoats be taken from rear-area soldiers to give to frontline troops.[13] The dead were stripped of virtually everything useful: all equipment, shoes, belts, and overcoats, leaving only the uniform and underwear on the body.[14] Rear-area soldiers wore secondhand clothing, given new clothes only when sent to the front.[15] It was usually impossible to tell anything about the previous owner, but occasionally an indelible imprint was left on the inherited garment, such as coarsely sewn repairs of overcoats rent by shrapnel.[16] Many garments outlived their wearers.

Soldiers on campaign traveled primarily on foot and spent most of their time outside or in structures that they built themselves with spades. Once soldiers were cold and wet, they might stay that way for days or weeks.[17] They had limited access to water, and soap was issued at the rate of 200 grams per month for men or 300 grams for women.[18] Opportunities to undress were rare.[19] Naturally, soldiers were often filthy. Nikolai Chekhovich wrote to his mother: "Today was the first time for a whole month that I could scrape the mud off my clothing."[20] Under these conditions, the uniform became an outer layer of the soldiers' bodies, a veritable second skin.

Underwear and Fellow Travelers

Underwear in the Red Army circulated even more frequently than other forms of clothing, as it was the most likely to be washed and changed, soldiers carrying a spare set in their packs.[21] Underwear had a particular place in the Soviet project as part of its modernizing and civilizing mission. Underwear, soap, and bathhouses had been central objects to raise the cultural level of workers and peasants via improved hygiene in the 1920s.[22] It was rare that anyone in the army had a chance to change their underwear. Some soldiers ritualistically changed theirs before battle, for good luck; others believed that changing underwear would bring bad luck.[23] Regular changing of underwear happened when soldiers went to a bathhouse, which could be as often as once every ten days or as rare as every couple months.[24] Predictably, lice and various skin ailments were a constant problem, and underwear tended to be dingy. One veteran mused: "Our shirts and drawers had become encrusted with salt, becoming gray-yellow. Our unwashed bodies stung from the bites of parasites."[25] Killing lice was a common pastime, as Lesin wrote, "If one asked what our company is doing, we would answer in two words: crushing lice."[26] Other soldiers recalled the lengthy process of popping lice with their teeth and nails or freezing them in the snow.[27] Some old soldiers believed that lice emerged due to homesickness.[28] In particularly desperate situations, lice could go from the personal banner to the division's, as Boris Suris wrote in his diary, "We managed to save the division's banner (the *starshina* of the Commandant's Company carried it out on his back, and it got lice)."[29] References to lice were censored from letters and newsprint.[30] Lice could not be ignored at the front, and battles with lice took place primarily in bathhouses. These could be set up by the soldiers themselves or be traveling sanitary stations, where soldiers would have their clothing disinfected and deloused in special ovens.[31] While they would get their uniforms back, chances were that they would get a different set of underwear.[32] When soldiers did get a chance to undress, it could be a moment of revelation as to how much their bodies had changed. Chekhovich wrote his mother about how tan his face and hands had become, while Irina Dunaevskaia recalled the shock she felt when confronted with how much weight she had lost.[33]

Except for sailors, who had striped shirts of which they were immensely proud, the underwear issued to soldiers came in one of three types: flannel boxers, long white cotton drawers with a long-sleeved undershirt from May to October, or white flannel drawers with matching shirt from October to May. Ties were often substituted for buttons for the sake of speed of manufacture. For men in the ranks, this would be familiar and utterly unremarkable clothing. However, for female soldiers these garments could be vexing, as one veteran recalled: "We almost burst into tears, when they gave us those long johns."[34] Aesthetics aside, women were not provided with any specialized underwear such as bras and had to make do on their own, often from scrounged materials such as bandages.[35] The army did not provide anything like sanitary napkins either. Women were left to informal measures to decide this problem themselves, as one sniper told the Mints Commission: "It's very difficult during menstruation. There are no bandages and nowhere to wash up. The girls told the divisional Komsomol organization about this, and they told the medic to give as much bandages and wadding as we needed." Others used the soldier's personal first aid packet.[36] Improvisation was a way of life in the army.

Underwear often became the victim of the regime's shoestring budget, as soldiers traded clothes for food, leading to a shortage of underwear at the front and increased surveillance on the road to the front.[37] Lacking a change of underwear meant disinfection was impossible.[38] At least one soldier remembered how he and his comrades purposefully allowed lousy clothing to burn in order to receive fresh replacements, a practice occasionally endorsed by high-ranking commanders.[39] Sometimes soldiers simply forgot to remove flammable items, such as lighters and grenades, from their pockets, emerging from the bathhouse to find the ashes of their clothing.[40] Underwear, like the bodies of soldiers, was de jure the property of the state, but de facto under the control of the soldier.

Bottoms

Pants

Pants in the Red Army were of the jodhpur style and underwent less change than most other parts of the uniform in the course of the war. While at the war's beginning it would be typical for commanders to have significantly higher-quality blue trousers and for everyone to be issued wool trousers in the winter and cotton in the summer, these practices largely gave way to more practical concerns. Rank-and-file soldiers received trousers with knee reinforcements. The army provided cotton pants in the summer and cotton padded pants in the winter. The increasingly diverse cadres of the army made sizing a major issue: one female soldier recalled that the padded trousers she was given fit "like a jumpsuit" and came up to her armpits. While men's clothing was a necessity for female medics and snipers crawling across no man's land, the desire to maintain a semblance of femininity was very strong; it was not uncommon for women to receive skirts or dresses for official functions and holidays or even to craft their own.[41]

Portianki and Boots

Red Army soldiers didn't wear socks. Instead, they wore an ancient article from the peasant wardrobe. *Portianki*, or foot wraps, which came in winter and summer weights, were simply rectangular strips of cloth held in place by tension. While familiar to most peasants, they were alien to many urbanites: "For a long time the biggest problem remained portianki. If there happened to be even one wrinkle, your feet could be rubbed to blood, and commanders instilled in us that it was a crime to have blistered feet in the army."[42] Foot wraps were quite comfortable when properly wrapped, did not wear out at the toes and heel like socks, and could be turned around to a dry corner for multiple uses.[43] A highly economical and traditional item, portianki were an exemplar of the Red Army's mastery of doing more with less. The feet of

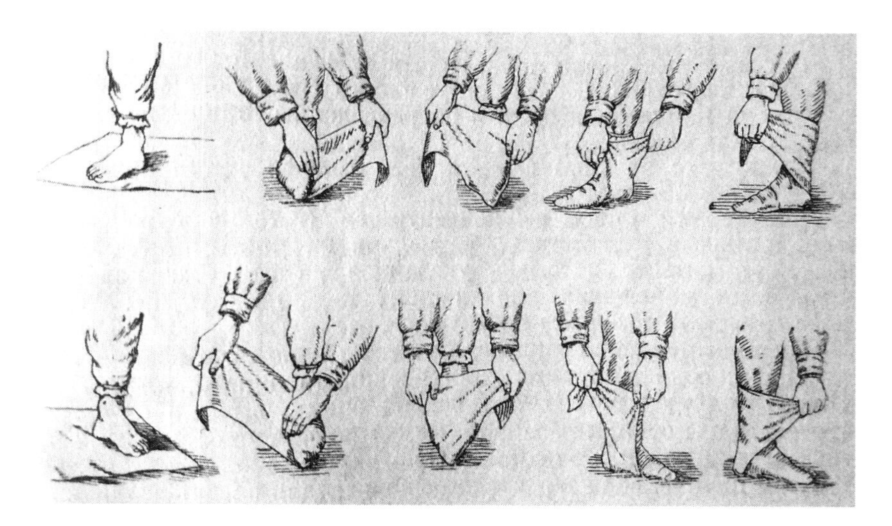

FIGURE 2.2 *Portianki* from a Red Army manual. *Posobie komandiru i boitsu strelkovogo otdeleniia* (Moscow: Voenizdat, 1943), 123.

soldiers who operated state-of-the-art tanks and planes were wrapped in the rags found in the bast shoes of medieval peasants.

Once portianki were wrapped around the feet, one had to slip them quickly into whatever footwear they had been issued. These could be low boots with puttees (*obmotki*), which replaced jackboots as a stopgap measure.[44] Some soldiers especially liked American-made boots received through Lend-Lease, though in the postwar period praising them became taboo.[45] Puttees, or "three-meter bootlegs," so-called because they were made of long strips of cloth, had a tendency to come undone at inopportune moments and seem to have been generally unpopular. Jackboots were not without their problems: "the top of the boot was made of artificial leather [*kirza*], the sole of rubber, and the nose of low-quality leather; even when oiled with blubber they would get wet in a hard rain."[46] *Kirza* (*kozhzamenitel' Kirovskogo zavoda*) was first used in 1936–1937 and authorized to replace leather goods in 1940, as a way of stretching resources for the expanding army.[47] Some soldiers tried to use captured German boots, but many complained that they were too tight at the ankle and too wide at the top, one soldier recalling that he had to cut himself out of a pair of captured German boots and retrieve his own worn-out boots.[48] Some soldiers sported boots made of green cotton tarpaulin for summer months.[49] Commanders were authorized to wear "chrome boots" of shiny leather. Finally, for extreme cold, Red Army personnel were issued *valenki*, traditional peasant boiled-felt boots.

Seasonal change was often the worst time for soldiers. As winter turned to spring, felt boots that were ideal for extreme cold became sponges absorbing frigid water, leading to increased cases of frostbite.[50] As fall turned to winter, people froze in thin *kirza* boots. The situation in spring and fall frequently aroused the scorn of inspectors, as they observed barefoot soldiers or soldiers wearing rubber antichemical stockings in the absence of adequate footwear.[51]

There was no single piece of the uniform that was more important from a practical perspective than boots. Not having proper footwear could lead to unimaginable suffering when one was constantly exposed to the elements and walked everywhere. Boots too large, too small, or too worn out led to suffering. Chekhovich wrote his mother about the declining fortunes of his boots, which "held, held, and then started leaking" despite "nearly daily repairs," until finally he lamented "[I] really rubbed my feet raw—till they were bloody . . . but kept going."[52] The army was deeply concerned with reserves of boots and their repair, dedicating considerable attention to them.[53] Boots were supposed to be cleaned to a bright shine at all times.[54] Boots became a significant part of the military's civilizing process, for they were the article most quick to get dirty, the hardest to keep clean, but also the most practically significant. As the emphasis on the soldier's appearance increased, boots were expected to be cleaned and shined, especially when soldiers were interacting with civilians.[55] Boots were constantly an issue, right up to the end of the war, and often replaced under combat conditions. One soldier, on receiving a new pair of boots in Berlin, autographed his old ones adding "got to Berlin, 1945" and threw them into a tree as a record of his achievement.[56] Like everything else a soldier wore, boots were ubiquitous yet personal. Boots were the part of one's uniform most likely to cause pain and the first thing taken off the dead to give the living.[57]

Tunics, the Personal Banner

Government Symbolism and the Tunic

Unlike boots, tunics were not recycled from the dead to give to the living. A tunic (*gimnastërka*) was expected to be worn by a soldier for six months, being replaced with the winter or summer issue of clothing, and was often salty from sweat and faded by the sun.[58] Tunics underwent the greatest amount of change during the war. As the primary uniform item, the tunic was the main text of a soldier's biography. It was the garment that carried insignia and to which all medals and orders were affixed, serving as a soldier's personal banner. The tunics themselves were among the simplest uniforms issued by any army to its combatants—simple pullover shirts made of cotton or wool, with elbow patches for enlisted men.

There was a specific technique to wearing the tunic. A belt was required to give the loose fitting shirt a smart appearance, and soldiers folded the garment in the back so as not to create unattractive wrinkles in the front that could cause blisters. Not a single button or hook could be undone without the commander's permission. A fresh, white collar liner (*podvorotnik*) was to be sewn in every day. Without it a soldier was out of uniform.[59]

The shortages brought about by the war in both men and material made being stylish difficult, because the army was forced to simplify uniforms, including the authorization of nonuniform buttons (often simply stamped steel) in November 1941.[60] A report from September 1941 lamented that uniforms were usually incorrectly sized, shrank one size after the first washing, and poorly fitted the growing number of older soldiers in the ranks.[61] As the war continued, uniforms would be recycled from the wounded, and soldiers in training wore exclusively secondhand clothes, fresh tunics being reserved for those at, or en route to, the front.[62]

FIGURE 2.3 Guards Senior Sergeant V. I. Panfilova, Eighth Guards "Panfilov" Rifle Division (named in honor of her father, I. V. Panfilov), 1942. RGAKFD 0-286566 (V. P. Grebnev).

In 1941 tunics still retained traces of the Civil War, bearing the insignia for branches of service (infantry, artillery, cavalry, etc.) devised during that conflict. Two chest pockets were to hold the soldier's documents, a first aid packet, and assorted sundries. Rank, in the form of red enameled geometric shapes, and branch of service were worn on the folding collar in vivid colors—raspberry red with black piping for infantry, medium blue with black piping for cavalry, black with red piping for artillery, and so on. Brass pips identified branch of service—a tank for armored troops, crossed shovels for engineers, and the like—making soldiers easily distinguishable. Commanders had additional piping on the cuff in their branch of service color and chevrons on their sleeves; political officers sported a gold-braided Red Star with hammer and sickle on their sleeve.

While quite attractive, in the immediate aftermath of the Winter War these insignia were found to be too conspicuous on the battlefield and were replaced in August 1941 by "defense collar tabs" of olive drab with rank and insignia in muted green.[63] The flashy insignia continued to be popular. V. E. Ardov wrote to Stalin explaining that commanders saw the colorful insignia as a privilege that "inspires people, arousing respect for them from the population, [and] decorates the difficulty and danger of battle service."[64] The desire to be stylish could overtake other concerns, including safety. Stalin underlined "insignia" in Ardov's letter and a dramatic change in soldiers' appearance would mark the Red Army's shift in fortune and turn westward.

Pogony and the "New-Old" Uniform

On January 6, 1943, less than a month before the German surrender at Stalingrad, the Red Army received an order fundamentally changing its uniform. The new uniform featured a standing collar with two buttons. Pockets would no longer be seen on soldiers' tunics and would be inset on officers' and sergeants' uniforms. This new model looked even more like a traditional peasant's tunic than the early war uniform, but this was not its primary association. The "new-old" uniform would be familiar to Soviet citizens, as it was a return to late tsarist uniforms, including the previously hated shoulder boards.[65] The return of pogony was announced at a moment when victory at Stalingrad was inevitable. However, the decision to introduce the new uniform had been made in October 1942, a period that General Vasilii Chuikov, commanding the Sixty-second Army at Stalingrad, described as "the most horrible period of the enemy's assault."[66] Their introduction diverted resources from civilian clothing and underwear production, which were already in short supply.[67] Since the outbreak of the war the army had been simplifying all of its equipment in order to stretch resources. Clearly, this change in symbolism took precedence over materialistically pragmatic concerns. Pogony would mark a new relationship to authority and the Russian past and become a symbol of a successful, reborn Red Army.

There had been several previous attempts to give the army a makeover. In 1940 the state looked to create more gallant peacetime uniforms (many of which echoed those of 1812), and in 1941 a distinctive uniform for elite Guards units was considered.[68] According to David Ortenberg, the editor in chief of *Krasnaia zvezda* for most of the war, Stalin vacillated on whether or not "to bring back the old regime" by introducing pogony. Eventually he was convinced that the new-old uniform would help motivate soldiers, many of whom had no memory of the old order. Stalin demanded that journalists promote pogony as a symbols of order and discipline, declaring that "we are the inheritors of Russian military glory." Top Soviet officials apparently even scoured museums to find old pogony.[69] Rank would be indicated by the number of stripes for soldiers and the size and number of stars for officers. The branch of service symbols moved from collar to shoulder board with trim around the pogony in the prewar color scheme. The new pogony would come in everyday (peacetime and rear-area) and field (combat) versions, the former of which would even revive golden pogony for officers!

A *Krasnaia zvevda* article from January 1943 explained the timing of the soldiers' makeover in terms both pragmatic and historical, dismissing the idea that "it would not seem to be the time to become preoccupied with the soldier's appearance": "The introduction of pogony, which clearly express the subordination of juniors to seniors in service, strengthening the authority of leaders, has a principal and important meaning . . . And the fact that they have appeared on the shoulders of Soviet warriors at this moment, at the climax of the struggle, makes them a doubly honorable sign, forever linked with the legendary battle for the honor and independence of the beloved fatherland."[70] Pogony were markers of both increased military effectiveness and historic change. They also happened to be one of the preeminent symbols of the old regime. Boris Kolonitskii, an expert on symbols in the Russian Revolution, has pointed out that in 1917 the degree of antipathy toward pogony was a reliable indicator of a unit's "degree of revolutionization." It was not uncommon to demand the removal of pogony in the early phases of the Revolution or to drive nails through the pogony of captured White officers during the Civil War. Admiral Aleksandr Kolchak, one of the leaders of the White movement, explained to his Bolshevik captors shortly before his execution that he saw pogony as a "purely Russian form of insignia." He also saw them as a sign of discipline and a guarantor of performance by the army, which "when it was in pogony, fought, and when the army changed its spirit, when it took off pogony, this was connected to a period of the greatest disintegration and shame."[71] Ironically, the Main Political Directorate of the Red Army (GlavPURKKA) and many Soviet commanders would come to agree with Kolchak a few decades after the Bolsheviks shot him. General Chuikov told a historian in Stalingrad that "The factor of ambition [*chestoliubie*] remains . . . the title 'Guards' and similar things, the titles, given to our heroes, pogony—you think that Stalin isn't taking this into account?"[72]

The day after the return of pogony was announced, Aleksandr Krivitskii published a lengthy article in *Krasnaia zvevda* concerning the sea change in the regime's symbolism. After a lengthy history of uniforms in Europe and Russia, Krivitskii explained that during the Civil War the Red Army refused the uniform that its enemy wore, but now, having matured, it could wear the "signs of military dignity" that decorated "the uniform of the Russian Army in 1812" all the way to World War I.[73] In one of the few references to the Civil War in propaganda about pogony, Krivitskii dismissed what had been a fundamental difference between Whites and Reds as "water under the bridge." Instead, he emphasized the Red Army as the descendant of *Russian* military traditions in a way consistent with propaganda centered on the Soviet edition of romantic Russian nationalism, continuing a trend started in the 1930s.[74] The preeminent event in that tradition, the Patriotic War of 1812 in which Russian forces defeated Napoleon, is invoked, with pogony providing a material link to history that was repeating itself. Propaganda portrayed Red Army soldiers as the descendants of traditional Russian heroes such as Aleksandr Nevskii, Aleksandr Suvorov, and Mikhail Kutuzov, while occasionally paying heed to (also long deceased) Civil War figures such as Vasilii Chapaev, Nikolai Shchors, and Mikhail Frunze. The Soviet Union was no longer portraying itself primarily as a young regime oriented toward world revolution but rather as something like a nation with ancient roots that had its potential unchained by the Revolution. At the same moment that the soldier's uniform became more national in form, the motto on military banners changed from "Workers of the World Unite!" to "For Our Soviet Motherland!"[75]

As the Soviet state began to resemble a nation-state more closely, its army followed suit. The authority of commanders was strengthened. Dual command, the practice by which orders given by a commander needed to be cosigned by commissars, was repealed, giving commanders full authority. The term "officer" was reintroduced with a corresponding emphasis on hierarchy. Before 1943 the army was divided into privates, junior commanders (which included both sergeants and lieutenants), supervisory cadres (*nachal'stvuiushchii sostav*) and command staff.[76] Beginning in 1943 the army was divided into privates, sergeants, officers, and generals.[77] More than a simple change in phrasing, classification had become more rigid and new divisions were made between classes of cadres.

The year 1943 is often cited as the time by which the Red Army had learned to fight.[78] The cadres who came into the army in 1941–1942 were often poorly trained because crises at the front led to minimal training. Many soldiers learned how to fight at the front. By 1943 collective experience had accumulated, and the situation at the front had stabilized to the point where soldiers could receive more training and increasingly impressive weapons, a process detailed in chapter 5. Liberated civilians and German prisoners noticed the new-old uniform, as General Andrei Mishchenko noted in February 1943, the entrance of troops wearing the new uniform made an extraordinary impression. German prisoners exclaimed: "this isn't the Red Army that we beat in 1941. Now that army is no more—now the tsarist Russian army has taken form."[79] The new uniforms were the mark of a new army, one that was mastering its trade and would be advancing west rather than retreating east.

Soldiers were permitted to wear out their old uniforms with attached pogony until they were issued new tunics, and a variety of hybrid uniforms appear in photographs. (See figures I.1, 4.3, 4.19, and 5.4.) [80] The issuing of pogony was supposed to be accompanied by an improvised ceremony, which served as a reaffirmation of the soldier's responsibilities before the state and became an unforgettable moment in the soldier's life. Aleksandr Lesin recorded that he was given his pogony at the Kalinin Front "not exactly in a grand manner, but not without words of encouragement."[81]

With the new uniforms came new responsibilities that went beyond rhetoric. The term "the honor of your uniform" (*chest' mundira*), which before had been associated with tsarist officers, entered Soviet rhetoric. At the front troops honored their uniform with daring deeds, and in the rear by being a model citizen.[82] Wearing pogony "any little thing" that a soldier did was "of serious importance." Soldiers' behavior and appearance was subjected to intense scrutiny.[83] This was more than simply lip service, as memoirs and diaries record how soldiers were harassed and arrested for improper appearances or had their pay docked for swearing from 1943 on.[84] The growing cult of the uniform was one of the most publicly recognizable aspects of the campaign to professionalize the Red Army and create a separate caste of officers with corresponding hierarchical relations.

Improving the appearance and bearing of officers was part of the perpetual Soviet concern with "raising the cultural level" of what was still an overwhelmingly peasant society. Officers were permitted to wear a special uniform (*kitel'*) similar to European uniforms (they buttoned all the way down the front) and were encouraged to order clothing from tailors.[85] Official propaganda stated that in the old army: "The authority of a Russian officer was based on his higher cultural level . . . In our Red Army officers are required to be of an even higher cultural level."[86] Soldiers and officers alike complained about the low cultural level of the army's commanders. Some officers took it on themselves to write tracts reminiscent of nineteenth-century etiquette manuals, including how best to blow one's nose, chew food, use knives and forks, drink, spend money, and interact with the wives of one's superiors. These books so enraged General Aleksandr Shcherbakov, head of the Red Army's Main Political Directorate, that he forbade their publication, stating these texts "interpret in a perverted and philistine spirit questions of morality and etiquette, soldierly education, [and] military honor and often propagandize views alien to the Red Army." Shcherbakov reminded his subordinates that the purpose of discipline was to fulfill goals of the military, not to recreate the officer corps of the old regime.[87] Confusion over the meaning of pogony was not limited to the authors of ill-fated etiquette manuals, as many soldiers themselves were at a loss in their attempts to understand these new "small, personal banners."

Soldiers React to Pogony: The Empire's New Clothes

"They are introducing pogony. We are perplexed by it all [*My vse nedoumevaem*]," Boris Suris laconically wrote in his diary in January 1943.[88] Suris was far from alone in not knowing what to make of the "new-old" form. Even top officials felt awkward at the sight of pogony in January–February 1943.[89] Pogony soon became ubiquitous, and everyone adjusted to them, but initial responses varied.

Guards Lieutenant Rafgat Akhtiamov, a Tatar from a remote village, was very enthusiastic about the new uniform, stating "I have such clothing now as no one in the village ever dreamed about" in one letter and "the government has dressed us very well, I only want to go visiting, that's how pretty the new uniform is" in the next.[90] Akhtiamov, whose father had served in the Russian Imperial Army, betrayed no sense of the political in his letters home and seemed genuinely enthusiastic about the new uniform from an aesthetic point of view. As the representative of a national minority, he could have had reason to interpret this shift as a return to tsarist norms that marked his people as second-class citizens, but he reveals no sense of worry. Lev Slëzkin, a Muscovite, remembered that soldiers were at first indifferent to the introduction of pogony, but that once they received them "many were quite satisfied . . . not so much due to the restoration of an external link with a tradition that had disappeared for a quarter-century as that pogony gave the wearers a certain swagger and bellicosity."[91] The ranking political officer of the Seventy-ninth Guards Rifle Division told the Mints Commission that soldiers enthusiastically awaited pogony, a sea change from 1918–1919: "Even before they got their pogony, Red Army men sewed straps for pogony on their uniforms. Everyone couldn't wait to get their pogony. Some joked that they felt like plucked chickens without them."[92] Pogony made soldiers' shoulders look broader, creating a more masculine appearance, but not everyone perceived them as simply attractive elements of their uniform.

Many soldiers saw the return of pogony exactly as Red Army propaganda presented it. Senior Lt. Bogomolov lauded them as "a great thing for the strengthening of discipline," that gave "an even more martial countenance" and recalled "our forefathers." Another officer thought it simply impossible that soldiers could remove such a uniform and dress in civilian clothes, as had been common in 1941. For some, the fact that the government could afford to undertake such a clearly expensive and not apparently necessary campaign in the midst of war was a positive sign.[93] David Samoilov saw "the appeal to tradition" in general as a sign of "the entry of our government into the moment of maturity, the classical stage of development."[94] For many the national and traditional was either more attractive than, or in no way contradictory to, the ostensibly socialist society they were fighting for.

Others were confused by or even opposed to pogony. According to reports by NKVD agents, several saw pogony as signs of the bourgeois world, as something that the capitalist allies of the Soviet Union had forced on the socialist state. Others were simply disgusted by the return of the old, musing that old forms of address such as "Lord" (*gospodin*) and symbols like the double-headed eagle would reappear. Some took this as a sign that churches would be reopened, old hierarchies would reappear, and the Soviet Union would become a capitalist country. One soldier even detected a hint of fascism: "the fascists wear pogony. Soon they will clip on pogony and we will be eternal soldiers." Others bemoaned the low cultural level of Soviet officers as compared to tsarist cadres, "the old officer of the Russian army was a most cultured person, and ours—simply a disgrace [*sramota*]."[95] A sense of inadequacy before the heroic past overtook a sense of inferiority before enemies for some, while others still found the taste of the old regime unpalatable. Some were disturbed by the reintroduction of pogony as a sign of growing nationalism, as Russian national tropes were increasingly becoming a standard of propaganda.[96] In the end, regardless of their attitudes toward pogony, as soldiers they had no choice but to wear them.

Despite new uniforms and rhetoric, soldiers continued to live in filthy conditions and wear their clothes for extended periods of time. While recovering from a wound in August 1943, Boris Suris reflected in his diary about the lack of respect shown by soldiers to officers. While there was much talk of "the honor of an officer, of the dignity of the commander," it was useless "when so many of our commanders go around in low boots with puttees or boots that have been torn to shreds and patched up a thousand times, in salty and sun-faded uniforms."[97] Even the new uniform would fade and become covered with mud.

The dramatic change in the army's fashion caught a variety of people—those on occupied territory, former White Guards, and Soviet laborers deported to work in the Third Reich—unaware. Liberated peoples did not always recognize Red Army troops, some even asking, "Are you Soviets or Germans?"[98] The confusion caused by pogony did not end at the moment of liberation. Some recently liberated civilians asked whether there was "a new aristocracy" consisting of generals, and if "after the introduction of pogony there are no longer Communists or Komsomol members in the USSR?"[99] While these men and women had been under occupation, a return

FIGURE 2.4 Guards Senior Sergeant Bato Damcheev, most likely in the winter or spring of 1945, www.ww2incolor.com/soviet-union/soviet_1.html.

to tradition that had begun before the war came to fruition, leading to a new banner, new uniforms, and eventually, in 1944, a new anthem. The new "Anthem of the Soviet Union" spoke of "Great Rus'" rather than the *Internationale*'s "whole world of the starving and enslaved."[100] Even some Russian émigrés who had fled the Revolution gazed in wonder at the Red Army's pogony.[101]

A new understanding of what it meant to be Soviet was emerging, one with a new relationship with the past and a clear ethnic hierarchy in which Russian was assumed to be not only the lingua franca but also the common culture, similar to German in the Hapsburg Empire. New forms of social differentiation also came to the fore in the new officer class, elite Guards units and highly decorated soldiers. The collections of decorations a soldier earned became the main text of their personal banner.

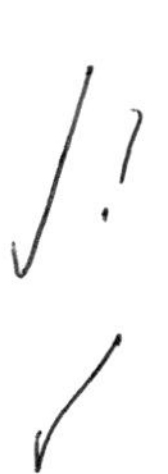

Decorations: The Emergence of New Icons

As a heroic past was increasingly made relevant and manifest, Red Army soldiers were encouraged to become heroes, modeling themselves on historical figures and repeating contemporary feats. The military press was filled with accounts of miraculous acts of heroism, much as the prewar press had lauded the exploits of shock workers, border guards, and polar explorers, and the Red Army used decorations as didactic tools to encourage bravery.[102] Both the actions of soldiers who had earned various decorations and the decorations themselves were to serve as means to raise the consciousness of soldiers: "We must teach others using the example of heroes; we need to tell the fighters more often about the best people, selflessly struggling for the Motherland, and drive these stories into the hearts of Red Army men."[103] This obsession with inspiration led Soviet leadership to research the tsarist system of awards. The military press later ran articles about the history of military decorations in Russia, going all the way back to medieval chronicles.[104] Alongside these articles, lists of those awarded with various decorations and stories of how soldiers received them were standard reading throughout the war. Inspiring soldiers to earn medals was one of the ways that the government tried to mitigate the lack of training and equipment it provided soldiers in the war's early days and eventually created a particular culture of decorations in the Red Army.

The Red Army's practice of wearing decorations at all times was unusual. US combat soldiers were famous for their apathy toward decorations. Several armies wore small ribbons in place of medals themselves and even those only on their dress uniforms, making their uniforms a text that few could read.[105] Although the Red Army officially went over to wearing just ribbons rather than medals in 1943, photographic and archaeological evidence shows that the wearing of medals was common until the final days of the war. In the Red Army decorations were part of the everyday uniform and a point of pride for soldiers.[106] People could be prosecuted for wearing medals they had not earned or for losing their medals, as German spies made extensive use of decorations in order to more convincingly pass themselves off as Red Army soldiers and officers.[107] Medals gave a soldier authority.

At the beginning of the war medals were quite rare. Decorations were most often issued on holidays and could take months or years to find their recipients.[108] As of September 29, 1942, 69,436 medals had not found their recipients, who had either been killed, wounded, or transferred to other units.[109] That month, Oskar Kurganov, a war correspondent for *Pravda*, noted that after waiting several months for a decoration "occasionally the person getting the award forgets for what exactly they were decorated, and those around the soldier don't know anything about it . . . It's clear that the power of such decorations to stimulate is highly insignificant."[110] As the war continued, the process of awarding medals was simplified and the government encouraged officers to award their subordinates more generously.[111]

By 1944 the army created standards for awarding some decorations simply for the number of years one had served, overwhelmingly benefiting career officers.[112] One Polish civilian observed in late 1944, "In no other army have I seen such a profusion of decorations. Every private had at least three, while the officers had at least a whole row, sometimes one on the left side and one on the right."[113] Decorations came in several genres and were awarded for everything from baking bread and surviving wounds to feats such as destroying enemy tanks or seizing enemy banners. There were a variety of badges, awarded before and during the war, that formed an ever expanding list. Paratroopers received a badge for learning how to jump. "Excellent" awards were given for professional competence. For example, one could earn the "Excellent Cook" badge for "systematically" doing such things as "quickly providing hot food and tea" and "using local resources to provide vitamins and greens," as well as properly camouflaging a field kitchen.[114] Within certain frontline communities, these decorations were highly prized, as they were given out for one's holistic talents and professionalism, rather than for a feat of heroism.[115]

The most common badge was issued to whole units and granted major privileges. This was the badge created to distinguish Guards units, elite military formations that had earned their status through combat exploits. The first Guards units were recognized in September 1941, and from May

1942 soldiers in Guards units earned double pay, Guards commanders received 150 percent of a commander's base pay, and all Guards personnel had access to better clothing, equipment, food, and weapons as well as the title "Guards" added to their rank. They also had the right to return to their unit after being wounded, enjoying a much more stable social world.[116] The ideological underpinnings of Guards units were not the Red Guards of the October Revolution, but the elite Guards Regiments of the Russian Imperial Army. Guards soldiers took pride in their privileged status, and Guards units received the most physically fit and politically mature replacements.[117] In a massive, largely anonymous army where cadres moved around constantly, the Guards badge gave soldiers a sense of corporate pride and belonging, as well as a means to make claims about their own worth.

Similar to Guards Badges, in that they were issued en masse, were campaign medals. These were received by every surviving participant in a battle: for example, for the Defense of Moscow (over 1,000,000 awarded) or the Defense of Leningrad (over 1,470,000 awarded). The most widely issued campaign medal, Victory over Germany, was issued to everyone in the army in 1945, over fifteen million people.[118] These medals could take months and sometimes years to reach soldiers and did not imply distinction on the battlefield, merely presence. Spotting a campaign medal could be a moment of fellowship, as two soldiers who served at the same battle swapped war stories and made newly arrived soldiers jealous. After the war, some veterans wore these medals on trips to the relevant cities to garner special treatment.[119] While Guards badges and campaign medals were about collective feats, most decorations noted personal accomplishment and sacrifice.

In July 1942 the Red Army introduced wound stripes to be awarded to all personnel who had been injured since the beginning of the war. Coming in two varieties—red for light (flesh) wounds or gold/yellow for serious wounds (broken bones or compromised arteries)—wound stripes served two purposes. They proved that people could struggle and survive despite injuries received at the front and celebrated these survivors as heroes to be emulated. Every wound had a story, and surviving the enemy's fire gave a soldier authority. In one propaganda piece a soldier uses his wound stripes to discuss his transformation from an enthusiastic but foolish greenhorn (saying of his first stripe "I don't respect this one") to an effective warrior respected by all.[120]

Several medals were issued in the Red Army for personal accomplishments, such as For Valor and For Combat Service. As the war continued and medals became more common, they seem to have become almost an expected accoutrement.[121] Medals were proof that people had done their duty, an individual mark that they had done something of note for the war effort, and something to write home about. Those without medals often felt a sense of shame. Lt. Vladimir Gel'fand confided to his diary in August 1944: "At war fiasco after fiasco. No orders. Not even a medal, at least for Stalingrad. I have no reputation . . . The earth crumbles beneath my feet."[122] Among the rank and file, the pressure to earn medals was perhaps less urgent, but still palpable. Mansur Abdulin recalled: "Here soldier psychology is simple: the war has been on for two years, and after Stalingrad it is somehow particularly shameful to not have at least one medal on your chest. 'What have you been doing there,' one wonders, 'if you haven't earned any decorations?!'"[123] This sentiment could be used to pressure soldiers directly. At the end of the war Lesin recorded an officer scolding a young lieutenant—"The war is ending, how are you going to remain without any orders?"—before sending him on a particularly dangerous assignment.[124] Interestingly, the culture of medals most similar to the Red Army's was that of the Third Reich, which also used medals as a stimulus that included peer pressure, created a complicated system of decorations, and called on its soldiers to wear their medals as part of their everyday uniform.[125] Soviet leadership researched both German and tsarist systems of decorations in late 1942, a period of expansion of both the system of decorations and the volume being issued.[126]

Medals proved you were a man, but orders made you a man among men (even if you were a woman). Orders were a higher form of decoration, reserved for the truly exceptional. Decorations existed in a hierarchy, and the medals that Red Army personnel received were to be worn in a manner that displayed this hierarchy with the highest decorations displayed most prominently.[127] Medals and chest badges were less significant than orders, which were worn above other decorations. Made of precious metals such as silver, gold, and platinum and studded with enamel or jewels, orders were an expensive item for the government to manufacture, yet some were made in mass quantities. The most commonly awarded order, the Order of the Red Star (cast of silver and enamel), was issued more than 2,860,000 times during the war.[128]

That the Soviet government was willing to expend such extensive resources on decorations speaks to the emphasis that they put on decorations as a stimulus. The statutes of these orders could be very specific and were published in the military press. For example, a soldier could earn the Order of the Patriotic War for such acts as destroying three to five enemy tanks or five enemy artillery batteries, being the first to enter and destroy the garrison of an enemy bunker that had pinned down an advance, and so on.[129] Many of the decorations conceived during the war were instituted in the spring and summer of 1942, when the Red Army was clearly losing.

The government continued to create new orders as the war continued, instituting a slew of them between 1942 and 1945. A major means of connecting the Red Army with the heroic Russian past, orders explicitly harkened back to the great men of Russian history, bearing the names of leaders such as Aleksandr Nevskii, Suvorov, Kutuzov, Pavel Nakhimov, Lenin, and even the Ukrainian hetman (Cossack military leader) Bogdan Khmel'nitskii. Each historic figure that had an order named after them also had a movie made about them. After being awarded an order, one became a "cavalier" of that order, something akin to an ancient knight. Orders also carried the concrete privileges of an *ordenonosets*—"order bearer." Every *ordenonosets* was entitled to a free yearly round trip by rail or boat to anywhere in the Soviet Union, free use of trams in cities, a reduction of taxes and utility bills, and a reduction of time before pension by a third. All of these privileges were announced to the public and would be enjoyed by their family after the bearer's death.[130] Stakhanovites, polar explorers, and marshals had commonly received orders in the prewar period; the new cadre of war veterans swelled the ranks of *ordenonostsy*.

Many orders were awarded exclusively by rank. For example, the Order of Victory—a diamond-studded platinum star—could be awarded only to front commanders and marshals for coordinating and prosecuting a massive offensive. The Order of Suvorov existed in three degrees: the first degree for officers commanding whole fronts or armies and their staff; the second for commanders of corps, divisions, and brigades and their staff; the third for regimental, battalion, and company commanders.[131] The Order of Glory was available only to soldiers, sergeants, and pilots and came with an immediate rise through several ranks, a 5–15 ruble monthly bonus in pay, a 50 percent reduction in time to pension, and free higher education for the bearer's children. This order was conspicuously modeled after the St. George Cross, a combat award from the Russian Empire that was often allowed to be worn in the Soviet period.[132]

Orders could also be awarded to military units, factories, ships, newspapers, republics, and cities, making them both corporate and personal awards. Any corporate entity would have its title read at official functions, and military units would often receive honorific titles for towns they had liberated or captured ("Fifty-second Riga-Berlin Guards, Order of Lenin, Orders of Suvorov and Kutuzov Rifle Division"). These decorations were inscribed on the unit's banner just as soldiers wore their individual medals on their tunics.[133]

The titles and decorations a division carried told a story of its feats that was used to inspire new soldiers in the unit.[134]

The highest award (technically a medal), Hero of the Soviet Union, could be won (often posthumously) by anyone of any rank and was routinely given to pilots for completing twenty-five missions. The Gold Star earned its recipients immortality, as they automatically received an Order of Lenin and their story became the subject of media attention. Should bearers win a second Hero title, their busts were to be cast in bronze and displayed in their hometowns.[135] During the war, more than 11,500 soldiers received this medal, 104 twice and 3 thrice.[136] People of humble origins, such as Aleksandr Matrosov and Zoia Kosmodem'ianskaia, who sacrificed themselves in the war came to share a place in the pantheon of the conflict's mythology with men of stature, such as Stalin and Marshal Georgii Zhukov.

Due to their transformative power and biographical importance, medals were rarely removed. Part of the punishment of being sent to a penal unit was the retraction of all decorations and rank, an act made public by reading the punishment and removing decorations in front of one's comrades.[137] When going on reconnaissance, medals and all forms of identification would be surrendered.[138] Medals and orders were to be taken from the dead and sent to their families or museums, as relics of the fallen.[139] When soldiers were buried far from home in unmarked graves, medals were often the only remains a family received. During the funeral procession of a high-ranking officer each of his orders would be carried by an officer on a separate pillow behind the hearse.[140]

Decorations soon entered the lexicon of "speaking Bolshevik," as people writing letters to various bureaucracies would be sure to mention medals they had been awarded to legitimize their demands.[141] For those who had been prisoners of war, a decoration could be "a pass to life," erasing their perceived guilt.[142] However, the manner of speaking Bolshevik was changing, as the language had come to incorporate references to a heroic past that had become much more ancient and that was repeating itself. The medals and orders themselves often made reference to a heroic past that posed Russians as an extraordinarily gifted nation, while their primary purpose was to identify and promote outstanding individuals as exemplary models.

Many desired further distinction. People wrote Stalin and Mikhail Kalinin with various suggestions for new awards. Before campaign medals were instituted, soldiers wrote in recommendations with elaborate schemes for Defender of Moscow and For the City of Lenin medals.[143] Snipers wrote Kalinin suggesting that they be given a chit to dangle from their Sniper badges, indicating their number of kills.[144] In May 1942 V. Markevich wrote Stalin with three suggestions: to devise medals for participants in the Russo-Japanese War and World War I; to devise medals for participation in the Civil War and Winter War; and to create some sort of distinctive mark for the wounded. Given the date of this letter, it is feasible that his letter served as the inspiration for adopting wound stripes. The failure to create medals for these other events implies that they paled in comparison to the Great Patriotic War that was then being waged.[145] Veterans were invested in making sure that their service was not forgotten, and many seemed to understand the language of hierarchy and distinction that medals offered, trying to shape its development.

Not everyone was enamored with the enameled language of power and recognition. Rashid Rafikov wrote home in 1944 after receiving an Order of the Red Star: "Of course, to get an order is a not a bad thing, if you survive this and keep your head on your shoulders. Otherwise I have no interest."[146] Even Tvardovskii's Vasilii Tërkin preferred going home to a medal.[147] The war correspondent Aleksandr Lesin talked to a man who, having been wounded, smoked the paper his recommendation was written on: "'I smoked up my order' . . . Try to interpret this fact. Wouldn't it be better to say that it not in order to get decorations that we fight the fascists!"[148]

Beyond these examples, it was not unheard of for soldiers to receive decorations for reasons other than combat service. One agitator received the For Valor medal for organizing a talent show, and many frontline veterans were infuriated that rear-echelon troops received medals while the services of those risking their lives often went undistinguished.[149] In the Second Guards Paratrooper Division, some officers wrote themselves up for fictitious feats of bravery and denied heroes decorations, while others promised medals to men and women for certain services, such as "sewing boots, giving out new costumes, for giving fuel, for living together [*sozhitel'stvo*]."[150] The final category was particularly painful for women in the ranks, as it cast a shadow on all of them. Iakov Aizenshtat recalled of 1943: "Medals and orders were quite a rarity then, even among those who had been at the front from the first day. I looked at the secretaries of the Front Tribunal, young women, and on their chests were orders and medals. It was clear to all, what they had received these medals for."[151] Soldiers' relationships with their commanders, who wrote recommendations for medals, were key to whether a soldier received decorations. In some situations, medals and orders became—or were perceived as—a sort of ersatz jewelry for officers' lovers, while women who refused officers' advances could be overlooked for decorations, regardless of their accomplishments.

Once recommendations were written, another set of factors came into play. Irina Dunaevskaia recalled that medals were usually conferred only when the army was advancing, as celebrating acts of heroism during a retreat brought attention to defeats. In general soldiers usually received a decoration a "step" or two lower in the official hierarchy of awards than they were recommended for.[152] Sometimes it was impossible to determine the names of those who should be awarded, as they spent so little time in the unit.[153]

Despite the often petty politics of medals, decorations gave the bearers a form of legitimacy as defenders of the Motherland that continues to resonate. Decorations were tied to concrete actions or moments in the bearers' lives and as such were an object pinned directly into their biography and put on public display. Medals were publicized and recognizable by soldiers and civilians. Soldiers could tell their stories via wound stripes and their decorations, lending a biographical element to the soldier's uniform, personalizing an otherwise anonymous article of clothing. For example, just by looking at the portrait of Guards Senior Sergeant Bato Damcheev (figure 2.4), we can tell that he is a Guardsman, was wounded twice, served as a scout and sniper, and fought at Stalingrad. We can also get a sense of the deeds he had to perform to earn two Combat Service medals and an Order of Glory.[154]

Nonetheless, this remained a language of power that the government controlled. The hierarchy of awards and the regulations on how they could be worn provided a sort of grammar of this language, but its real meaning was always subject to the state's control. Once medals started becoming more prevalent, concerns that soldiers might rest on their laurels mounted. One war correspondent for *Pravda* wrote from the Steppe Front in November 1943: "The generous awarding of decorations has gone to many people's heads . . . Brave soldiers need to be given orders, while those who have shamed themselves, who put on airs, need to have their orders taken away."[155] In July 1944, Shcherbakov warned that on the Second Ukrainian Front: "people live in the past, live yesterday's victories."[156] After the war this logic took on a menacing dimension. On January 1, 1948, all benefits tied to decorations were cut. While this had a significant financial component, as the value of benefits to veterans reached over 3.4 billion rubles by 1947, many veterans saw the loss of benefits as a betrayal.[157] Veteran Grigorii Pomerants recalled his reaction: "all of us with our orders, medals, and wound stripes became nothing."[158] Nonetheless, during the war, and, later, during the revival of the war as a foundational myth under Brezhnev, medals continued to be a point of pride, embodying the story of one's life at the front.[159]

Outerwear

Soldiers' overcoats (*shineli*) were single-breasted, sewn from coarse wool with hook and eye closures, while commanders had more ostentatious double-breasted models with buttons. Like the tunic, the overcoat bore rank and branch of service insignia, large diamond-shaped collar tabs in bright colors and chevrons, then muted defense collar tabs, and finally pogony with new model collar tabs, rectangular and showing reintroduced branch of service colors, from 1943 on. The basic outline of the overcoat had not changed since the time of the war with Napoleon.[160] In 1945 Aleksandr Lesin overheard a German woman say, that "all the Russians have left their land and put on overcoats; this is our punishment."[161] The overcoat was one of the iconic symbols of the Red Army soldier, engraved on the Order of the Red Star and prewar sculptures. The phrases "to put on the overcoat" and "in overcoats" are ubiquitous allusions to military service.[162]

Despite their symbolic power, overcoats were heavy and often among the items soldiers discarded, and one quartermaster suggested that they be withdrawn during summer.[163] Such losses would generally be temporary at the front, as one could usually receive secondhand outerwear from the dead.[164] Red Army soldiers were not issued blankets, instead overcoats were carefully folded and tightly wrapped into a blanket roll, which could provide a psychological sense of protection.[165] When improperly rolled, the rough wool of the overcoat would rub soldiers' faces raw and impede their movement.[166] Like most of the rest of a soldier's kit, the overcoat required practice and special knowhow to be used in a way that would not cause suffering.

Soldiers in the Red Army gave their overcoats mixed reviews. A veteran of the Finnish campaign remembered that the gray overcoats provided no concealment and hampered movement: "in the snow, we were like flies in sour cream."[167] Later the army shifted to a dark brown dye that blended with the earth. Many soldiers shortened the bottom of the overcoat to improve movement. As with all other items, size was key. Lesin recalled in 1942: "we were issued uniforms, overcoats. I got a shorty. It barely covers my knees, and the back belt is practically at my armpits. I won't be going to have my picture taken."[168] Genatulin rated the overcoat the ideal garment for soldiers: "Not very pretty, not always the right size, for soldiers it served as featherbed, mattress, and blanket on the bare ground, in the trench, and even in the snow."[169] Soldiers curled up together under overcoats: "There is no brotherhood that binds people closer than the brotherhood that is shared in the lines, and a shared greatcoat is one of its symbols. You feel warm and secure with a friend close by." Soldiers used their knapsacks or mittens as pillows, their rain-cape/half-tents as mattresses, the "shabbier" of the two overcoats to cover their feet, and the better of the two to cover their heads in a sort of "sleeping bag, warm and cozy."[170] Those in service became bedfellows via the overcoat, as the cold and the military's practicality and frugality forced people into greater intimacy. The overcoat could be improvised into a stretcher and its hems became a wick for improvised lanterns.[171]

The overcoat was one of several items issued for winter wear. A padded cotton jacket called a *vatnik, fufaika,* or *telogreika* (body warmer) was a traditional cold-weather garment initially considered an undergarment by the army, but it had become outerwear during the Winter War. (See figure 1.4.) It was considered to be less desirable than the overcoat, authorized for rear-area personnel to provide enough overcoats at the front and regarded as not dignified enough for soldiers to appear in public wearing them from 1943 on. More stylish were *polushubki*, lambskin coats initially authorized for officers only, but which more and more soldiers wore as the conflict continued.

Red Army personnel suffered from frostbite and the elements as much as any humans, and supplying sufficient cold-weather wear was a constant concern. The situation improved as the war continued.[172] The government and soldiers worked to define themselves in contradistinction to their enemies, who were portrayed as constantly suffering at the hands of "General Winter." "Winter Fritzes" were stock characters of propaganda, appearing in documentary films, cartoons, and the soldiers' own drawings. Consistently shown wrapped in women's shawls and bast shoes that gave them a pathetic, unmartial, and feminine appearance, they emphasized that winter was the period of the Red Army's initial victories and the humiliation of the enemy.[173] Pathetic groups of these beggar-like figures were often seen being led by a single Red Army man, proud and erect in his overcoat.

Belts

In order to stand proud and erect, the soldier needed a belt that would draw in the waist of the loose-fitting overcoat (which had buttons placed as a marker for the belt) or tunic.[174] Belts served many of the same functions that a tie serves in the professional world, as a soldier was not in uniform without one.[175] The belt was one of the first things taken from a soldier under arrest.[176] Soldiers were instructed in how to make tourniquets using their belts and how to properly wear them so that their equipment didn't flop and in regulation alignment. In full kit a soldier carried cartridge boxes, grenade pouches, a spade, a water bottle, and sometimes a bread bag on their belt.[177] Belts were important as both functional and decorative items, and as with everything else, a hierarchy of belts existed in the army. Before the massive expansion of the army in the late 1930s, all soldiers' belts were thin leather with roller buckles, with officers having a variety of wide, high-quality belts decorated with ostentatious buckles of the Sam Brown variety or a hammer and sickle. With the expansion of the army, and particularly once the war began, most soldiers' belts were thick cloth of various colors with rough leather binding (see figure 1.3). Leather belts became premium items, borrowed for special occasions such as dates, stolen from comrades, or used to mark a privileged position.[178] Generals, a rare sight, surprised some soldiers by the fact that they did not wear belts, having finely tailored clothing.[179]

Headgear

Headgear changed with the seasons, and like belts it acted as a clear indicator of hierarchy. The earliest hats of the Red Army had been peaked caps and the *budënovka*, or officially *shlem*, most popularly known for General Semën Budënnyi, famed cavalry commander of the Civil War. Having a large enameled red star and a large cloth star in branch of service color, the *budënovka* looked like something from another era. One version gives authorship to the artist Boris Kustodiev, another to the imperial regime.[180] According to Isaak Mints, the preeminent Soviet historian during the war, the Bolsheviks consciously chose headgear that resembled the legendary knights of ancient Rus'.[181] The outline of the *shlem* recalled ancient Russian warriors and onion-domed churches, many of which would be destroyed by men wearing these hats during the First Five-Year Plan. This most distinctive uniform item of the Red Army, coming in both summer and winter weights, was officially discarded in 1940, as it was found to be too cold, but continued to be worn through 1942 and could still be seen on surviving POWs in 1945.

The *budënovka* was replaced by the *shapka-ushanka*, a militarized version of a traditional winter hat. The standard issue *ushanka* was made of cotton, sometimes called "fish-fur," and although it was quite warm, some soldiers found it to be particularly ugly. (See image 4.14.)[182] For many soldiers, their hats became talismans, a long-lasting item that became molded to the wearer's head.[183] Headgear was said to retain the distinctive smell of its wearer—usually a mix of sweat, hair, and soap. Hats were also the place where soldiers tended to store a needle or two with thread to make repairs to their clothing.[184] Some claim that soldiers in hospitals could not sleep without their hats, as soldiers did everything—even faced death—wearing their hats.[185] Despite attachment, as winter turned to spring, *budënevki* and *ushanki* were replaced by *pilotki* (wedge caps), *furazhki* (forage caps), and sometimes berets for female soldiers. Every change of headgear was a marker that one had survived another season.[186] The *pilotka* was similar to headgear worn by most armies at the time, as were forage caps. All headgear carried a star with hammer and sickle that was at first red enamel and later changed to defense green, like collar tabs and rank pips. Officers wore a finer-quality *pilotka* with edging in branch of service color or forage caps with a bright band of material in their branch of service color until the turnover to olive drab (though the earlier caps remained popular). (See figure 6.1.) Generals had distinctive headgear: in winter the *papakha*, which recalled revolutionary heroes such as Chapaev; and in summer forage caps with ostentatious insignia.[187]

In elite formations—such as armor troops, cavalry, border guards, and NKVD units—personnel were authorized to wear forage caps, and sailors either black *pilotki* or sailor's caps with their ship's or fleet's title. As the war continued, many cavalry and scout units came to wear fur *kubanka* caps year round.[188] These flourishes contributed to a distinct corporate identity that made its members stand out in an otherwise anonymous mass institution and many of these symbols still evoke strong associations.

By regulations, *ushanki* and peaked caps were worn straight, and *pilotki* at a slight angle, two fingers above the eyebrow, with the star over the center of the face.[189] However, as Alexandra Orme observed, soldiers wore their caps every which way and their angle could speak volumes: "For you must know that a Russian soldier's cap is far from being just a covering for his head. If I had to deal with a dumb Russian soldier, I could guess his innermost feelings simply from the movements of his cap. The cap serves to express those feelings for which there are no words in imperfect human speech. The Red Army cap, as we came to realize has a language of its own."[190] Orme proceeds to describe how every emotion has its own angle, particularly once soldiers start drinking. Photographic evidence shows soldiers wearing their caps at a variety of angles, yet another example of the tiny ways in which soldiers made their mass-produced, standard-issue uniforms their own.

Uniformity and Individuality

Red Army uniforms were humble and required special knowledge to be worn properly—the proper fold of portianki, a tunic, or an overcoat was the difference not only between looking sharp or like an unmade bed but also between being comfortable or blistered and bloody. Soldiers learned how to wear and feel at home in these garments, which reflected both the shoestring budget of the Red Army and the peasant origins of its cadres. The Soviet Union skimped on uniforms and gear but not on weapons and medals. Many soldiers who donned the uniform earned distinction, transforming the anonymous cloth the army issued them into a personal banner showcasing their exploits on the battlefield. These exploits were translated into medals and orders devised by the Soviet leadership that spoke an official language dictated by the government.

Millions of people from different backgrounds passed through the ranks of the Red Army. When people stood in ranks, with tunics buttoned tight, boots shined and belt properly centered, it was not necessarily clear who was a peasant, a student, a worker, or a former criminal. Underwear would circulate among all of them. A Ukrainian would share his overcoat with an Uzbek, a convict and a Komsomol member could receive the same medals, and "backward" peasants would command "conscious" students.[191] Donning the uniform, men and women were exposed to a peculiar military modernity that forced them first to master proper wear and etiquette, and then the art of killing, relying on each other for survival. Officially, they would no longer be distinguished by their class, age, ethnicity, or education but rather by their rank, specialization, and the decorations they had earned via personal achievement on the battlefield.

Yet the army could never completely transfigure all of the men and women who filled its ranks. The juxtaposition of yesterday's peasants and ultramodern technology sometimes created visual dissonance: "In surprising contrast to the armored column appeared a little mule, quietly grazing by the side of the road. An old Uzbek in a *pilotka*, stretched like a *tiubeteika* [skull cap] on his head, with a rifle over his shoulder, rode astride, half asleep, along the ditch."[192]

The process of becoming a Red Army soldier consisted of both adjusting to uniforms and making them personal. Even as the army effaced many aspects of earlier identities, there were still times and places when one's prewar biography and habits bled through khaki tunics and gray overcoats. Every soldier wore the cap a little differently, despite regulations. Everyone knew which pair of boots, which belt and which tunic were theirs, despite the fact that all these items were standard issue. Soldiers found ways to individualize their humble belongings, using both the government's language of distinction and their own simple means—whether through tailoring or simply wearing something in their own particular way. They left their mark

on boots, hats, and tunics that all wore out in ways specific to their wearer's body and came to have their particular smell. Many of them also earned decorations that told of their exploits in the Soviet idiom of enamel and metal. However, doling out medals was not the only way the government showed its appreciation for soldiers. A much more humble, but vital measure was also used to mark their worth—the calorie.

CHAPTER 3

The State's Pot and the Soldier's Spoon

Rations in the Red Army

> Without a spoon, just as without a rifle, it is impossible to wage war.
>
> —Aleksandr Lesin

Aleksandr Lesin, who wrote the above lines, came to understand all too well how important being fed was to being able to fight. He served on the benighted Kalinin Front. In the spring of 1942, Lesin participated in an offensive that bogged down as starving and exhausted soldiers failed to take their objectives.[1] The Kalinin Front eventually became a lightning rod attracting Moscow's attention to the needs of soldiers' stomachs.

On May 31, 1943, Stalin signed an order underlining the failure of the Kalinin Front to properly feed its troops. Stalin found a "criminally irresponsible, un-Soviet attitude toward soldier's food" among those responsible for feeding the army: "Apparently our commanders have forgotten the best traditions of the Russian Army, of such eminent commanders as Suvorov and Kutuzov" who "demonstrated fatherly care about the everyday life and rations of soldiers." Concern for soldiers' well-being was posited as a "sacred duty" that commanders often ignored. Stalin drew on national and revolutionary traditions to shame officers into fulfilling their duties. Soldiers had been assured that the government was capable of providing for them. But failure was everywhere. "The Kalinin Front is not an exception," Stalin noted; "similar conditions occur on other fronts." Stalin's order was distributed to all fronts as a warning.[2] Alongside highlighting failure and prescribing punishments, the document provided extensive corrective prescriptions, reinforcing and setting norms that would remain fundamental through 1945. This document, as Stalin's word, marked the culmination of a flurry of similar inspections in 1942–1943. It reflected and constructed Soviet norms and expectations of nourishment during a total war that called on citizens to make great sacrifices.[3]

In the first years of the war the Soviet Union lost its bread basket, making hunger inevitable. Under these conditions the government's dedication to provide was reaffirmed to soldiers, who were promised ample provisions in return for their service. This ideological commitment and the very real consequences of fighting a war on an empty stomach made breakdowns in provisioning deeply disturbing, prompting the government to reaffirm its role as provider.

This chapter explores the quotidian details of provisioning, a material embodiment of the bond between Red Army soldiers to the Soviet government during the war. It was difficult to imagine such a key resource as food outside of the horizontal bonds between citizens and the vertical relationship to the state. The very term used for rations, *paëk*, implied mutual obligations. Paëk could be seen as the physical embodiment of the socialist adage "to each according to his work," as its etymological root implied an earned share in a common cause.[4] We will see how rations were assembled by the government and later received and used by soldiers at the front—how paëk functioned, was experienced, and occasionally transformed by soldiers.[5]

This chapter is divided into three sections. The first describes how the government thought of rationing, where it drew its resources, and what it sought to provide. The second examines failures in the provisioning system and how standards improved as the war continued. The third describes how soldiers used rations and responded to failures. Rations were a key resource in setting the army apart as a separate class of citizens and in creating clear hierarchies within the army; they both brought together and divided those in the ranks.

The State Provides: What and How

In 1941–1942, the Soviet Union lost vast resources to the rapacious Wehrmacht. Even by 1945, after the war had shifted to enemy territory, the gross production of the Soviet food industry stood at half of the level of 1940.[6] When Stalin issued his famous "Not One Step Backward" order (227) in the summer of 1942, he pointed out that if the army retreated any farther, it would be dooming itself to starvation: "The territory of the USSR, which the enemy has seized and strives to seize, is bread and other foodstuffs for the army and rear, metal and fuel for industry, factories and plants, supplying the army with weapons and ammunition, rail roads . . . Every new scrap of land we leave to the enemy will in every way possible strengthen the enemy and in every way possible weaken our defense and our Motherland."[7] Food was a resource that could mean the difference between victory and defeat, one that would feed you or the enemy in a zero-sum equation. As a result, both sides used scorched-earth tactics whenever they were forced to retreat, leaving civilians in a dire situation.[8] Forced to wage war regardless of a catastrophic loss of material, Soviet leaders strived to establish total control over food distribution under the chaotic conditions of a war it was clearly losing, as well as refining a hierarchy around what was arguably its most vital resource.

Despite these immense losses, in the course of the war the Soviets were able to provide more and more adequately for the military. In 1941 as many resources as possible were evacuated to the east, and agricultural production shifted there as well. The full-scale development of agriculture in the east, American Lend-Lease aid, and the recapture of resources led to palpable improvements in 1943.[9]

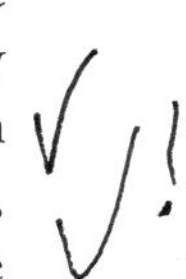

The Rationale of Rationing

Paëk, a common term for rations, revealed a certain moral economy of provisioning.[10] Paëk came from the Turkic root *pai,* which meant "share, part in a common cause, coming through mutual agreement to every individual, in the paying or receiving of a monetary sum or other form of personal property."[11] The root had close associations with an individual's "part, fate, destiny, and happiness," and a participant in a common enterprise such as a cooperative was often called a *paishchik*.[12] The root itself presupposed the necessity of a common cause and mutual obligations in the circulation of rations; it is not a form of welfare but part of a bargain based on who earned what.[13] Yet the term also spoke to a certain ambiguity about who owned the paëk—the government or the soldier. This was never fully resolved.

Pai-based understandings of state-citizen relations had been key to how both the Imperial Army in 1914 and the Bolsheviks in 1918 apportioned resources. The *paika*, a special ration issued to soldiers' families in time of war, was instituted by the tsarist government in 1912 and later adopted by the Bolsheviks during the Civil War. Joshua Sanborn notes that paika was a manifestation of the government's understanding that it had a reciprocal relationship with soldiers and owed both them and their families more than other citizens. It was also a powerful tool to control soldiers' actions—cutting off the paika could leave a family to starve.[14]

Rations had to keep soldiers functioning while taking up minimal space and weight in their packs or on supply wagons.[15] This logic gave fats and carbohydrates a privileged place. Both provided a high number of calories per volume and a feeling of satiation. Meat (especially salt pork and sausage) and potatoes were looked upon as ideal ingredients, and bread the fundamental foodstuff.[16] The war forced a new set of norms on the army, established in September 1941, that would remain essential until 1945.[17]

The range of calories guaranteed to men under arms varied depending on position. A manual for *fel'dshers* (medical assistants) published on the eve of the war stated that a person in a state of total relaxation needed 1,700 calories, a tractor driver 3,000.[18] A soldier could receive between 3,088 and 4,692 according to different estimates. An official history of the rear-area services claimed that soldiers received between 2,659 calories (in the rear) to 3,450 (for soldiers at the front) to 4,712 (for airmen) according to the September 1941 norms.[19] Frontline soldiers were to receive 3,505 calories. Control over calories became a common part of frontline inspections, which were also supposed to insure balanced nutrition.[20]

Any system of mass catering imagines a more or less generic body that it will be feeding, ignoring to some degree differences in age, sex, and mass that might warrant special attention, not to mention culturally constructed differences that could also be of significance.[21] Red Army provisioning practically ignored these differences, which was in line with prewar provisioning in the Soviet Union. But in other armies—such as the British and French—adherence to ethnic and religious tradition was used as a way to discipline soldiers drawn from the colonies. Utilizing differences as a form of rule was a hallmark of imperial thinking. Red Army soldiers were seen as citizens of a modern state, and as such their paëk was supposed to be universal and rationalized.[22] The Red Army redefined its soldiers' identities by specialization and rank: who a person was did not depended on where they were from or which God they prayed to, but rather on their rank, specialization, and location, all of which impacted what kind of food they received.

Rations followed several trajectories: up the ranks, from the rear to the front, the changing of seasons and climate zones, and finally, soldiers' specializations. Commanders, as the *heads* of the military units, received a supplemental ration (*doppaëk*), which included extra meat, cookies, and higher-quality tobacco—an extra 450 calories, totaling 3,490–4,000 calories per day.[23] The hierarchy of rations had an impact on soldiers' language—strong tea was referred to as "general's tea."[24] Those serving in frigid climates, such as the Karelian Isthmus or Far North, received special rations of extra vodka, salt pork, and vitamin C (154 additional calories).[25] In winter (from October until

April) soldiers were given one hundred extra grams (3.5 oz.) of bread. Those recovering from wounds had a different set of norms and specific food.[26] The closer soldiers were to the front, the more rations they were entitled to. Simply put, in risking their lives and thus most directly contributing to the war effort, frontline soldiers earned more resources.[27] Soldiers at the front received roughly twice the amount of bread as those in training, and roughly three times the amount of bread as those in hospitals. This logic translated into civilian rations, where the more directly civilians' jobs contributed to the war effort, the more calories they were given.[28] Specialized troops received particular kinds of rations. Pilots were given a highly portable ration that included condensed milk and chocolate in case of fatigue or an emergency landing. Reconnaissance troops took special rations with them onto enemy territory, while elite formations received additional rations, including much-coveted white bread.[29]

Some privileges seemed based on biological needs, others on status. Soldiers at the front needed more energy because they were engaged in strenuous combat, but they also deserved more because they were risking their lives. Among the items included in rations were tobacco, rolling paper, matches, and vodka—none of which were necessary for physical survival, but all of which were deficit items that carried important social weight.[30]

As a result of multivector norms, the volume and composition of rations could vary dramatically.[31] Soldiers in the rear were nearly unanimous in their complaints of hunger while undergoing training, and kitchen duty was a much sought-after assignment.[32] Grigorii Baklanov recalled of the rear-area diet: "you'll live, but even in your sleep you won't have sinful thoughts."[33] Many looked forward to going to the front as a place where they could finally get enough to eat.[34]

Locavores, Pillagers, and Boxed Lunches

As the war progressed, responsibility for feeding soldiers became increasingly localized, some claiming that the army provided itself with 65 percent of its provisions.[35] In the first months, it was typical to appeal to higher ranks and invest them with sole control over both food and transport.[36] In the first two years of the war, the authority and competencies of rear-area officers were expanded and their personal responsibility clarified. While central reserves provided necessities that could not be produced locally, subsidiary agriculture (*podsobnoe khoziaistvo*) accounted for more and more of people's diets. Unlike their US or Commonwealth allies, Red Army soldiers often produced the rations they were eating and knew who prepared them.

Agricultural work became a common duty of men and women in uniform, as military units began to tend their own rear-area farms and soldiers were sent to assist local collective farms.[37]

Red Army personnel were, to a great extent, *locavores*, in contrast to their US and British allies and similar to their Wehrmacht foe. The United States had taken pains to develop its famous C and K Rations—prepackaged, ready-to-eat, standardized, and self-sufficient meals, containing everything from a can opener, wooden spoon, entrée, and dessert to gum, cigarettes, matches, and toilet paper, all prominently displaying brand names.[38] However, US soldiers quickly grew tired of quartermaster officers' over-reliance on a limited assortment of canned goods, which became the subject of postwar inquiries.[39] The British were similar to the Americans, being primarily an expeditionary force, but their rations were simpler. The Wehrmacht combined ready-to-eat items with those needing preparation and were notorious foragers, often living off what they pillaged from Soviet peasants.[40] Indeed the Reich's strategy called for the extermination by famine of millions of Soviet citizens.[41] The Red Army relied heavily on whatever was available locally, drawing from central reserves when local reserves failed.[42] Red Army forces planned on feeding soldiers whenever possible with hot, fresh food from field kitchens located just behind the front line, with mobile bakeries, kitchens, and even herds of livestock in tow.

The provisioning methods of the Wehrmacht and the Red Army relied on similar logic: a preference for field kitchens, the issuance of a variety of ready-to-eat items in the event that a hot meal was unfeasible, and the extensive use of local resources. There was, of course, a significant difference between their provisioning strategies. The Red Army, for most of the war, was taking from its own citizens in an economy that treated all resources as "the people's," and thus constituting a horizontal connection between provider and defender. This was in sharp contrast to the Nazi strategy of exploiting racial "others," most of whom were slotted for extermination or enslavement. Wherever they were provisioning, the Red Army showed concern for the feeding of local civilians. Local provisioning, a key aspect of the Soviet ration system, obscured the borders between military and civilian. Both combatants and wide swaths of the civilian population received rations during the war, but soldiers were much better fed.

The Menu

The menu at the front often impressed those who had been wasting away in the rear.[43] Whenever possible, soldiers at the front were to be provided with hot, fresh food by field kitchens twice a day.[44] Hot food was to be brought up just before dawn and just after dark with bread and meat for lunch.[45] The soldier's meal was supposed to consist of two dishes—a soup and a porridge—and tea, brought up in twelve-liter thermoses.[46] In practice this was often combined into one dish.[47] Cooks were supposed to divide the ration so that every soldier received the same portion of meat in their soup or kasha.[48] Frontline menus varied greatly, but could become monotonous as one dish became constant.[49]

The Red Army veteran and food historian Vil'iam Pokhlëbkin noted that priorities during the war ignored assortment in favor of sufficiency and practicality in provisioning, centering on "basic products, without which not one person in the rear or at the front can exist . . . first of all, bread and salt . . . [then] meat and fish, fats and vegetables . . . What kind of meat, which kind of fats—this is all unimportant."[50] An official table of exchange existed to ensure that each soldier received the proper number of calories. The government saw categories in terms of meat, bread, grains, vegetables, and so on, without attention to whether the meat was pork or beef or rabbit or fish, or whether it was canned or smoked or fresh or even took the form of powdered eggs—which concentrated calories but were unpalatable.[51] Meat was meat (or anything with fat and protein), bread was bread (whether dried into crouton-like *sukhari* or fresh-baked). Provisioning, at least initially and primarily, was concerned with caloric, not culinary, value. Provisioning officers made concessions to taste by the continued use and inclusion of basic spices (bay leaf, salt, pepper, onions, sometimes garlic) in rations.[52]

Field kitchens were supposed to service no more than 180 soldiers but often served 300 or more. The kitchens consisted of three pots (soup, kasha, and tea) on wheels, with an oven. They could be so close to the front as to endanger the cook's life or so far that food would be doled out cold.[53] Cooks were supposed to provide nourishment to their comrades a few hundred meters or a few kilometers in front of them, but their ability to do so varied depending on their skills, the resources available to the rear area, and conditions at the front.

In battle, in echelon, and whenever troops found themselves too far from a field kitchen, two types of ration provided them with sustenance: the NZ and dry rations (*sukhpaëk*). The *neprikosnovennyi zapas* (literally "untouchable reserve"; in the British and German armies referred to as "Iron Rations"), or NZ (sometimes referred to as *nosimyi zapas*—"portable reserve"), was carried in a soldier's knapsack at all times, consumed only with a commander's authorization.[54] However, experience showed that soldiers could eat them surreptitiously.[55] The NZ typically consisted of canned or smoked meat, tea, sugar, salt, and dried bread.[56] The dry rations, often distinguishable only by not being labeled "NZ," were slightly more generous. They consisted partially of things soldiers could prepare themselves, such as concentrated soups and grains, and partially of ready-to-eat items, such as dried bread and canned food. Preparation was difficult, as the army failed to provide enough dry spirits (alcohol that burned without smoke) to allow soldiers to cook these concentrates, while starting a fire could draw enemy fire and prove fatal.[57] Eventually concentrated soups and grains were reserved for field kitchens.[58]

The contents of NZ and dry rations varied dramatically: they could be freshly slaughtered boiled lamb, salt pork, lard, compressed animal fats, sardines, sausage, or American Spam (sarcastically referred to as "Second Front"), or a variety of goods, depending on what was available.[59] Canned goods were often unlabeled.[60] Pokhlëbkin remembered different kinds of NZ and sukhpaëk that had been provided to him from the "hard, dingy, yellowish-gray chunks" of *kombizhir* (combined fats) in 1942 to American lard in 1943 to a beautifully smoked piece of Hungarian or German salt pork in 1944. He mused that the NZ's "character changed in relation to historical and military conditions."[61] The food that soldiers received depended on a variety of factors largely beyond their control, factors that treated fat and meat as the same thing and where calories were king.

During the war, food was *the* deficit resource, something that everyone needed. By creating a hierarchy of distribution, the government directly ranked whose contribution was most significant to its continued survival and made providing for those who were risking their lives for the Soviet Union their first priority. Soldiers were keenly aware of being better fed than their families. Ibragim Gazi wrote his wife and child that "As soon as we get a chance to eat something good, I say: this is for kids, and our children probably don't have this."[62] While soldiers might feel guilt, they generally felt the pangs of hunger less acutely than their families in the rear or under German occupation.

An Inviolable Camp: Rhetoric, Realities, and Explaining Failures

Boris Slutskii reflected on the realpolitik of the stomach just after the war: "The cruel antitheft laws of war, executions of chauffeurs for two packs of concentrates, were necessitated by the famished convulsions of a country that robbed its own rear to fatten its front."[63] While such a policy left children, dependents, and the elderly with the smallest rations, in some cases condemning them to death, it fit with the logic of a sovereign power struggling for survival during total war.[64]

One of the clear messages sent by the Soviet state at the front was that the USSR could provide for its men under arms only so long as they stopped retreating. In his November 6, 1941, address to the Red Army (when German troops had reached the Moscow suburbs), Stalin described how the hardships of the war "converted the family of peoples of the U.S.S.R. into a single and inviolable camp, which is selflessly supporting its Red Army and Red Navy."[65] As a result of this rhetoric of an "inviolable camp" making sacrifices for the Red Army, any shortages at the front were not the failure of the Soviet system or the people but rather of identifiable individuals to which the government had entrusted the sacred task of feeding. Stalin declared that "the government allocates enough varied and nourishing foodstuffs" for the army and it was "only . . . a negligent, dishonest, and sometimes criminal attitude on the part of commanders" that "degraded" soldiers' rations.[66] This tracking of failure as the result of corrupt individuals was nothing new and would remain a continuous trope of Soviet discussions of provisioning, even as systematic problems became obvious.

Breakdowns: Their Consequences and Their Culprits

The physical impact of hunger was impossible to ignore. Marshal Zhukov reportedly declared: "A full soldier is worth five hungry ones!"[67] Failures in provisioning were cited by Red Army officers as the direct causes of desertion, illness, and sometimes the breakdown of combat operations.[68] Soldiers died from digestive maladies at or on their way to the front, and night blindness due to lack of vitamins was very common.[69] In a meeting of top political personnel of the army, one officer exclaimed that when soldiers were not fed, "What kind of combat effectiveness can you expect from them?"[70] Wherever breakdowns occurred, culprits needed to be found and punished.

Stalin's 1943 warning from the Kalinin Front noted that some commanders were "using their authority, disposing of ration stocks as if they were their personal property, illegally expending foodstuffs," which had a disastrous effect on the army.[71] Early in the war draconian laws concerning the theft of socialist property developed during collectivization were reiterated.[72] Despite the consequences, commanders having more or less total control over resources on the ground often pilfered. As one provisioning officer told his colleagues in January 1943: "The fighter could be full. But why doesn't he get all of his food? We came to a definite conclusion: starting from the DOP [Divisional Exchange Point] people steal, and when food gets to the kitchen, they steal there too."[73] This ubiquitous theft, while considered insignificant in comparison to graft under the old regime, included cases of illegal trade in foodstuffs as well as of officers throwing unsanctioned feasts using the soldiers' rations; the latter problem became worse around holidays.[74] Theft by commanders was considered such a scourge that there was even talk of separating commanders from general provisioning. However, it was decided that feeding commanders separately would mean "they would just stop looking in on the troops."[75]

A few cases of theft (or scapegoating) within a unit could have a ripple effect and send men at the top and bottom of the rear area into eminent peril for treating communal property as personal property. At best, a tribunal or punishment battalion meant humiliation; at worst—death. For example, in May 1944, on the Third Belorussian Front, one Private M., a cook, was sent to a punishment battalion for two months for hiding 5.25 kilograms of meat and 4.9 kilograms of flour; a Lt. L. went before a military tribunal for the illegal use of a variety of luxury items (including sugar, meat, and fish); the head provisioning officer for their army, a Guards major general, was removed from his position for allowing these abuses under his command.[76]

Theft aside, provisioning was a challenge. Under the difficult conditions of armies on the move, provisioning officers were forced to find secure storage during every advance and retreat, often in places utterly ravaged by war.[77] Incompetence could make a difficult situation catastrophic.[78] Food was left to rot or to be consumed by rats or left unguarded and stolen. A report from the Transcaucasian Front in January 1943 noted the "extreme carelessness" and "unsanitary conditions" in which "grain is stored in heaps on a dirty floor" and "four hundred tons of potatoes were ruined," yet no one was brought to answer. Inspections frequently found both field kitchens and canteens serving military personnel (including Moscow canteens that fed the staff of the People's Commissariat of Defense) dirty and undersupplied.[79] Vegetables were boiled without being peeled, creating muck.[80] Soldiers could be given raw food with no way to prepare it or, worst of all, simply given nothing.[81]

All of this spoke to a violation of the government's obligation to its soldiers, who were quite conscious of their duties and those of the state. Wherever Soviet leadership noted that the paëk was not being received, Soviet power was quick to find the culprits and ameliorate the situation. In Stalin's admonition to the Kalinin Front on May 31, 1943, the army was to retroactively make good what it had failed to give frontline soldiers for up to five days of foodstuffs and up to fifteen days of luxuries (tobacco, soap, vodka, etc.).[82] These obligations took on a wider scope as the war reached its turning point, as the government promised not only to provide calories but to emphasize taste.

Improvement

By 1943 Soviet leadership demanded very high-quality rations, and standards sometimes contradicted the logic of provisioning more generally. Vil'iam Pokhlëbkin noted that the categories used by the army to apportion foodstuffs were dramatically simplified and made no appeal to variety. As a result there "came the 'era' of the potato, or pea, and suddenly the 'macaroni period' or continuously only oats or pearl barley"—whatever was on hand was whatever was going to be served.[83] As the war progressed and the Red Army's fortunes changed, these "eras" became inexcusable.[84] The army began placing greater emphasis on who was cooking. Alongside the call for better and more varied ingredients, the army sought to improve the skills of cadres doing the cooking by finding professional chefs who were already serving in the army and replacing men with women. The military press took aim at bad cooks and praised good ones, while skilled cooks received medals.[85] A special badge was created for "excellent cooks" in 1943, and intensive training courses were held in 1943 to train new (mainly female) cooks.[86] According to Pokhlëbkin, this led to a period of experimentation among military cooks that altered the face of postwar Soviet culinary traditions.[87]

This spirit of innovation was not merely a phenomenon of the front line. Lend-Lease food from the United States required cooks to come up with new ways to use unfamiliar products, such as Spam, Vienna sausages, and deviled ham. In 1944 a special manual was published, translating the labels of Lend-Lease products and explaining how to prepare them.[88] Provisioning officers experimented with "vegetarian days," specialized foods for those in hospitals, wild herbs, frozen foods, and various dishes that could be prepared in the rear and given to troops at the front. In one particularly innovative moment, a provisioning officer described how frozen meat dumplings (*pel'meni*) were air-dropped to troops caught in encirclement.[89]

Food was rhetoric made substance. Soldiers could literally *feel* when the state was not holding up its end of the bargain. The government had promised to feed its soldiers and to punish those responsible for any failures in provisioning. This was a promise that the Soviet leadership intended to keep despite tremendous losses in every type of resource imaginable. As battlefield successes began to show the army's worth and liberation gave access to resources, a new set of expectations emerged, leading to greater demands from the soldiers. Political and provisioning officers encouraged soldiers to speak honestly about how they were being fed. At a conference for propagandists and agitators in 1943, Shcherbakov declared: "We should have taught agitators to start with *makhorka* [soldier's tobacco]. Is there enough of it? Have you eaten today? Every agitator—the new and old—should start with this."[90] Military psychologists discovered strong correlations between morale and being well-provisioned.[91] One commander told the war correspondent Vasilii Grossman: "The worse the front, the more food reminds you of peacetime."[92] In conversation with soldiers, agitators and provisioning and political officers learned how important hot food, tea, spices, and a smoke could be to men risking their lives in defense of the Soviet Union.[93]

Pots and Spoons: Eating and Drinking in the Red Army

Conditions and Improvisation

Red Army soldiers were supposed to receive hot, fresh food. Mikhail Loginov, a platoon commander on the Kalinin Front, recalled how after the ten-kilometer round trip to the field kitchen, his soldiers brought back "cold soup, cold kasha, and cold tea. There is nothing and nowhere to heat up the food—neither dry spirits, nor firewood, and anyway, to start a fire at the front is forbidden. The enemy would notice and immediately bombard us."[94] Hot food was often an unrealizable goal, as field kitchens could service three hundred or more men scattered over a wide front at each meal. Posting kitchens close to the front endangered them with bombardment and capture. In the chaos of the front, some field kitchens ended up delivering themselves to the enemy.[95] Soldiers carrying food could be killed, thermoses destroyed.[96] During successful offensives, troops could outrun their provisioning services and were sometimes forced to live on what they could capture.[97]

Improvisation was a way of life. Making do was a necessity in a world without chairs, tables, napkins, and other trappings of civility. Tvardovskii's hero Tërkin recalls how soldiers' new habits are inexcusable in "heaven"—the civilian world as exemplified by a rear-area hospital, where "you can't eat off your knee/ Only from the table" nor "mangle bread with a bayonet."[98] Eating at the front was something that was done wherever the food found its consumers—in bunkers, mud-filled trenches, forests, bombed-out cities, and along dusty roads.[99]

The calculations done in the rear concentrated on the body, not the psyche. Even when food was ample, soldiers suffered from nerves and exhaustion. The milieu could be inimical to eating, as Loginov recalled: "From no man's land a little wind blows, bringing the slightly sweet smell of corpses, filling

FIGURE 3.1 Red Army soldiers eating, Second Baltic Front, 1944. Note the variety of mess tins they are eating from and the shovels hanging off their belts. RGAKFD 0-291527 (V. P. Grebnev).

the trench. We have trouble breathing, and a few get nauseous and throw up. Dinner is brought up in thermoses, but I can't look at the meat or kasha. I give my portion to the soldiers, and myself have only bread and cold tea from my canteen."[100] Conversely, animal carrion near the trenches could become food for soldiers.

During the first two years of the war, when the situation with meat in the army was critical, soldiers and resourceful cooks found a solution on the battlefield.[101] Lesin recalled how in April 1942, horsemeat became the main source of food for him and his comrades, and a variety of sources show this was common practice. At first this idea disgusted him, as something alien to his culture: "To him, a Tatar, *makhan* [horsemeat] is the same as pork to a Russian. We all have a taste for horsemeat now."[102] The military translator Irina Dunaevskaia initially described soldiers mocking Kazakhs who ate horse but later noted that she and a comrade were "lucky" when a shell killed a horse, and they ate "makhan (there is no other way to refer to this at the front than by this Tatar word)."[103] Less fresh horse carrion (*propastina*), was, of course, of questionable quality, but also common.[104] Some units earned a bad reputation for their love of horseflesh. A report filed after an inspection of the Fiftieth Army (on the Kalinin Front, where F. S. Saushin, Lesin, and Loginov served) noted that a certain Colonel Samsonov admitted his division had eaten 175 of its horses, leading others to snicker "be careful; don't leave your horses standing around, because the 'Samsons' will eat them right away."[105] Horses were not part of rations, and their consumption could be both demoralizing and counterproductive. But in the darkest days of hunger, they soon found their way into the soldier's pot as an expedient way to make up for what the state could not provide.

Troops sometimes resorted to theft as a means of ensuring survival.[106] Theft, or the perception of thievery, could destroy bonds within a unit, so a fair distribution of rations was key to morale. At the front, exact measurements of food proved impractical. Some *starshiny* found their own way out of this, using magazines and discs from weapons as ersatz weights, a practice condemned by *Krasnaia zvezda*.[107] However, the most common arrangement to ensure fairness in the distribution of rations was a system found in many armies throughout history, in which rations were doled out into piles and "one of the soldiers turns to the side, and the one who divided the rations points to a portion and asks, 'Whose?' The soldier turned to the side names any name."[108] Such a system ensured that any inequality in rations was pure chance. This maintained a sense of fairness at the lowest level of ration distribution and kept disputes over what was probably the most valuable commodity to a minimum.

Eating from the Same Pot

Eating in the Red Army was a collective activity that could strengthen bonds between the diverse people in the ranks. It was a time of rest, when soldiers took stock of their situation, remembered home, got to know each other, and replenished their physical strength. While soldiers seemed to always want more to eat, situations where food was ample were not necessarily occasions for celebration. The strict ratios of products to soldiers proved difficult to fulfill as casualties mounted on the front. The head of provisioning of the First Belorussian Front, N. A. Antipenko, admitted that, since the calculation and reporting of casualties always came with delay and the "higher authorities continue to send food for the entire unit . . . a soldier in the course of an offensive received unlimited food."[109] Moments of rest and feeding underlined the losses that a unit had suffered. A passage from an autobiographical novel by Grigorii Baklanov captures this moment eloquently: "Only after he swallowed did he look at what he was eating. In his mess pot was thick, yellow pea soup. And with this spoon, with his eyes closed, he mentally held a funerary feast for those who today were no longer with them. They were still here, all the same; they could stumble into the kitchen at any moment, sit in the sun."[110] Eating was when you realized that you were alive, a visceral moment that separated the living and the dead.[111] As a result, army food could evoke strong emotions and potent memories, and, as we see from this quotation, could create the sense that those who had fallen were near.

A sense of communality was supported by the most quotidian details of provisioning. Soldiers were supposed to receive two dishes, yet they were issued only one mess pot, and given that shortage was a general rule, there were often fewer pots than soldiers.[112] Red Army pots came in two styles. One was a copy of the German mess tin issued in both world wars, which was a kidney-shaped aluminum pot with a bail-like handle and a shallow top that doubled as a cup.[113] The other was a simple round pot of varying depths with a bail handle but no top. The mess pot was not entirely the soldier's and not entirely the government's, much like the food that was consumed in it. The pot was issued by the army, but it was one of the few items of a soldier's kit that seemed to belong specifically to him or her. It was not uncommon for soldiers to decorate their pots with their name, a place where they had served, or the name of a friend or random acquaintance. Lev Slëzkin carved the names of two Estonian women he met before the war onto the side of his pot "in memory of a pleasant, romantic meeting."[114] By carving names, initials, places, and dates into government-issued items, soldiers turned an anonymous piece of metal into a personal item that recorded parts of their biography.

Mess pots served not only for consuming and occasionally preparing food. One female soldier recalled, "Mess tins! We had them for food, to wash our clothes in, to wash up ourselves with—everywhere mess tins!"[115] They served as desks.[116] Many soldiers lacking vitamin A suffered from night blindness, which created serious problems on long marches under the cover of darkness. In this case, banging a rod on the mess pot of the man in front enabled the blinded soldier to complete night marches.[117] A soldier's mess pot was something like a room in a portable home, serving as dining room and sometimes as kitchen and shower. However, as all activities in the army took place in the company of others, soldiers shared their pots with their comrades.

When food was doled out to soldiers, often one pot was filled with soup, the other with kasha.[118] Soldiers would eat in pairs, as Gabriel Temkin remembers: "We ate from one *kotelok* (mess tin), using approximately the same size wooden spoons. We would eat by turns, I a spoonful and then he a spoonful, slowly, as becoming among comrades. Having finished the soup or kasha, we would lick clean our personal spoons and put them back in place, where they were customarily kept—behind the top of the right or left boot." Such an arrangement helped to build a sense of comradeship, as soldiers of different ages and ethnicities found themselves eating from the same pot. The spoons that they dipped carried special meaning, as Temkin noted: "Frontline soldiers would sometimes, in panicky retreats, throw away their heavy rifles but never their spoons."[119]

Virtually nothing that soldiers carried belonged to them. Their clothes were the property of the government. When they went to a bathhouse to wash up, they were not guaranteed to get their own set of underwear back. Their weapons belonged to the army, as did the food they ate. However, the spoon was something that the individual soldier *owned*. Draft notices instructed inductees to bring a spoon, cup, towel, and change of underwear.[120] Given that the towel and underwear would soon be worn out, the spoon and cup were among the few items from the civilian world that soldiers would carry throughout their service. Spoons were frequently individualized with initials and artwork and are often the only way to identify soldiers whose remains are found today. The spoon could be wooden or metal, a traditional Russian triangular spoon or an oval soupspoon. German and Finnish folding spoons were also popular, as they were easily carried and their handles doubled as forks. Some soldiers made their own spoons out of scrap found on the battlefield, such as downed planes.[121] In one case, an officer found craftsmen from among his soldiers, took them from the front line, and put them to work carving spoons for soldiers in need.[122] Spoons were a frequent item in government supply orders throughout the war: in the third quarter of 1942 alone, 1.9 million wooden spoons were ordered.[123]

The spoon was the only utensil soldiers were expected to have—all of their food was designed to be eaten either with a spoon or bare hands. The spoon became a mark of a real soldier. Vera Malakhova, a frontline surgeon, recalled an embarrassing moment near Odessa. While joining a group of soldiers sitting down to a meal, she realized that she lacked something the men around her all possessed: "'What sort of a blankety-blank are you? Just what sort of soldier are you? Why don't you have a spoon?'"[124] Even dry rations could not be consumed without a spoon. A soldier reduced to a minimum carried a spoon and a rifle. The soldier's spoon separated the military and civilian worlds. In a letter home in 1939, Lev Slëzkin describes how soldiers confronted silverware in a café "like troglodytes, looking with tender emotion at knives and forks (in the barracks we eat only with spoons)."[125] Spoons were *the* implement of individual consumption and a deeply prized, rare piece of personal property. Yet every aspect of the soldier's rations could be treated as if it were personal property.

Currencies, Rituals, Substitutes, and Valuables

Food became a tradable commodity under conditions of extreme scarcity. People receiving rations throughout the country were often willing to part with durable goods (such as clothing and jewelry) for consumables (such as bread, meat, and vodka). As one war correspondent recorded in his diary in January 1943, "The modern form of payment is vodka and bread."[126] Boris Slutskii recalled: "In the trenches there was a lively exchange business! Tobacco for sugar, a portion of vodka for two portions of sugar. The prosecutor struggled with this barter in vain."[127] Some soldiers traded frontline trophies for food. These exchanges both highlighted the rituals of consumption that took place in the army and allowed those who did not drink or smoke to participate in or profit from them by either exchanging or giving away their portions of tobacco and alcohol.[128] These coveted items were not only potential commodities but also consumables that were used collectively.

"Let's smoke one, comrade!" was the chorus of a popular wartime song.[129] Tobacco was considered to be so important that the provisioning officer of the Kalinin Front was flown to Moscow to procure it in the spring of 1944 and ordered not to return without *makhorka*. He did this despite orders not to send delegations from the front to beg from manufacturers.[130] Tobacco was such an integral part of military culture that the state was dedicated to providing its soldiers with smokes despite a union-wide reduction to 25 percent of prewar production.[131] Smoking, a communal activity experienced as a different form of time, brought soldiers together in moments of rest and was often accompanied by sugary black tea.[132]

Tea—which, according to a nutrition textbook from 1940, was "almost without nutritional value"—was to be given to soldiers *hot*, twice a day, and manuals reminded soldiers that it was preferable to water.[133] Some aspects of the soldier's ration were clearly aimed at psychological rather than nutritional benefits and were invested with important social meaning. Tobacco and tea were useful stimulants; the latter, served warm, could save those dying of frostbite. Tea was a particularly good delivery system for sugar and quick calories. Both caffeine and nicotine could enliven people psychologically numbed by lack of sleep, endless manual labor, fighting, and long marches. They also leant themselves to ritualized, habitual use.[134] Vodka, a depressant, could calm the nerves of soldiers who had seen ghastly sights.

Vodka had only recently returned to the Red Army soldier's ration, the experience of the Finnish War having shown its value in staving off frostbite and death by exposure. Beginning with the Finnish campaign, one hundred grams (3.4 fl. oz.) of vodka were issued in winter. In the course of the war, vodka was used as a reward for frontline service and became a necessary part of revolutionary holidays. The amount soldiers were given and conditions under which soldiers earned their vodka ration fluctuated, but it was clear that vodka was seen as a privilege earned by those actively defending the Motherland. Soldiers engaged in offensive operations were always allotted a vodka ration, sometimes a double ration.[135]

The army's approach to rationing vodka rested on the notion that it could manage the delicate balance between calming nerves and inducing drunkenness. Soldiers, however, disposed of their rations in various ways and often found means to acquire more than their allotment. Some female soldiers reported never having received a vodka ration; others that they gave theirs away.[136] Many Muslim soldiers gave their vodka to their comrades.[137] Trade and gift giving disrupted the army's attempt to manage soldiers' use of vodka, as did theft by officers, and the lag between tallying losses and computing the quantity of necessary supplies. Commanders wrote home to their local representatives begging not to send alcohol in care packages.[138] Mansur Abdulin recalled the catastrophe that ensued when Red Army soldiers discovered an intact distillery abandoned by the Germans in retreat: many "tied one on," and when thirty enemy tanks appeared, drunken soldiers who had fought bravely elsewhere met a ghastly end.[139] Access to alcohol only increased as the war continued, as Red Army men gained access to the wine cellars of East Central Europe.[140] Once this happened, it became increasingly difficult to control consumption habits.[141]

Drink offered one of the few escapes for men under severe stress but could lead to disaster. The ambiguities of vodka as doled out by the state had a peculiar effect, according to Pokhlëbkin: "By 1945 the use of vodka, which had been low-class and forbidden, suddenly became very prestigious among the mid-level leadership . . . and refusing your allotted portion of spirits was already understood as an element of opposition and disloyalty." Who, after all, would refuse what the government had provided?[142]

While Red Army soldiers were provided with vodka, they were on their own when it came to finding water. Soldiers were issued half-liter canteens and were supposed to bring their own mugs upon mobilization. The canteen, however, often suffered from several shortcomings. To economize on precious aluminum, used for both canteens and airplanes, the army began manufacturing glass canteens. A report concerning equipment in the first three months of the war concluded: "The canteen in and of itself is convenient, but the glass ones are very fragile, and the aluminum ones are too few and expensive to make."[143] Glass canteens continued to be manufactured as a stopgap measure. Even though metal canteens were supposed to become the norm, over four million glass ones were ordered in the third quarter of 1942 and five million in the third quarter of 1943.[144]

The army published norms for hydration as well as recommendations on what, when, and how to drink.[145] It was estimated that every soldier consumed 10–15 liters of water a day, drinking 3–4.5 liters and using the rest for food preparation and cleaning. Medics were responsible for testing and marking all water sources.[146] Soldiers were officially tasked with finding or digging their own wells and building their own filtration systems.[147] The army discouraged soldiers from drinking water, as there was no way to ensure that water found would not prove harmful or lethal, particularly given the presence of the dead and tendency of retreating Germans to poison wells. Much of the time, soldiers just took their chances, drinking from ditches or wherever else they could find water.[148] This was another aspect of the importance of tea: providing tea ensured that soldiers would be drinking water that had at least been boiled.[149]

Water was scavenged but not trusted, preferably converted into something else. Nonetheless, this nonissue liquid became a way to dull hunger, as Boris Slutskii recalled: "Not just Kazakhs and Uzbeks, but heads and commanders of MPVO [Local Anti-Aircraft Defense] in the artillery regiment added many liters of water to their kasha—so that at least something would slosh around in the belly."[150] Inspection reports also mention this practice with alarm, as it was assumed that diluted food would not be properly digested.[151] Despite the army's attempt to control completely what soldiers consumed, at times what they ate and drank was entirely beyond its control and often a reaction to failed attempts at provisioning. Water all too often took the place of a soldier's "daily bread."

Bread was the most important component of a soldier's ration. A regiment (at full strength just over three thousand soldiers) would eat 2.6 tons of bread a day. Whether freshly baked in mobile field ovens or dried for long-term storage, bread made up half of the calories in a soldier's ration, was officially considered "the primary foodstuff," and, at 500–800 grams (17.6–28.2 oz.), was the largest portion of rations by weight. The wide gap in bread rations was one of the most palpable examples of the hierarchy of foodstuffs between the front and rear. In Russia bread occupied a psychological and cultural space symbolizing sustenance writ large.[152]

At the front, soldiers' obsession with bread could be extreme. Saushin, a provisioning officer on the Volkhov Front, recalled two instances of the close relationship soldiers had to bread. The first came from the dark times of 1941, when, after a prolonged period of being cut off from supply, soldiers received their rations. While crouching under fire, one man "held his rifle in one hand and a half loaf of bread in the other. It was uncomfortable for him to bend to the earth, and when necessary lie down and rise again . . . 'Drop the loaf, you'll get yourself killed!' I yelled to him . . . The Red Army man stopped for a second, and with surprise and fear looked at me. 'But it is bread! Don't you understand Comrade Commissar, bread' . . . It seems that for him it was easier to take death" than part with his bread.[153] During an inspection, General Shcherbakov was disturbed by how thinly the men sliced their bread. A soldier responded: "The thinner you slice it, the more there is. You see it's worth its weight in gold."[154] On the starving Leningrad Front, there was reluctance to give the men their bread ration in one lump sum—they ate it immediately.[155] Gabriel Temkin recalled that the young soldiers in his platoon were glad to be in the army as it was the place one could find ample bread. They would even save it for last as "bread is good by itself."[156]

Discontent and Subversion

Soldiers often complained about their paëk. The soldier with whom Temkin shared his mess tin grumbled: "Two things . . . bread and tobacco, should be distributed according to needs, and not according to the silly equal stomach principle. Take bread, the food most important for a human being. Is it fair to give somebody, a big guy like myself and a small guy like you—no offense, Gavriusha—the same daily paëk?"[157] Appetites, metabolisms, and differences in body mass were outside the scope of rationing, to the resentment of some.

Station was a key factor determining what soldiers received. The paëk did not always seem fair, and interest in how comrades of other ranks or branches of service ate speaks to the moral economy of provisioning. Boris Slutskii recalled how enlisted men envied the rations received by officers.[158] Resentment targeted not only those of higher rank but also elite branches of service. When the army was approaching Berlin, a soldier in Rakhimzhan Koshkarbaev's platoon described pilots as "devilish aristocrats" for receiving cookies and chocolate while infantrymen had "forgotten the taste of sugar": "I am thinking about the future, Commander. When the war ends, and they start to write its history, some good-for-nothing descendant will put it into their head to define the extent of participation of a branch of service in battles by how well they were fed. And it turns out that the poor infantry didn't play any role. Just try and prove later that you trudged through half of Europe with your stomach."[159] Was a pilot risking his life any more than an infantryman? Why did a "devilish aristocrat" deserve more and better rations than cannon fodder? Even if he needed more calories to fulfill his task, why did a pilot get them in the form of scarce cookies and chocolate? The fact that the government clearly used calories and scarce goods as a measure of worth made these questions all the more sensitive.

Hierarchies and sympathies could create a situation that reinterpreted paëk. Machine gunners often received extra vodka and rations, commanders could send newly arrived soldiers to the field kitchen to fatten up and might share their additional officer's rations with an old friend under their command.[160] Commanders learned to send the soldier who knew how to flirt with the (female) cook to get their rations, as she would pour them a thicker soup.[161] Interaction between the sexes was just one of many ways in which understandings of food as something more than calories interfered with the state's mission of nutrition.

Early in the war soldiers began to challenge the calorie principle of provisioning. As the war dragged on and they were forced to live through the above-mentioned "eras" of one or another foodstuff that had been stockpiled, they complained. Pearl barley porridge was known as *shrapnel'*, and one prosecutor mused that the common expression denigrating female soldiers who lived with commanders, PPZh, was allowed to enter into common usage because it distracted soldiers from a more demoralizing phenomenon—PPS, *postoiannyi perlovyi sup* or "eternal pearl barley soup."[162]

National Difference and Military Cuisine

Strong reactions to rations occasionally arose from the way in which provisioning disregarded the cultural traditions of some of the men in the ranks. The war was the first time large numbers of several traditionally Muslim nationalities were mobilized into the Red Army. Culinary practices within the army often varied dramatically from what these men had eaten in the prewar world. The meeting of different nationalities at the front could lead to an expansion of culinary horizons, as Uzbeks ate borsch for the first time and Ukrainians ate *plov* (pilaf). Pokhlëbkin claims that the war introduced many people from east of the Ural Mountains to the potato, which Lizzie Collingham declared "became the food of the Second World War."[163] One Azerbaijani draftee (who would die defending the Brest Fortress) complained on the eve of the war that he couldn't eat the local food.[164] Nikolai Inozemtsev made several references to the *chebureki* (fried pies common in the North Caucasus and Crimea) that his comrade Akhmetov made on special occasions.[165] An article from the newspaper *Za Rodinu* describes how a Yakut, Ukrainian, and Russian all prepared national dishes for their comrades—variations of dumplings.[166]

Consuming rations could bring together or alienate soldiers from different ethnic backgrounds. The Red Army's Political Department was particularly disturbed when some soldiers refused to share tobacco with anyone other than their co-ethnics, interpreting this act as a threat to the "Friendship of the Peoples," the rhetoric of harmonious coexistence among the many nationalities of the Soviet Union, which served as a cornerstone of the Soviet system.[167] Top political officers also discussed the importance of tea for some nationalities in 1943: "Things are bad with hot tea. This question is particularly sharp in non-Russian units. Uzbeks and Kazakhs especially love tea. If one of them gets a medal they all go to drink tea with him. But here we hit the question—where can they drink tea?"[168] Creative commanders provided these spaces. Some soldiers greeted reinforcements from Central Asia by carving out a *chaikhana* [tea house], getting some *pialy* [Central Asian-style teacups], and cooking plov with horsemeat for them.[169] Others resented the refusal of many Muslim soldiers to eat pork.[170] Given, the ubiquity of hunger, such behavior could seem criminal. In the army everyone was forced to eat things that they found less than appetizing, but for some, the food available challenged fundamental conceptions of themselves, which could occasionally lead to choosing hunger over betraying deeply held beliefs or to eating unfamiliar foods that their bodies did not always accept.[171] Even as provisioning improved and the army began to emphasize variety and such amenities as tea houses, offering an alternative to pork was not something that interested the government. The war's deprivations instead "taught" people how to eat anything.[172]

From Hunger to Feast

By mid-1943 the organization of the rear area became noticeably better and the resources available to the army richer. Aleksandr Lesin's diary is marked by constant references to food and hunger in 1942, but by the summer of 1943 food is not what is on his mind and is rarely mentioned through the end of the war. Rafgat Akhtiamov, who had written his parents several times in 1941 and 1942 asking them to send food, wrote home in 1943: "Don't worry about me. Now all is well with food."[173] When soldiers mention food in interviews and memoirs, 1943 is generally remembered as the year in which quality and quantity noticeably improved.[174] This trend continued, raising expectations among the troops. An artillery officer interviewed in March 1945 boasted: "We are fed very well, as Guardsmen . . . People have become so finicky that they say: 'I don't want a pig, I want suckling pigs, goose.' There is enough of it there. People have gotten so fat that they are like peaches."[175] Vasilii Grossman noted the same trend: "Among the infantry rosy, plump faces have appeared, which never happened before."[176]

With this new abundance came new responsibilities. Lesin called for the public execution of anyone stealing from the local population in Latvia, specifically citing the ample food available.[177] By the end of the war food had become sufficient enough that it could be wasted, as Boris Slutskii recalled: "all around the infantry overran kitchens, knocking mountains of kasha into the dirty snow—even though in the kasha they heaped 600 grams of meat per person, and not 37.5 grams of noble egg powder."[178]

Mobilizing Calories

The way that belligerents imagined the war could not be divorced from resources, especially food. Nazi planning imagined the Soviet Union as a space of extraction; occupation policies turned these imaginings into reality. Placing food near the center of its concerns, the Soviet government reexamined its relationship with its citizens, categorizing those defending it on a higher plane under conditions in which the possibility of starvation was very real. The implementation of this relationship created hierarchies, which stated in quantifiable terms whose life the state valued above others. While many aspects of provisioning would be reconsidered, these hierarchies remained intact and were indeed refined in the course of the war as a variety of elite formations and specializations saw privileges added to their status. In addition to creating new hierarchies, invested with real benefits, this system had the potential to efface identities that had existed before being drafted into the army.

The army as an institution was not interested in accommodating the culinary norms of the variety of peoples who composed its ranks; it was concerned with the much more vital function of keeping people fed. Muslims were fed pork alongside atheists and Orthodox Christians. In dire straits, Russians learned to eat horse from their Turkic comrades. If the cook of a unit happened to be Uzbek, men from European Russia might find themselves eating *plov* (likely with horse) for the first time. The army became a place where large numbers of men and women from a variety of ethnic and regional backgrounds came to share something like a common culinary culture. Despite the fact that provisioning was so localized, everyone in the army was likely to have received similar portions of borsch and kasha. Everyone experienced the same periods of feast and famine, shared while dipping their spoons into the government's pot. They would use similar tactics to survive when the army failed to provide and use their rations in ways that suited them. It would be nearly impossible for these soldiers not to appreciate how much better their rations were than those of their families in the rear. The shared experience of suffering and improvisation, followed by feasting and victory, is part of what made the Great Patriotic War such a central event in Soviet history. Food could unite and divide men and women in the ranks.

By the war's end, the abundance enjoyed by the army came from a much better-organized apparatus with access to more and more resources.[179] Everywhere it went, the army established a monopoly on foodstuffs, and in areas ravaged by war the army was often the only source of provisions for both civilian and military personnel. In Berlin, in the course of May 1945, the Red Army was feeding two million of its own soldiers and four million German civilians.[180] The army fed entire enemy cities, incorporating their populations into military provisioning via ration cards.[181] Once the provisioning system was fully functioning, the mutual obligations implied by the word *paëk* came to encompass former enemy civilians and prisoners of war. In return for recognizing the Soviet Union's monopoly of sovereignty, former enemies were provided with sustenance.[182] The army had come a long way from the dark days of 1941–1942, and its ability to provide for an organization of such scale moving so quickly was truly phenomenal. Of course, these calories were not given for free but as fuel for deadly and exhausting work.

PART TWO

Violence

FIGURE P.2 Red Army soldiers in battle near Kursk, 1943. Note the dead German soldier and DP-28 machinegun carried by the soldier on the right. RGAKFD 0-56024.

CHAPTER 4

Cities of Earth, Cities of Rubble

The Spade and Red Army Landscaping

> The trenches are like an underground city, with signs pointing to platoons, companies, Lenin Rooms, etc.
>
> —Oles′ Gonchar

Arriving at the front in 1941, Valentina Chudakova, who would later command a machine-gun platoon, was pleasantly surprised by the visible presence of order:

> I had no idea what the front was like. Or rather, I thought that everyone shoots at each other day and night, comes together in hand-to-hand combat, runs and hides wherever they find a spot . . . I was pleasantly surprised: here there was absolute order. What can I say—a real earthen city—like that of some ancient settlers. The main trench, that's the central street, and from there to the front and the rear are side streets and dead ends. On the side streets, leading in the direction of the enemy, there were all sorts of things—pillboxes, wood and earth fire positions and caponiers, covered foxholes . . . In the side streets headed toward the rear dugouts were hidden under snow-covered roofs. No commotion, no fighting—silence . . . I whistled: "Well, see how they've worked! Mama-infantry sure has dug in!"[1]

Her intuition that the front would be a space of chaotic violence was correct—it was the potential for destruction that led to the creation of the organized, essentially urban space that so impressed her. These cities of earth would be built everywhere along the front in response to the cities of rubble that modern weapons could create.

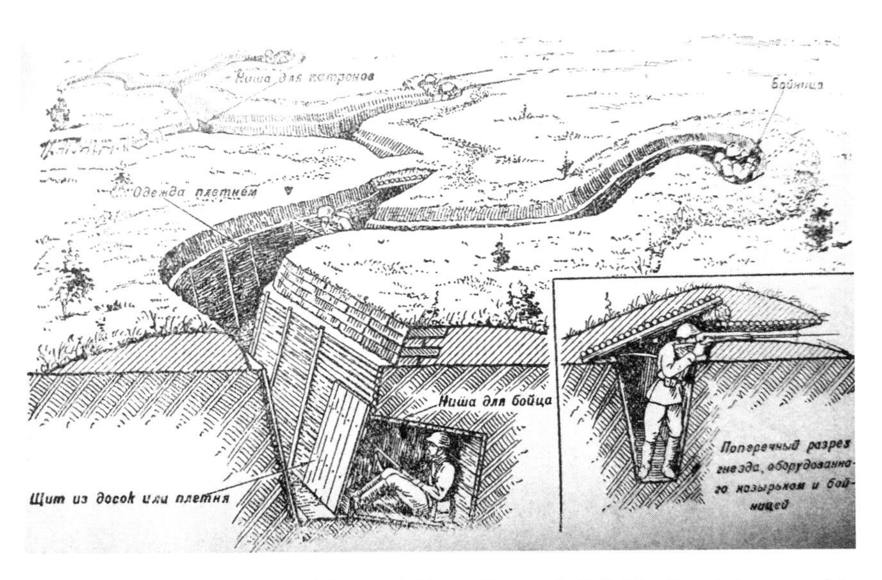

Figure 4.1 The city-like space of the trench. The caption reads "A finished trench for a squad." Ushakov, *Voenno-inzhenernoe delo*, 42.

During the Great Patriotic War, survival at the front was virtually impossible without the help of an ancient hand tool: the spade. The idea of the spade as a loyal friend became a major trope of military propaganda.[2] That such a humble object received so much attention is telling. The spade was the key to soldiers' reading, shaping, and using the landscape. Soldiers in the Great Patriotic War were reported to move about twice as much earth as soldiers in the trenches of World War I.[3] The soldier's small spade was an anonymous object, standard to everyone, and manuals taught soldiers to dig trenches according to a regular plan, yet Red Army soldiers excavated highly personalized spaces.

Red Army soldiers turned a wide, open field of fire into a narrow, constricted, and sometimes even cozy system of trenches, where they could more effectively kill without being killed. Soldiers marched from trenches they had dug to spaces they would turn into trenches. Excavations were to be carefully camouflaged, even as they dramatically altered landscapes. Those at the front attempted to recreate aspects of normal life amid death in the cities they created with spades. A small, relatively primitive object and the labor it facilitated were central to the experience of millions of people in one of the most technologically advanced conflicts in world history.

This chapter examines the deadly environment soldiers inhabited and the endless work that allowed them to survive. The first section details the

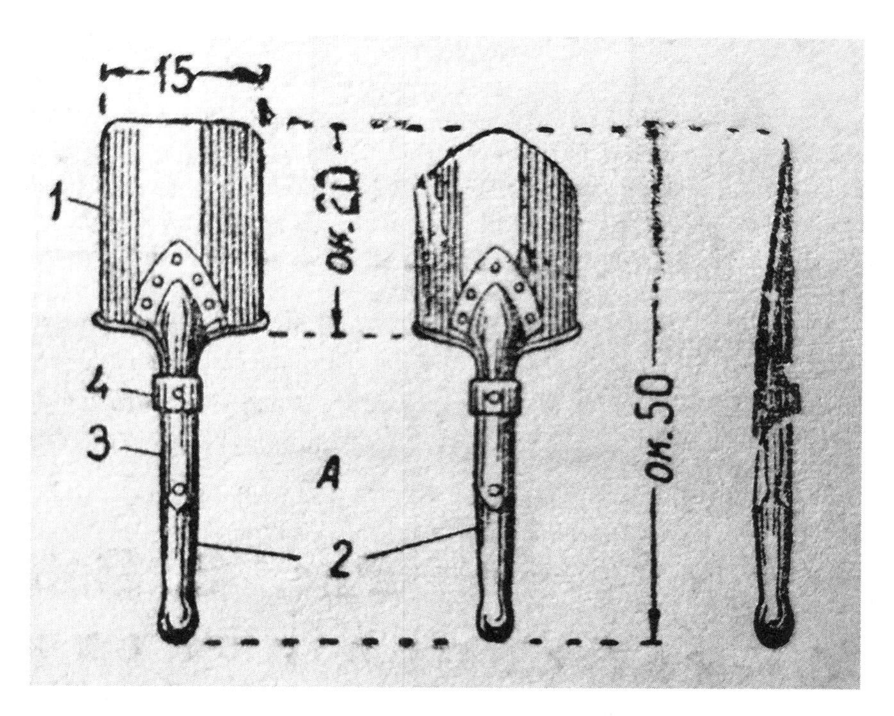

FIGURE 4.2 The soldier's spade, a commonly reproduced image from manuals. *Nastavlenie po inzhenernomu delu dlia pekhoty (Inzh. P-39)*, 8.

destructive powers soldiers faced and how exposure to massive devastation changed the way they moved, read the landscape, and understood the war. The second explores how soldiers used their spades to survive. The last two sections investigate attempts to recreate normal life in the trenches and soldiers' reactions to the deaths of their comrades.

The cities of earth that soldiers excavated mirrored Soviet cities in significant ways. Defined in contradistinction to both the village and capitalism, Soviet cities were to serve as key sites in the forging of new Soviet people. This transformation was most often from backward, unwashed peasant to conscious, cultured worker. Institutions such as libraries, clubs, museums, and bathhouses were important vehicles of this change, but so was communally organized living and an organization of space that directly connected workers to workplace. The factories around which socialist cities were built gave them their meaning, with a green band of parks to separate residences from workplace. In practice, socialist cities were often haphazard realizations of their rationalized blueprints with a marked lack of infrastructure. The spaces of production were built before permanent housing and the new urbanites had to improvise living space, plumbing, and various comforts. The emphasis on extreme exertion in labor, communal living that created larger collectives, and improvisational practices would all be reproduced as soldiers excavated cities at the front. The transformation from civilian to soldier would be realized by the soldiers' labor and tweaked by the state's attempts to provide the front with the hallmarks of urban infrastructure. Massive green spaces were preserved for the sake of camouflage. Foxholes and bunkers became managed spaces of transformation just as factories and communal apartments had been before the war.[4] In short, "the cities of earth" described here were a militarized iteration of general practices of Soviet urbanization—except now the sense of urgency and space had changed dramatically, as the enemy reduced Soviet cities and villages to rubble.

The cities of earth were dynamic border cities of a society fighting off a deadly foe. On this massive front, all territory—both around the globe and around a soldier's foxhole—was divided into "friendly," "enemy," and the liminal "neutral." The consequences of this division were very real for the soldier and formed the understanding of space by which tactical and strategic decisions were made. At the front being on enemy or neutral territory was to be directly in danger, while being on friendly territory provided a (very limited) degree of safety.[5] Commanders envisioned territory in terms of maps on which they often manipulated objects or drew where they thought their troops were positioned. Reliance on telephones meant that this was often conjecture, as artillery fire cut lines or troops became cut off from their headquarters. The crisp space of the map, although enhanced by constant reports and smaller, detailed maps provided by subordinates, could reflect the landscape faced by soldiers but captured little of the experience of those occupying the real terrain that it represented. The dirt and confusion of the battlefield was not part of these abstractions. Many commanders took tours of frontline positions to understand what exactly their soldiers were up against, and top-ranking officers often made use of aerial photography and other forms of reconnaissance.[6] On tour at the immediate front line and indeed more often than not in the bunker where he made decisions, the commander was confronted with realities that were not expressed on the two-dimensional world of the strategic map with red and blue lines dividing the real estate into "ours" and "theirs."

Despite the massive, impersonal nature of World War II, it was experienced as something intimate. Grigorii Baklanov explained that "for a soldier, the front is whatever faces his foxhole."[7] Although soldiers received rumors and official information from all over the front and rear, the immediate front, which they viewed through the narrow perspective of the trenches they built, was the world that they experienced. Yet soldiers' individual foxholes added up to an enormous front.

Cities of Rubble

In the 1940s it was much easier to envision and move through vast expanses. Aerial reconnaissance had given opposing armies the ability to see far behind the front lines. Movement that would have taken weeks even in World War I could now be a matter of days or even hours. In the interwar years technologies of the Great War became even deadlier. Tanks and planes were faster and better armored and packed heavier weapons, dramatically expanding the space of the battlefield. Hand grenades and automatic weapons had become more efficient, portable, and plentiful. Soldiers could be observed and injured more efficiently than ever before. Anything that could be seen could be destroyed.

The new potential for killing dramatically changed the way that soldiers experienced the landscape. Soldiers quickly came face to face with the capacities of new technology to kill and maim. Corpses were rendered unrecognizable, as one female soldier wrote her friend: "I lost my beloved black-eyed Sashka. I couldn't find his head. I gathered the remaining pieces of meat and buried them with my own hands."[8] A traumatized veteran of the Leningrad Front recalled: "when a person next to you is ripped to shreds, when you are soused in his blood, his insides and brains are strewn over you—this is enough in peacetime conditions to make you lose your mind."[9] The terrible experience of shelling could place unbearable strain on soldiers. Viktor Astaf'ev described a bombardment as follows: "it seemed as if now, this minute, the earth was shifting or had just shifted its axis."[10] Another veteran confided in his diary that during prolonged shelling: "it doesn't so much frighten as much as oppresses, and you think: 'Eh, the devil take it. Wouldn't it be better if they hit us, so that our suffering would be over!'"[11] Soldiers understood how vulnerable they were at the front; it was graphically demonstrated on the bodies of the fallen and carved into the landscape. The Germans dropped 148,478 bombs on the city of Leningrad alone. Numbers of this scale are abstract, but the effect of a few bombs on a small piece of real estate held by a few soldiers was made explicit by shell-pocked landscapes and dismembered bodies.[12] While this experience was universally unnerving, the ways in which people experienced this destruction changed depending on the direction in which the army was traveling.

Early in the war, soldiers underwent the shock of destruction as they retreated through towns and cities, often harassed by German aviation, which enjoyed total domination. Soldiers felt powerlessness and a sense of shame in front of civilians, whom they could not protect and left to an unknown fate. When they stopped, it was often in areas drained of inhabitants, as the Red Army frequently evacuated civilians from the "front strip" (*prifrontovaia polosa*), a space five or more kilometers behind its lines.[13] As the fortunes of war changed, so did the ways in which Red Army soldiers witnessed the destruction of the landscape and its inhabitants. When an area became the center of a protracted battle, undamaged structures became captivating. A division commander recalled at Stalingrad that after 156 days of combat, he saw an intact house for the first time: "We were so used to ruins—that seemed so normal to us—that an intact little house was a remarkable phenomenon and attracted our attention. We even stopped, looking at this surviving house."[14] During the retreat soldiers often felt themselves helpless victims and bystanders. When the Red Army began to retake territory, as early as December 1941, soldiers became both liberators and witnesses to much more systematic destruction than shrapnel and strafing could accomplish. As the army moved westward, destroyed villages became an everyday sight. The consequences of failing to defend territory became sickeningly clear. Diaries, letters home, and memoirs noted the inconceivable destruction that the retreating Wehrmacht left in its wake, as Vasilii Chekalov, an artillery officer, recorded early in 1943: "I've had my fill of wandering through other people's destroyed huts."[15] Lt. V. P. Kiselëv, also an artillerist, wrote in his diary: "I say 'villages,' but you see they remain only as geographical notions; they've all been burnt to the ground by shot and shell."[16]

Soviet authorities were keen to use the ruins of former cities and villages as a pedagogical tool to convince Red Army soldiers that their cause was just. In 1942 an article in the widely circulated *Propagandist Krasnoi Armii* discussed how soldiers came to realize the level of destruction visited by the enemy. When confronted with ruins, they asked an agitator: "Where did all the cows and sheep go? Why were the houses and mangers burnt? Where did all the people who lived here go? The Germans ate the livestock. The Germans burned the village. Germans drove the women, children, and teenagers into backbreaking slavery."[17] In the spring of the following year, when large areas occupied by the Germans were liberated, the entire Red Army held meetings dedicated to showcasing Fascist villainy. These "Meetings of Vengeance" became a common genre of agitation where soldiers were gathered around graves or destroyed villages and told of the horrors that the Germans had visited on Soviet people. By May 1943 the main propaganda publication of the army reported that 80 percent of personnel in the active army had attended at least one of more than a thousand such meetings.[18] In units far from the

FIGURE 4.3 Red Army soldier reunited with his family in a destroyed village near Belgorod, 1943. RGAKFD 1-105157 (Ia. I. Riumkin).

FIGURE 4.4 Kiev, city of rubble, view of Kreshchatik (the central street), November 1943. RGAKFD 0-271941 (A.S. Shaikhet).

front visitors from afflicted regions, including children who had been mutilated and veterans who had witnessed atrocities, gave harrowing accounts of what occupation meant. Letters from recently liberated areas were read aloud. Soldiers also discussed their homes under occupation: "Each of us has his or her own score to settle with the Germans. I for the destroyed and burned city of Briansk, for the tears of my mother and sister, who have already languished seventeen months in fascist captivity, and may no longer be among the living. Starshina Pakhomov for Rzhev. Dorozhchenko—for Khar'kov, and all of us together—for our Motherland, for our wives and children, for our brothers and sisters, fathers and mothers, factories and plants, for everything Russian, Soviet."[19] Hatred and rage toward the enemy and anxiety about the fate of loved ones could bring together people from all over the Soviet Union.

To appreciate how important such propaganda could be, it's worth remembering the fact that many soldiers were hostile to the regime and/or came from remote regions that did not feel the direct threat of destruction by the invaders.[20] Face-to-face meetings with survivors of destruction and witnesses of Nazi atrocities, set in the ruins of towns and villages and by the graves or exposed corpses of victims of the occupational regime, had a tremendous effect on soldiers, as the political agitator Guards Major Akai Neuspbekov recalled: "The reinforcements we received that summer from Kazakhstan saw such ruins, German villainy, that this meeting impressed in them hatred of the German barbarians."[21] The letter of Tatar soldiers from the First Ukrainian Front to their faraway homeland told of the horrors they had seen and the rage they felt: "We have fought our way through the lands of Ukraine and seen with our own eyes the monstrous acts carried out by the Germans in the occupied zones. Ukrainian soil was drenched in blood and tears. It freezes the heart and causes one's blood to run cold when we see thousands of old men, children, and women torn to pieces and thrown in ditches and wells, when we see mountains of ash where wonderful Ukrainian cities and towns once stood."[22] Ritualized vows, using similar language, often followed meetings of vengeance and turned the tragedy of death and destruction into an epic event, motivating soldiers to carry the struggle forward and move all the more quickly forward. While group letters were often scripted by political officers, their sentiments were probably genuine. Even if expressed in the terms of the Communist Party, these letters were written in response to real horrors soldiers had witnessed. A resolution from the North-Caucasian Front declared: "Dear brothers and sisters on occupied territory. We hear your moans and calls for help, your prayers for liberation from German slavery. We, warriors of the Red Army, are coming to free you.

No difficulties, hardships, or obstacles can stop us. We swear to the Soviet People, to great Stalin, to fight until the last German on our land is destroyed . . . Not one drop of blood, not one bitter tear of innocent Soviet people will go without vengeance."[23] Every day of occupation meant more women raped, towns destroyed, and fellow citizens humiliated and murdered. Quick movement westward would save lives.[24] A letter from children to their father read aloud at a meeting on the Southern Front put it plainly: "Papa, we were put on list No. 475 by the Germans—of people to be shot. The only thing that saved us was the quick advance of the Red Army."[25]

Meetings of Vengeance would continue throughout the war, and in addition to vows by those present to seek vengeance, signs narrating German atrocities to passersby were also planted. Boris Komskii recorded in his diary on October 1, 1943, that in the village of Shablykino, where he saw only graves and chimneys, the following sign was posted: "Here was a regional center. It was thoroughly looted and burned by the Germans. Fighter, remember and avenge!"[26] The Political Department of the army reported that soldiers were often bolder and more disciplined and worked harder after Meetings of Vengeance, and the evidence of what occupation meant became ever more abundant as soldiers marched west.[27]

FIGURE 4.5 Meeting of Red Army troops in honor of the victory at Stalingrad. The following spring, Meetings of Vengeance, using ruined cities as a backdrop, would be held throughout the army. RGAKFD 0-347117 (Ia. I. Riumkin).

Retreating east or advancing west, soldiers' lives in the cities of rubble and cities of earth were interspersed with nights in peasant huts. These encounters could be tense, with peasants angry at being abandoned early in the war or suspicious of soldiers as Soviet power returned. However, liberated peasants were often immensely grateful, thirsty for news, and ready to share horrendous stories of what occupation had meant. These could also be fleeting moments of romance, as men and women facing an uncertain fate comforted each other. Many mothers worried about leaving their young daughters alone with soldiers, while others offered food and shelter to soldiers who resembled their children, hoping somewhere others were doing the same for their kin. Often these were moments of coming together around a humble feast cobbled together from soldiers' rations, the harvest of peasant's gardens, and moonshine. These meetings are a motif of much of the fiction inspired by the war and pepper the diaries, letters, and memoirs of soldiers. They were a glimpse of what life might be like if those present survived the war and a moment of interaction with the people soldiers were defending and liberating.[28] However, these were but brief pauses in a life that consisted of constant movement and excavation.

The way soldiers moved was idiosyncratic. Every move made had to be carefully concealed. A Red Army manual on marching reminded soldiers that due to aviation all movements had to take place either at night or in areas of limited visibility. Soldiers were also instructed to leave no trace of their presence.[29] Night became a time of increased activity for soldiers not only on the march but while in the trenches as well, and the routes soldiers traveled had many peculiarities.

While soldiers would cover long distances by trains or trucks, and certain types of troops (cavalrymen, artillerists, tankers, and motorized riflemen) rode rather than marched, most soldiers traveled on foot. Vasilii Grossman wrote in his notebook during the spring thaw: "Certainly no one has seen such filth: rain; snow; grain; a liquid, bottomless swamp; a black dough kneaded by thousands of boots, wheels, and tracks."[30] In dry weather roads were plagued by dust clouds, an occurrence so common that scouts were trained to judge the size and type of enemy formations by the dust cloud a column raised.[31] The soldier-artist Kharis Iakupov recalled how dust covered everything on the march: "Not only people but all military hardware looked as if it had been painted one color—a whitish ocher. The sandy dust covered everything with a thick coat: the clothing and face of a soldier. Only the eyes shined. The sand got into the mouth and crunched between teeth."[32] In winter thousands of soldiers trudging the same path could turn it into an icy slough. The mere presence of so many soldiers turned roads into highways, and these paths came to resemble the thoroughfares of major cities, as Nikolai Nikulin, a veteran of the Leningrad Front recalled: "I observed strange, remarkable vignettes on the road near the front. Lively like an avenue, it had moved in two directions. Toward the front reinforcements traveled, food and weapons were delivered, tanks moved. The wounded were hauled away from the front. And along either side there was hustle and bustle."[33] The "hustle and bustle" included the issuing of rations, trade among soldiers, and burials of the dead. As the war progressed, the army became more motorized and its highways increasingly organized.[34]

However, highways were not the only means of travel for soldiers. An old saying by General Suvorov, repopularized during the war, declared, "Wherever a deer can pass, so can a soldier."[35] Soldiers were expected to be able to cross any obstacle: mountains, rivers, swamps, and even minefields.[36] Fronts stopped wherever two armies could not dislodge each other from their positions, something that often depended on climate and terrain. This meant that if the front stopped in a swamp, among mountains, or deep in the woods, soldiers would have to travel to them. If roads did not exist, it was the job of soldiers to build them or to carry additional supplies (bullets, shells, grenades, rations) to remote positions. Yet roads could also be very dangerous, leaving soldiers exposed to fire and often sewn with booby traps.

The Wehrmacht was known to generously sow mines, particularly in retreat. Manuals warned soldiers of the tricks the Wehrmacht used and explained how to disarm mines.[37] Chekalov recorded in his diary: "On the paths of retreat the Germans are trying to sow death: they mine roads and paths, bunkers and surviving buildings. Machinery doesn't enter Zaluch'e, and people move with caution. Time and again—just like some dry grass—the treacherous whiskers of mines fall underfoot."[38] Mines were utilized not only in retreat but also on the battlefield; combined with barbed wire, they dramatically limited soldiers' ability to move and avoid fire.[39]

On the battlefield every move soldiers made was determined to avoid enemy detection and fire to stay alive. Red Army personnel mastered the arts of crawling *po-plastunski* (an extremely low belly crawl) so as to keep their rear ends below machine-gun fire and their weapon out of the dirt, running in short bursts and navigating rows of wire and mines. Soldiers had to avoid the natural instinct to crowd together under fire, keeping a distance of six to eight paces from their neighbors.[40] Orders to move on the battlefield gave landmarks for soldiers to use as orientation.[41]

Landscapes were instantly divided into sectors of fire with clear points of orientation. Trees and churches were no longer features of rural life but points for zeroing in on the enemy.[42] The war correspondent and famed author Evgenii Petrov noted how a picturesque hill became simply "height number so-and-so . . . and that Russian birch, standing by the road—it's not a birch at all but a lone tree. That's how its marked on the map. And the creek—not a river but a front line. And the edge of the forest—not a glade but an excellent position to set up a fire base."[43] Viktor Nekrasov's avatar Iurii explained to his sweetheart while on an idyllic stroll that all he could think about was where he could best place machine guns. When she took offense, he responded: "It's just habit. I now look at the moon from the perspective of its expediency and usefulness."[44]

The need to use the landscape to kill turned soldiers into cartographers. "Fire maps," "fire reference cards," or "fire cards"—a small map using the landmarks chosen by the commander of a unit or a weapon crew—were a regular part of the duties of junior commanders, machine gunners, artillerists, antitank gunners, and mortar men.[45] These maps were a graphic representation of the soldiers' need to make rational use of the landscape to kill with maximum efficiency, and every soldier was obliged to know the points of orientation. A particular feature of the landscape could be used as a mark to open fire.[46] These maps were arranged with a clearly marked north-south orientation, indicated landmarks from right to left with range clearly indicated, and were scaled as accurately as possible. Neighboring units, the disposition of the enemy, and the arc of fire from the flanks of the unit were all indicated. These documents were signed and dated down to the hour and served as a key for anyone to use the stretch of ground in front of them to cause maximum damage to the enemy. They represented a sort of procedural expertise where junior commanders graphically demonstrated their technical mastery, often including azimuths necessary to hit their targets. The reference card allowed anyone to exploit weapons available to great effect and were copied to be viewed by their author's superiors. Through fire cards soldiers and commanders reduced an unmanageably complex landscape into a set of clear, rationalized vectors for killing.[47]

Junior commanders, scouts, and observers also constantly recorded and often mapped changes in the enemy's disposition, the planting of mines, and so on and copied their own fire maps to be sent to their superiors. Scouts engaged in nighttime raids to capture enemy soldiers for interrogation. The possibility of capture could make soldiers' knowledge about their trench system a liability for their comrades. Whether they were recording maps or simply traveling through the maze of tunnels they had dug, soldiers learned quickly to orient themselves. They came to know their section of the line expertly, but lack of knowledge of their vicinity or dramatic changes brought about by weather or bombardment could have catastrophic effects,

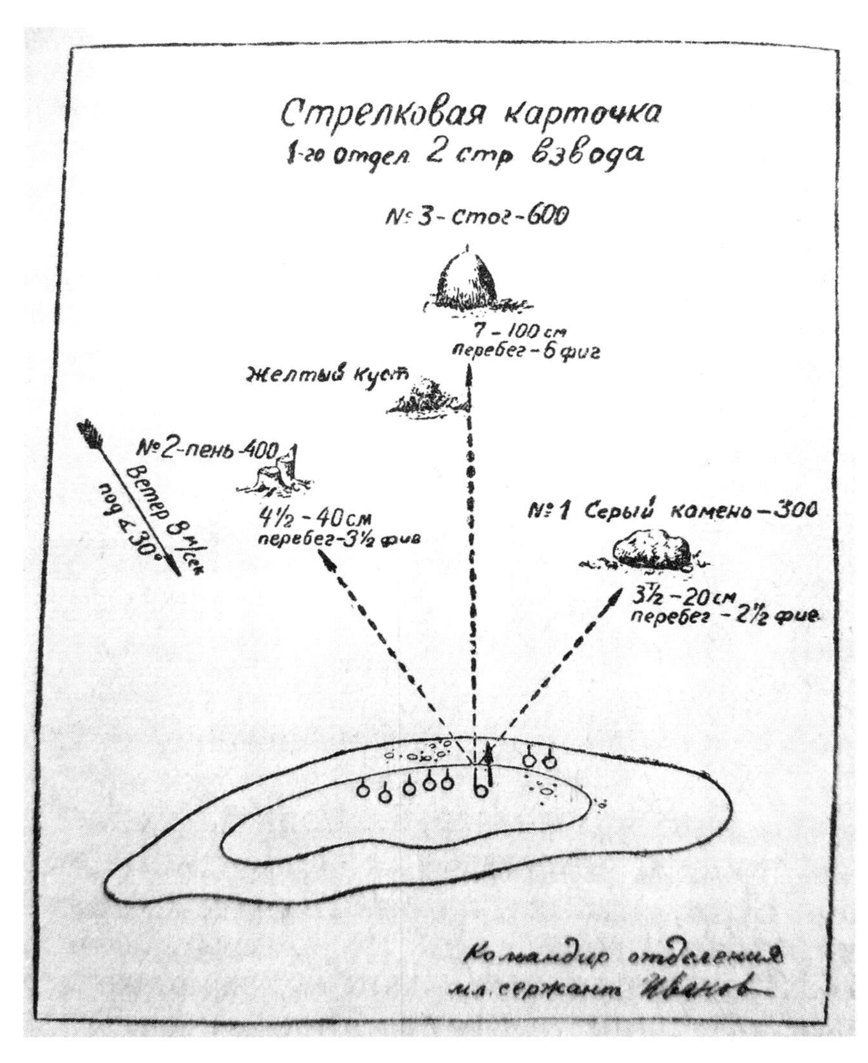

FIGURE 4.6 Fire card for a squad (roughly eight soldiers) using a rock (1), stump (2), haystack (3), and yellow bush as orienteers with gun sight markers, signed "squad leader jr. sergeant Ivanov." V. V. Glazatov, *Pamiatka komandiru strelkovogo otdeleniia po vvedeniiu ognem v boiu* (Moscow: Voenizdat, 1942), 14.

as soldiers could wander into enemy territory or be mistaken by their own soldiers for the enemy.[48]

Every piece of real estate gave advantages to one side and hindrances to the other. Commanding heights allowed an army to use artillery and machine guns more effectively. Controlling a road kept supplies and reinforcements rolling freely. Access to a body of water could provide transport and drinking water. In battle, any hill or remnant of a building could become a strong point that would cost lives to take. Every advantage in topography or adept usage of the landscape meant better chances of survival and more chances to kill the enemy. Navigating and surviving this environment required very specific skills. Yet even proficient soldiers inevitably suffered in this habitat.

Toward the end of the war, the military translator Boris Suris reflected on the awful conditions faced by a soldier across the Oder River from his bunker, summing up the twin miseries of danger and bad weather: "He sits in a shallow foxhole, which he only just carved into the cursed dam; exploding bullets keep clicking against the parapet. To the right of the brick factory two machine guns cut ceaselessly, artillery beats with a hurricane of blasts. Cold and wet as a dog since last night. They didn't bring up anything to eat all day—just nibble on a crouton [*sukhar'*]. Or doze off, if you can, crouched in a heap, switching off with your partner."[49]

Survival at the front meant killing and not being killed yourself while waging a parallel struggle against the elements. In such a deadly environment soldiers of all branches of service became heavily reliant on their spades, which allowed them to craft safety in the earth.

Cities of Earth: What the Spade Built

Soldiers' spades, often referred to as "the soldier's loyal friend," dangled off the belt near their right buttocks.[50] It was, after their weapons, the most important item to their survival. A thin sheet of metal 20cm long and 15cm wide, drawn over a 30cm handle, the spade could have a square or a sharp head. It was the soldier's task to wipe condensation from, sharpen, clean, and oil it as soon as his work was done.[51] Shortages of spades plagued the army, and even in the third quarter of 1944, when supply was at its height, only 80 percent of the active army had spades.[52] Due to wartime shortages, both the spades and the covers that held them were simplified; spade covers became entirely skeletal or multipurpose carriers (e.g., with a pocket for carrying grenades), and spades themselves were redesigned to use less metal.[53] Despite the spade's modest appearance, soldiers created whole cities with it.

Kliment Voroshilov, the people's commissar of defense before the war, declared that a soldier was "not prepared, if he has not mastered the spade, has not become accustomed to use it with the same skill as a spoon at the table."[54] Soldiers were expected to begin digging immediately if they stopped anywhere where they might have to fight. Digging was a way of making a claim to any territory, even a few meters taken from the enemy. This was of particular importance to an army that instructed its soldiers to be very aggressive and had retreated deep into its own territory. Red Army soldiers were trained to dig under fire, lying down while observing the enemy with their rifles an arm's length away. They were instructed to find a space that provided good fields of fire and that would be easy to conceal. In less than ten minutes soldiers dug berms behind which they could lie down.[55] The spades that soldiers carried were impossible to use standing until one had dug themselves into the earth. The position one had to dig in was unnatural and uncomfortable: in full kit, the soldier was to keep his body and head as close to the ground as possible and thrust his legs into the ground to give him force.[56] When soldiers were lucky, they might be issued larger implements that were carried with their unit's headquarters, but this was largely the province of combat engineers and artillerists.[57] Under extreme circumstances, civilian labor could be mobilized to dig massive networks of anti-tank ditches, a practice that was common near major cities, but the cities of earth were overwhelmingly of the soldiers' own construction.[58] Outside enemy observation, soldiers could dig while kneeling, but those training them assumed that they would be digging under fire.

Army manuals stated the time necessary to dig a breastwork out of everything from sand to snow and the thickness of breastwork needed to stop a bullet from a machine gun at a hundred meters. They also measured how many hours it would take for one man to dig such a barrier.[59] In an abstracted

FIGURE 4.7 Red Army Soldier entrenching while observing the enemy, Central Front, 1943. RGAKFD 0-312319.

Figure 4.8 Soldier removing sod while digging in full kit (m.1936 knapsack). *Inzh. P-39*, 32.

Figure 4.9 The deepening of a foxhole from lying down to standing: fighting from behind a berm. *Inzh. P-39*, 33.

Figure 4.10 The deepening of a foxhole from lying down to standing: narrow one-man foxholes, 1941. RGAKFD 0-94214 (A. S. Shaikhet and Garanin).

"average" soil (probably black earth), a soldier was expected to dig a camouflaged hole in which they could fire while standing within an hour.[60] This structure had the advantage of providing 360 degrees of fire, while "protecting the soldier against the fire of aviation and tanks passing over the trench." It was extraordinarily narrow, providing just enough space to stand and wield a weapon. These were the basic unit of the cities of earth—the place of work for soldiers, built to a standardized plan.

Manuals gave detailed instructions on how to dig an ergonomically informed, user-friendly foxhole. This included a special side step for getting out of the trench and throwing grenades, as well as a niche to fire from.[61] Specialized foxholes for machine guns, mortars, artillery, and so on were all part of the repertoire. All of these specific structures were given with dimensions in manuals, and soldiers were supposed to use their spades (conveniently 50cm long, with elements that measured 15 and 20cm) as a means to measure their excavations.[62] Manuals provided a boiled-down essence of necessary survival skills that required interpretation, personal experience, and constant revision (often provided by military newspapers and journals). They were universal in their prescriptions, but flexibility was, ironically, part of this universality.

If the army couldn't make the landscape uniform, it could hold its soldiers to a uniform standard. While specialized tips were given as to how to construct a fighting position in mountains, swamps, the steppe, and snow, these structures posited an average person and ignored differences in height, age, and gender that could be significant. Mansur Abdulin reflected: "Even I, a nineteen-year-old miner used to heavy labor from a young age, have moments when I can't go on. What is it like for those who have just finished ninth or tenth grade? Or those who never did manual labor?"[63] For the tall or the short, trenches could prove too deep or too shallow. However, as one article that highlighted a short soldier's endeavors to make a trench his own stated: "blame your parents for giving birth to one so small as you. But the trench is as it should be, dug by one measure, by regulations. You are the boss here now, make yourself at home."[64]

Soldiers' shovels moved the sand of the Karelian Isthmus, stones of the Caucasus Mountains, black soil of Ukraine, marshes around Leningrad, permafrost near Murmansk, and shattered concrete of Stalingrad and Berlin.[65] They helped forge the fencing and log work that would take the place of a trench in swamps, which Dunaevskaia described in her diary: "The ground is swampy, and we take our places in a log framework, which neither defends nor saves from either shrapnel or the cold. Our bunks are made of the same logs."[66] Spring mud was virtually impossible to dig into, but with enough wood a drainage system could be built to make such a situation livable.[67] Manuals gave timeframes for each type of soil but failed to mention rocks, roots, and all the little things that could make digging hell. There was also no discussion of how soldiers were supposed to keep themselves from becoming utterly grimy while creating and then navigating a claustrophobic space made of soil.[68]

Many soldiers had received minimal training before arriving at the front, and although most recruits came from a peasant background and were accustomed to working the earth, the specific types of labor required to build cities of earth differed dramatically from traditional peasant landscaping and building. Sod houses were sometimes built in emergency situations by peasants and workers, but the demands that everything be invisible and impervious to enemy fire made frontline structures unlike anything that most had ever built. Regardless of their background, men and women in the service came to share in the ubiquitous tasks of excavation.

Once soldiers finished their fighting positions, they were supposed to dig at least three more "reserve positions," then lines of communication connecting foxholes. These were the streets of their frontline cities. They had to be zigzag to provide protection from shrapnel and fighting positions in case the line was overrun by the enemy. Their depth depended on the amount of time soldiers had to dig them, so soldiers might crawl on their bellies, hands and knees, or walk fully upright between positions.[69] Finally, soldiers dug false trench systems. They were also responsible for building obstacles to stop enemy infantry and tanks, including barbed wire, "hedgehogs," antitank ditches, and so on.[70] Still later, they built bunkers, latrines, warehouses, wells, observation points, antitank ditches, and more advanced infrastructure.[71] Tankers and drivers dug earthworks to conceal their vehicles or to rest in. Every soldier was "not just a warrior but the builder of his own fortress,"[72] who "should transform any locality into an ally."[73]

A soldier was never finished digging and building. All structures had to be impervious to both bullets and bombs. This meant reinforcing trenches and dugouts with several layers of wood or other materials improvised from whatever was at hand: trees, munitions boxes, peasant huts, or destroyed apartment blocks. Manuals included detailed descriptions of which thicknesses of wood to use and how to manufacture standard components for trench construction.[74]

All of this effort worked only if soldiers mastered the rhythms and rules of behavior at the front. They needed to be vigilant. By regulation, most soldiers would be relaxing in their bunkers, coming out only to fight, while a few stood watch.[75] Often snipers or groups of observers provided this cover

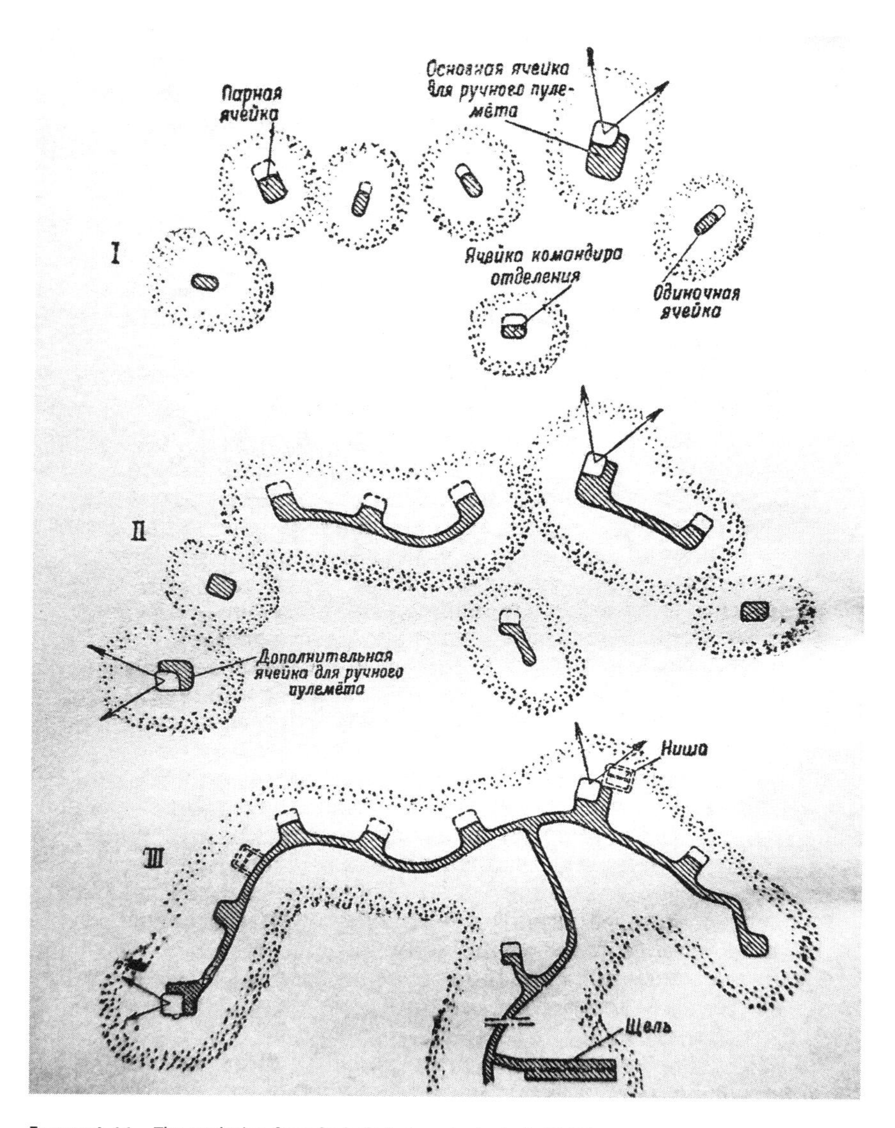

FIGURE 4.11 The evolution from foxhole to trench. *Inzh. P-43*, 27.

during the day. While allowing the majority of soldiers to rest, this meant that one soldier's drowsiness or inattention could get his or her comrades killed, wounded, or captured. Soldiers caught sleeping were often humiliated or severely punished, including by execution.[76] However, those who were on guard were often poorly trained or lax in discipline. Grigorii Pomerants recalled that soldiers almost never acted according to regulations and

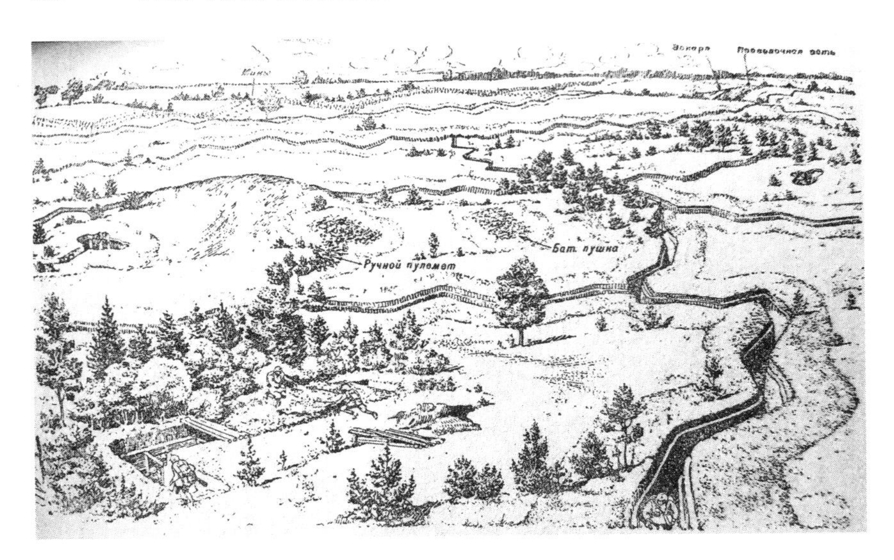

FIGURE 4.12 The end result, a fully formed network of trenches. Ushakov, *Voenno-inzhenernoe delo*, 33.

seldom knew the password.[77] A report from April 1944 showed that the rear area was rife with people who didn't fulfill regulations, talking with strangers and letting civilians walk freely through the areas they guarded.[78] Calls for vigilance remained a constant aspect of military propaganda throughout the war.

It was assumed that the denizens of the cities of earth would wear helmets, as its streets were often exposed. Propaganda often linked the spade and the helmet as equipment that could save a soldier's life, yet Red Army soldiers were often found at the front without helmets.[79] This was due in part to shortage, as large numbers of helmets were lost in 1941. As with everything else, the army recycled helmets, taking from the wounded and dead and from rear to front, to give to the living directly in combat. Yet a significant number of soldiers remained without helmets. One 1942 study from the Leningrad Front found that 83.7 percent of a sample of soldiers with head wounds were not wearing helmets.[80] As Anatolii Genatulin recalled, many soldiers refused to wear helmets: "Many fling their helmets into the grass as if they were useless. When it's hot these helmets heat up as if you slapped a just-forged cauldron on your head, and if you take it off and put it on your pack its still heavy. And later it turns out, that a helmet is a most necessary thing in battle."[81] Genatulin uses this passage to introduce an autobiographical story about how almost everyone in his squad was killed by shrapnel, their brains being splattered all over him. He alone survived due to his helmet. So many helmets were thrown away that the government even offered a bounty to civilians who collected discarded helmets from the battlefield.[82] Articles encouraging soldiers to wear helmets abounded in the military press, which often linked helmets and spades as objects that could be the difference between life and death. A June 1944 article described how a tiny fragment could kill a man and how a helmet could deflect a bullet.[83] Despite regulations and the pragmatic desire to increase one's chances of survival in an endeavor centered on killing, many seemed to believe in talismans more than in the uncomfortable steel shell the army had provided. Nikolai Nikulin, who never wore a helmet, recalled, "We usually defecated into helmets, then threw them over the parapet of the trench, and shock waves from explosions tossed everything back, onto our heads."[84] The desire for comfort that led soldiers to discard helmets could also lead to laziness, as soldiers avoided the manual labor so key to their survival.

Mikhail Kalinin recalled a common retort in an article for agitators: "Why dig foxholes, when in half an hour we won't need them?"[85] Since soldiers moved often and dug in almost everywhere they stopped, digging became an endless rhythm of their lives. Despite its ubiquity, soldiers often did not appreciate just how important this work was until they had been in combat. The perils of laziness were not always clear to soldiers at the front, as we see from official reports and propaganda. Marshal Voroshilov noted in a report from May 1942, that soldiers often built fortifications only "to clear their conscience," neglecting to dig in or creating positions that were difficult to fire from.[86] Veterans learned quickly how key digging was to survival, as the sapper A. V. Luzhbin wrote from his hospital bed: "If during peaceful training in the rear the parapet doesn't have a tangible meaning, at the front its thickness is a question of life and death."[87] Failure could be used as a tool to teach others, as one political officer recalled using dead soldiers and their shabby bunker as "a form of visual propaganda," which "had a very positive result—they started digging in quite well."[88] Soldiers without spades felt cheated, as one censored letter during the Battle of Kursk complained: "the discrepancies have chewed us to death. For example, they say 'the spade is the soldier's friend,' but we went into battle without spades."[89] It became clear to everyone that survival meant digging.

The simple soldier could survive mechanized warfare only with the help of the spade. Soldiers wrote about their landscaping in their diaries, as one chauffeur complained in the last winter of the war: "The ground had thoroughly frozen; even crowbars break. But we gnaw through it unremittingly. That is a soldier's duty. And if you don't bury yourself in the earth, you'll die

FIGURE 4.13 Soldiers, including Sergeants I. F. Vas′kin and T. D. Osipov, eating in a well-appointed trench, Kalinin Front, 1943. Most carry SVT-40 Rifles. RGAKFD 0-286688 (V. P. Grebnev).

like a fly."[90] *Krasnoarmeets*, the most widely circulated journal in the army, ran a story in which Mother Earth promised to protect a brave soldier.[91] But this promise was made good only through the soldier's near-endless work of not only digging but woodworking and other crafts. Creative excavations also helped offset a lack of manpower.

Norms were supposed to dictate the density of manpower per square kilometer of front. However, losses often meant that only the skeletal remains of a unit held territory meant for significantly more people. At the front the combat medic Tat'iana Atabek recorded: "In our trenches the squads [about eight soldiers] are 500 meters from each other, and we were taught that by regulations they should be no more than 30–40 meters."[92] One veteran recalled the soldiers' ironic phrase "two Russians make a front."[93] It was imperative that the enemy have no idea that such a shadow force was in front of them, and hiding their numbers was a major task that provided hours of extra work for soldiers and led to the preservation of green space throughout the cities of rubble.

The progression of trenches from unconnected dots to a massive network resembling a primitive city was supposed to happen without the enemy catching a glimpse of any of it. Camouflage was a key part of this endeavor, especially given the overwhelming air superiority enjoyed by the Germans in 1941–1942. Enemy planes and artillery could rain death on anything they could see. As was repeated in manuals, "the task of military camouflage is to conceal that which is true and show that which is false."[94] Camouflage was often meant not to entirely conceal the presence of soldiers, as it was felt that this would attract the enemy's suspicion, but rather to deceive the enemy as to the intentions of those in the trenches.[95] The landscape was to be filled with false clues about what the army was up to.

Being seen meant being shot at. Trenches were supposed to seamlessly blend into the landscape, preserving its profile. Soldiers were supposed to carry a personal (150 x 100cm) net to help camouflage themselves at all times.[96] Every foxhole was to disappear behind some aspect of the local flora, which meant soldiers could find themselves hidden by roses, raspberries, corn, rye, sod, or whatever sprang from the ground.[97] From the first slice of earth a soldier lifted, every effort was to be taken to maintain concealment. This meant that soldiers vigilantly cleared their fields of fire, gathered brush, and cut into sod gingerly to preserve the network of roots that kept the grass alive and gave cover. Commanders carefully chose the ground and, when possible, watched their soldiers work from the direction of the enemy. Soldiers eventually dug false positions to attract enemy fire (often using soot to give the impression of depth) and produced fake machine guns and artillery pieces from wood that needed to be both realistic enough to draw attention but also camouflaged enough to seem like the actual object—a real-looking machine gun left out in the open might seem too incredible to waste a mortar shell on.[98] Nets and screens were constructed to place over communication trenches, foxholes, and vehicles. When Red Army men opened fire, it was often through a loophole invisible from the enemy's lines, which could feature a camouflaged trapdoor.[99] As Vasilii Grossman recorded in his notebook, the Red Army soldier had "become so crafty that not even a professor could think of it. He builds such a foxhole that you can step on his head and not notice him."[100]

Camouflage required not just labor but changes in the way soldiers acted. Every move made was to take the landscape into account and use its features to provide cover. Soldiers were encouraged to choose paths based on which plants best matched the color of their uniform or to move in short bursts from tree to tree or rock to rock. They would also smear their helmets in mud or flour to match the clay or snow of their trenches.[101] Noise and light discipline became key to survival. This meant concealing smoke and flashes, particularly at night. Nothing could shine or jingle in the soldier's kit, and conversation had to be kept quiet and to a minimum.[102] In the steppe soldiers were instructed to "strictly observe the regime on the battle line, don't show yourself, sit in the ground and secretly observe the enemy. Leave the trench only with permission and only at night."[103]

Of course, the enemy was also trying to hide his every move. As an article on observation put it: "The modern battlefield often seems entirely empty. Locating a fortified and camouflaged enemy requires ceaseless and in-depth observation of the smallest signs of the life in the enemy's order of battle." The trenches were dotted with observation points, in which soldiers kept journals of every change in the landscape and enemies' disposition. Observers tried to locate minefields, artillery, and machine guns, reading the landscape more fluently than other soldiers, either with the naked eye or with the help of binoculars, periscopes and stereoscopes.[104]

The Germans noticed how rapidly and skillfully Red Army soldiers used their spades, finding a racial explanation: "instinctual use of location, exemplary construction of positions, and skillful camouflage strikingly characterize every Red Army soldier," stemming from their "inborn, mindless submission to fate, lack of initiative, blind obedience from fear of being shot."[105]

Soviet propaganda declared: "You could say that every blow with a shovel on the battlefield is equal to a well aimed shot. The spade makes a soldier invulnerable, and thus terrifying to the enemy."[106] The spade was a soldier's shield; often carried tucked into the belt in combat, it could be a weapon.

Spades even occasionally deflected lethal fire, and some soldiers went into attacks covering their faces with spades.[107] The spade was indispensable to soldiers on the battlefield, but the world created by the soldier's spade consisted of more than just violence. As Mikhail Loginov put it: "We don't just wage war at the front, we also live there. The trenches are our home."[108]

Domesticating the Front

Once soldiers settled into a position for any period of time, they began to concentrate on comfort, starting with the soldier's foxhole. Mansur Abdulin recalled:

> Every burrow is different in shape and volume, since it is dug to the taste and complexion of the builder . . . A foxhole is a soldier's place of work, his fire position. But it's also his home . . . The passion for improvement is inflamed: he starts to dig out a niche for grenades, another for ammunition, a third for his submachine gun, so it will be at hand. And you want to find a place for your mess pot . . . A soldier has already made himself at home.[109]

Niches for all manner of items—weapons, gasmasks, food, and water—came to mark the walls of the trench. After soldiers finished their fighting positions, they began constructing more complicated places to live with the same attention to detail. First they built a small burrow to escape the elements and enemy fire, a place large enough to lie down in, directly in the trench wall. These could be for one or two soldiers, with walls of bare earth or lined with branches or wooden boards.[110]

The first door for any structure a soldier made was the *plashch-palatka*, an ingeniously designed tent portion and rain cape that was both clothing and architectural element in one. (See figures 1.4 and 4.16.) A veteran of the Leningrad Front praised this humble cloth: "the indispensable accessory of the soldier. It defends him from rain and blizzard, covers him from the sun, and serves as both bedding and tent. And they'll bury you in it, when your time comes."[111] The plashch-palatka could be used as a very basic shelter or combined in any number of combinations from a simple one- or two-person tent to dozens put together into a yurt. When not being used as a tent, it could be worn as a rain cape, with one corner buttoning up so as not to be dragged on the ground and another being drawn into a hood via two drawstrings. It even featured an opening that soldiers could put their shooting arm through. Lightweight and multifunctional, the plashch-palatka was an all-purpose item that served as a blanket, a method of carrying wounded and supplies and makeshift door, roof, camouflage, and, finally, as a shroud for the dead.

If the plashch-palatka was a portable house, an element of both architecture and clothing, then the spade allowed soldiers to build elaborate shelter wherever they found themselves. Living space was often separated from the space of fighting by tens of meters, echoing in miniature Soviet city planning that sought to connect workers with their factories but also provide a belt of greenery that separated the spaces of labor and rest. The bunkers that soldiers built and inhabited were spaces for rest and recovery, sparsely decorated, but often with a few touches of home. As one correspondent with the Mints Commission wrote: "The present communication was written in a small bunker—the dugout of the commander of the Chemical Company and his political deputy. The décor: a window the width of two logs without glass on the level of the earth. Two board beds; a small iron oven; a log floor; the door is hung with a double sheet of chemical paper. The only decoration on the wall—two sheets of *Frontovaia illiustratsiia* [a popular illustrated magazine]. On the other wall—a submachine gun and a gasmask bag."[112] Structures like this became ubiquitous at the front, as soldiers adapted to their environment and built customized spaces.

This world had its own rhythms and eccentricities to which soldiers had to adapt; a process that mirrored country folks' adaptation to city ways. Soldiers often became largely nocturnal, as Boris Marchenko wrote home to his wife in September 1942: "I try not to sleep at night, in general, and sleep during the day—it's better that way. In general it is impossible to answer the question of when I sleep, day and night and neither day nor night. In a word, I've gotten used to sleeping in snatches and feel no inconvenience whatsoever."[113] Loginov recalled similar patterns, but his soldiers had more trouble adjusting: "We don't know plates, forgot what hot tea, good books, and music are . . . The most simple everyday details of life are beyond our reach. We have a special schedule—there is no difference between day and night. We sleep, not knowing sheets, not getting undressed, at odd moments. And so we never get enough rest and always want to sleep."[114]

Beyond sleep, other conditions were dramatically different. Once soldiers dug in, they could become acclimated to the technology that shaped their landscape. They learned to tell enemy and friendly planes from the sound their engines made or the direction of artillery fire by its shriek.[115] Others learned simply to ignore these sounds, as Nikolai Chekhovich wrote home to his mother: "the constant whistle of bullet and mortar fire, and even the creak of 'Katiusha' and the German 'mule' [rocket mortars] have become normal." He later boasted to his mother that he could drink tea

FIGURE 4.14 Soldiers in a bunker with an improvised stove. RGAKFD 0-312479 (O. B. Knorring).

under bombardment.[116] Soldiers bragged of their ability to sleep through bombardments or even battle as well as to sleep fully clothed, with grenades and spades hanging from their belts.[117]

While danger could become unremarkable, other conditions could prove unbearable. Lice were a frequent companion in the trenches, as were other vermin such as mice. The trenches were subject to the elements, as one female sniper recalled: "awful rain began . . . I arrived at a dugout, dampness; the dugout is dripping, uninviting, cold, and dirty . . . even though we were on high ground, the water came up to our knees, and we stood in water all the time."[118] A certain level of discomfort was inevitable under these conditions, but both soldiers and the government strived to create the best living conditions possible.

The environment in which soldiers lived was always of interest to the state, who frequently sent agents to investigate conditions at the front. In early 1943 interest in the everyday life (*byt*) of soldiers became a major campaign. A discussion among top political officers led by A. S. Shcherbakov, head of the Main Red Army Political Directorate, revealed that soldiers were living under unacceptable conditions. Food was pilfered and poorly prepared, warm clothing wasn't finding its way to the front, and the bunkers in which soldiers lived had major shortcomings. Political officers at the front were deemed not to understand the importance of looking after soldiers' comfort. Those present understood that "where the barest needs of the fighters are not satisfied—the number of extraordinary events [desertions, self-mutilations, etc.] grows. Enemy agents take advantage of this."[119] Keeping cities of earth up to code was key to securing soldiers' loyalty and health.

Just as in Soviet cities, infrastructure was sought but often found wanting. Political officers noted that in some units, bunkers were dug too shallow, forcing soldiers to constantly hunch over.[120] In others, the bare clay floors were cold and miserable, while windows remained without glass, and a thin rain cape served as the only door. Finally, lighting was cited as a major problem. Soldiers spent a lot of time in their dugouts—the space in which they were to sleep, write and read letters, clean their weapons, shave, and repair their uniforms and equipment. Some form of artificial lighting was key, and in a few bunkers, soldiers burned telephone cable to provide light. One soldier colorfully described how when burning telephone cable there was so much soot that "in the course of an evening so much crawls into your nose that you can't pick it all out," while a report to Moscow found that the practice could lead to death from smoke inhalation.[121]

Both soldiers and political officers came to the same conclusion: soldiers should simply improvise lamps from whatever could be found at hand. Most often the bodies of these makeshift creations were either the casings of artillery shells (the 45mm cannon being particularly common) or recycled ration cans. Soldiers used a variety of fuels—kerosene, animal fat, or a mix of gasoline and salt. For a wick they often cut strips from the end of their woolen overcoats.[122]

Improvisation was key to survival and comfort at the front, just as it often was in Soviet cities. Indeed manuals officially encouraged a "do-it-yourself" ethic. In contrast to the US and German armies (both drawn from consumer-oriented societies)—which provided a wide array of field stoves, flashlights, and other appliances for their soldiers in the trenches—comforts and necessities in the Red Army had to be improvised from whatever was at hand. Materials could include artillery shell casings, cans, abandoned civilian houses, or whatever else could be salvaged.[123] The military surgeon Vishnevskii described one instance of ingenuity: "There's a highly original wash-stand in the MSB [Medical Sanitary Battalion]. It is made out of a ration box of American sausages. In the bottom of the box there is a hole and a rifle cartridge is soldered in place."[124] Other soldiers described making ovens out of whatever they could find: "Nearby a 76mm gun was positioned. We asked the troops for a can that had been used for lube and made an oven out of it. We found a piece of old iron, turned it into a tube, and put it outside, not too high—level with the ground. It got dark, we struck up the oven, it got warm, you could even heat up soup in a mess tin."[125] Everything had to be scrounged and had to function in such a way as to be invisible from enemy lines—the trail of smoke of even a cigarette could draw enemy fire.[126]

The infrastructure of the cities of earth was improvised but also expansive. The political department encouraged units to organize workshops that could repair clothing, and commanders would often try to find various craftsmen—tailors, barbers, or woodworkers, for example—to make their lives easier.[127] Self-sufficiency became a priority, touching all aspects of a soldier's life. Soldiers were expected to find their own wells. They were supposed to bathe every ten days, to have their clothing deloused in special rooms. These bathhouses took the form of either a specially constructed bunker with running water and an oven or traveling tents.[128] Being able to wash was a real joy, as it rescued soldiers from lice and the possibility of contracting typhus. However, female soldiers often found themselves in an awkward situation, being forced to bathe with men or wait last in line, when there might not be any hot water.[129]

Latrines were standardized, coming in two camouflaged models, and soldiers had to travel 30–40 meters from the main trench line through a communication trench to answer the call of nature.[130] References to urination and defecation among soldiers highlight the lack of privacy in the army. Boris Suris confided to his diary that in his training camp the bathroom was the only place where one could get any privacy.[131] In the field such privacy was rare. Irina Dunaevskaia recorded an entire squad, in ranks, peeing in one puddle in the middle of the road.[132] Vera Malakhova recalled the particular suffering of women on the march: "You'd be marching along, exhausted, worn out beyond belief. Suddenly you'd have the urge to go, but how could you? And they [older soldiers] 'saved' us . . . It was dangerous to go off somewhere, because sometimes there were mines. So the three of them would stand up, turn their backs to us, open their greatcoats wide, and say: 'Dear daughters, go ahead, don't be bashful. We can see that you can't march any further.'" Until then, she had to hold it in, with an aching stomach and bursting bladder.[133] Whatever you did in the bunkers, in trenches, or on the road, chances are you wouldn't be doing it alone and unobserved.

Red Army units had a special section, from April 1943 on dubbed SMERSh (Smert' shpionam—Death to Spies), dedicated to finding spies and traitors in the active army, rear, and newly liberated territories. Known colloquially as *osobisty* ("specials" or as the "Special Section"—*osobyi otdel*), these often unpopular men were part of the NKVD, their presence a continuation of prewar surveillance organically embedded in any Soviet institution. *Osobisty* and sometimes political officers recruited soldiers to observe and report back on their comrades, and occasionally to create provocations.[134] These men listened carefully to conversations and reported back in detail any anti-Soviet sentiments, conspiracies to desert to the enemy, or cases of panic and cowardice. Commanders sent regular reports of "extraordinary occurrences" to the Political Department of the army. Griping is a norm in any army, but in the Red Army this could lead to serious results, as the veteran Boris Slutskii recorded in a poem about a machine gunner: "For three jokes, facts of three, he will never see tomorrow."[135] However, soldiers who seemed to be on the verge of surrendering to the enemy could redeem themselves in battle, as the story of the Red Army man "Chechen" reveals: Under investigation for reportedly declaring that the food was so bad he was ready to desert, the Germans too strong to defeat, and the Allies disinterested, he was pardoned when he inspired a group of cowering soldiers to open fire with his "heroic example." Redemption was possible but difficult to achieve, and some informants hunted those who would desert to the enemy.[136] This war was in many ways a battle for souls, and the army's Political Department sought to reveal to soldiers what was at stake but also to feed the soul.

A variety of institutions providing spiritual sustenance that had marked Soviet cities also found their place at the front, whether in dedicated bunkers that served as libraries, clubs, or tea houses, many of which were portable. The creativity shown by some political officers in arranging the free time of soldiers and attempting to influence them could be truly impressive. For example Izer Aizenberg, a regimental agitator, devised an *agitkul'tchemodan* (cultural agitation suitcase) that he compared to an "illusionist's case." Aizenberg described the contents of and scuttlebutt around his case:

> It works out like this: one group takes a map, hangs it up, and starts to trace their fingers around cities the Germans have bombed and where our pilots are bombing. They take an interest in other theaters of the war, ask what is happening in Tunisia, and so on. Another group plays checkers, a third reads brochures—riddles and songs—and laugh jollily. Serious brochures are read in the corner. In the case there is paper and envelopes; they take the paper and write a letter home or turn out a combat sheet [a short form of propaganda produced by soldiers themselves]. There is also a mirror. Sometimes a line forms when you take out this mirror: one comes up to it—"Let me have a look at myself," another "I'm overgrown, let me have a look." At the height of this work the agitator asks for everyone's attention and holds a ten- to fifteen-minute discussion or reads an interesting article.[137]

Captivating props allowed political officers to influence soldiers and explain events on their front and around the world. The state also sent musical instruments (particularly accordions and harmonicas), traveling theaters, and musicians and even organized talent shows among the soldiers themselves. The military press encouraged the creation of mobile libraries, and some agitators gathered collections of literature in the languages of "non-Russians" as part of a larger effort to help integrate these soldiers into the largely Slavic culture of the army.[138] The Soviet project of enlightenment and shaping the individual continued even under the primitive conditions of a world constructed by the soldier's small spade. The bunkers, no less than workers' barracks and urban clubs, were spaces of transformation.

Even in the chaotic conditions of Stalingrad, with its near constant combat, a political officer explained that soldiers "sitting in foxholes and dugouts in constant battle, need some relaxation, they need time to relax and let it all out, to get dry, warm, write a letter somewhere, bandy a few words with their comrades . . . share their impressions. This was also very important: through the exchange of experiences from mouth to mouth fighters helped each other."[139] In these tight quarters the sharing of experiences went beyond veteran soldiers instructing newly arrived reinforcements in the tricks to

FIGURE 4.15 A frontline club, Fourth Guards Rifle Division, Volkhov Front, 1942. RGAKFD 0-144470 (Garanin).

FIGURE 4.16 Frontline post office, 1942. Note the soldier on the left wearing a plashch-palatka. RGAKFD 0-101371 (Chernov).

staying alive and becoming a hero. The space was a veritable melting pot. Many urban youths were first introduced to the peasantry that still demographically dominated the country. They might hear different versions of the Civil War,[140] folk tales, or entirely different concepts of biology.[141] Educated people from good families became friends with professional criminals, sharing stories about their prewar lives.[142] Kazakhs or Uzbeks might sing their traditional songs, to the amusement or bemusement of their comrades.[143] All of the diversity of the country came together in these bunkers, and in the need to while away the sometimes interminable hours between battle and training, there was little else to do but talk, read, and write. The world built by the spade had little place for privacy—the battlefield exploits, opinions, love affairs, and hygienic practices of a soldier were all exposed to the surveillance and judgment of his or her comrades.[144]

The particular ambience of the dugout—that intimate, relatively safe space so close to death—was recorded in songs, interviews, and memoirs. None more famous than "In a dugout" (V zemlianke), which begins with a description of a lamp flickering in a tight bunker and climaxes with the singer proclaiming that his love is far away but that "Death is only four steps away."[145] As one female scout remembered: "The dugout was very dimly lit with oil wick lamps, filled to the brim with people. You can't imagine how

FIGURE 4.17 A camouflaged bathhouse, 1942. RGAKFD 0-121639 (Kolesnikov).

loud it was, and so much smoke that you could hang an axe."[146] The bunker was the space where soldiers wrote their loved ones, thought of home, and socialized. As Boris Suris recorded in his diary: "It's always jolly and lively at our place. Sometimes the operations guys come, saving themselves from the boredom that rules their section; sometimes scouts in white camouflage suits bring a report; sometimes we suddenly get crowded with guests from the neighboring outfit."[147] Boris Slutskii used the bunker as a motif in his poetry, musing on how officers imagined themselves in a Moscow restaurant while sitting in their bunker, reflecting that "No one has forbidden us to live well!" A typical line reads as follows:

And we're alive. We wait for lunch.
And in the meantime, we argue, banter, fool around.
We are glad that we aren't under the rain.
And are used to the fact that we are under fire.[148]

One veteran recalled the homey smell of Red Army bunkers as opposed to the stench of captured German dugouts: "Our bunkers were filled with the tart fumes of tobacco and the aroma of bread."[149] Unlike the trenches, where total concentration was needed, the bunker was a domesticated space separated, albeit by a slim margin, from the space of killing and dying.

Bunkers could be the loneliest places or replace a lost sense of family for the soldier. Viktor Kiselëv complained of an every-man-for-himself attitude in his rear-area training camp.[150] Irina Dunaevskaia (a Leningrad University student turned army translator) was kicked out of a bunker because her (male) roommate "didn't want to acquire cultured habits," which was a relief, as she was afraid of acquiring her comrade's coarse manners.[151] Her tendency toward quiet reflection would later draw the negative attention of other comrades sharing her bunker.[152] For other soldiers, the closeness of the front was a balm to their souls. Major A. Luzhbin wrote in his diary: "Every soldier and commander of our army carries in their heart a burning love for their family, their mother and father. But their families are not at the front and they shift this onto each other . . . this earnest, gigantic strength of love bonds together a unit with a firm, militant friendship and works true miracles—it brought us victory."[153] Intimacy, while not easy to come by, served as both a substitute for families left at home and as a guarantee of combat effectiveness.

However, not all intimacy at the front was welcome. As noted previously, the bodies of female soldiers were often viewed as the property of commanders. The intimate space of the bunker was where by either mutual affection, continual harassment, or force, men and women entered into relations. Many soldiers recorded in their diaries the primacy of sex in soldiers' discussions and that many men doubted that a woman at the front could stay single or faithful to a distant partner.[154] The closed quarters of men and women often led to uncomfortable moments for female soldiers, as Tat'iana Atabek confided to her diary: "At night, as if an insane dream, I hear impassioned whisperings: 'I love you and will never leave you alone.' Attempts to embrace and kiss and that moaning prayer to give him my lips. I felt that I had no strength to stop this person . . . and felt so insulted—I had never been so wronged in all my life—that I wept uncontrollably."[155] Other women also reported being sexually harassed in bunkers.[156] Occasionally female soldiers, particularly those with elite status, were given special, separate accommodations, in what could jokingly be called "the harem."[157] Sex at the front could lead to serious indiscipline, such as soldiers who should be on watch having sex. One wry chauffeur noted: "Such a case, it seems, has not been foreseen even by our ubiquitous military regulations. It is necessary, apparently, to add to it: 'it is forbidden to f*ck sentries.'"[158] "Girls" in the army were often forced to regulate their actions very strictly or risk being seen as sexually available, and some women found it necessary to set stringent boundaries immediately.[159]

While talk about women and promiscuity was common, many soldiers noted that the presence of women was looked on fondly as a way of softening the harsh conditions of the front. One commander told the Mints Commission in 1944, "At the front, at headquarters, in the hospital, and in the trenches our girls are the only pure souls."[160] Women in the army often fulfilled traditional gender roles, something that could lead to correctives from the military press: "In a number of units a wholly unnecessary 'division of labor' has been established. Girls do the laundry for all soldiers, wash the floors of bunkers and garrisons of men, and men-soldiers start up the stoves in the dormitories of girls. Such a relationship to girl-soldiers weakens military discipline."[161] Despite being a breach of discipline, it was precisely these hints of civilian life that many soldiers cherished.[162] Any reminder of home could distract soldiers from the hardships and constant danger of the front. Nonetheless, those comrades who had made today bearable could be gone tomorrow without a trace or rendered into the earth.

Living with the Dead

The dead became part of the landscape, haunting the memories of friends and leaving unavoidable traces of their existence. Under combat conditions

FIGURE 4.18 Dmitrii Baltermants, *Na voennoi doroge*, 1941. © Dmitrii Baltermants. Courtesy of Paul Harbaugh and Tat′iana Baltermants.

it was often impractical to remove the dead. Nikulin noted a macabre experience on the Leningrad Front in his 1942 diary:

> A corpse smells intolerably. There are many of them here, old and new. Some are dried and black, like a mummy with shining teeth. Others have swelled as if they are about to burst. They lie in different poses. Some inexperienced soldiers dug themselves a niche in the sandy walls of the trench, and the earth, crumbling from a nearby explosion, buried them alive. And so they lie, all curled up, as if sleeping under a thick layer of sand. The picture looks like a grave cut down the middle. Here and there in the trench parts of bodies trampled into the clay stick out—a spine, a flattened face, a hand—all brown in color like the earth. We walk directly on them.[163]

Reports found that soldiers, particularly in the winter months, did not rush to bury their friends, which could lead to fears of epidemics during the spring thaw.[164] Bodies left on the battlefield could attest to the length of battles that traversed several seasons and bore witness to different units that had passed through, particularly if they had distinctive uniforms (such as the black pea coats of Naval Infantry).[165] These men and women decomposed into the landscape yet left traces of themselves everywhere.

Many observers found the attitude of soldiers toward the dead unconscionable. Vasilii Grossman, writing about the fate of a corpse that had lain frozen for two days in the winter of 1942, exclaimed: "No one wants to bury him—laziness. He has bad comrades! . . . Too often one is forced to observe the approach of reserves to the front, reinforcements passing recent battlefields among unburied dead, scattered all over. Who knows what is happening in the souls of those people going to take the place of those lying on the snow?"[166] Shortly before Grossman put his rancor to paper, an army-wide order had called for the standardization and organization of burial of soldiers, noting: "The burial of the dead in battle often takes place not in mass graves, but in foxholes, fissures, and dugouts. Individual and mass graves are not registered, not marked on maps, and not properly formalized." The order noted that by not providing a proper burial, the army missed an opportunity to "mobilize the masses of Red Army soldiers for the decisive battle with German-Fascist invaders, to engender in the soldiers hatred of the enemy and the drive to avenge the deaths of their comrades," and that no record existed to identify the dead or where they were buried.[167] Despite orders to the contrary, many soldiers would come to their final rest in the foxholes and dugouts they had built, unmarked and unknown. Under these conditions, the spade became the tool that committed soldiers' bodies to the earth.

Improper burial continued to be an issue throughout the war, with Mikhail Kalinin instructing agitators in 1943 to "cultivate respect for the dead among Red Army men, to honor them." Soldiers were supposed to bury their comrades and turn maintenance of the graves over to members of the Young Pioneers youth organization.[168] An article in *Krasnaia zvezda*, the daily newspaper of the army, pointed out that some municipalities had allowed grave sites to become overgrown as early as 1943.[169] The war swallowed whole villages, and with them all knowledge of nearby graves.[170] Even when the desire to provide a proper burial was apparent, the logistics of burying so many soldiers so quickly could prove daunting. A report from the Karelian Front described how burial teams (*pokhoronnaia komanda*) during a successful offensive dug properly formatted graves with markings rather than in simple pencil, leading to these men falling 20–40 kilometers behind their units.[171] All valuables belonging to soldiers were supposed to be sent to their families, including medals, which often took the place of remains as kin mourned. Exact locations of place of burial, often a mass grave, were to be provided to relatives.[172] However, even orders from 1945 sought to correct the failures of those committing soldiers to the earth to properly fulfill their task.[173]

In contrast, Red Army soldiers had seen—and later often witnessed the destruction of—neatly formatted German cemeteries.[174] Some claimed that this disparity inspired them to pay more attention to how they buried their own.[175] On January 26, 1945, the artillerist Viktor Kiselëv noted in his diary:

> to an inexperienced person, it could seem that the losses of the German were minuscule, because their corpses are much fewer than ours. The thing of it is that the Germans have a custom of taking away their dead to bury them in the Motherland. Their soldiers are severely punished if they retreat without taking the body of their commander. And what do we do? Russians have nothing like this respect for the dead, and it is not for nothing that you see so many neglected, uncared-for bodies in the fields and on the roads. Everyone who walks by wrinkles his nose and expresses indignation ("why haven't they been buried yet?") but himself doesn't lift a finger, covers his nose, and walks by. And only among *natsmen* (Uzbeks, Kazakhs, people from the Caucasus) do we see this noble custom.[176]

However, soldiers might be forgiven for being overwhelmed by the number of their dead comrades surrounding them and wanting to avoid such pungent reminders of their own possible fate.

The constant proximity of death and the dead could have a dramatic effect on soldiers' sense of self. Anatolii Genatulin wrote that he couldn't feel the same depth of feeling for dead at the front, having become used to death: "my boyish soul didn't so much harden as learned to live with it or became numb." He explained that it was perhaps his youth that didn't allow him to feel for the dead or "because I was one of them, I was shell-shocked and survived by some miracle" that his joy at being alive overtook his pity for those dying around him.[177] Chekalov recorded a similar feeling: "Death!? I have become so close to it that its blood-freezing breath seems normal. It repeats itself every day in hundreds of variations and tiresomely reminds us of itself. And perhaps this is exactly why you appreciate not only that which is alive but even the dream of life, of the past."[178] The simple ubiquity of death helps explain the indifference of soldiers to the bodies of their comrades, particularly to anonymous bodies littering roads and battlefields. Only one's closest friends warranted attention. As Chekhovich wrote home to his mother and fiancée: "Many times I have seen a person reduced to a bloody mess . . . It's hard to force myself to pull out the documents of a dead person if I didn't know him well."[179]

However, many soldiers put considerable effort into the graves they provided for their friends. A do-it-yourself ethos was common here as elsewhere. Often only one's closest comrades knew where a soldier had been buried and provided details of burial to a comrade's family.[180] Vasilii Chekalov ordered his soldiers to build a monument to a close friend who had been killed.[181] Tat'iana Atabek noted in her diary the substantial effort made by troops in technical branches in March 1944: "I have noticed that tankers have their own manner of burying their comrades (they use tank tracks as a fence) as opposed to artillerists, who cordon off their graves with casings from artillery shells. Not far from us two Guards tankers are buried. So that no one could step a foot on the ground near their grave, it is surrounded by the track of their once fearsome tank."[182] Graves such as this were much more conducive to the pedagogical role the dead were supposed to play in soldiers' lives. Loginov recalled a "lesson of bravery" when his soldiers stopped for a moment of somber ceremonial contemplation in front of the graves of soldiers who had perished the year before.[183] Nikolai Inozemtsev recorded that at the burial of his close friend (the third to die in the course of a single operation), three salvos were fired at the German lines, realizing Soviet propaganda mobilizing grief as fuel for vengeance.[184]

In this potentially indifferent milieu, many soldiers feared disappearing into nothingness, as if their lives had never existed.[185] Perhaps this is why reflections on the eternal nature of monuments to soldiers were common both in wartime propaganda and postwar memoir literature. The government would in many ways bind its legitimacy to a cult of the war dead and

FIGURE 4.19 Soldiers of the Eighth Guards "Panfilov" Rifle Division vow vengeance over the grave of a fallen commander before going into combat, April 17, 1943. RGAKFD 0-285341 (V. S. Kinelovskii).

raise generations of children to venerate mass graves at battlefields that were often a short walk from their homes or schools.[186] However, during the war itself, the soldier's spade was often all that was available to commit remains to the earth. The scale of events could render both the landscape and human form unrecognizable and turn people into the landscape.

Cities of Soldiers

The unimaginable scale of the Great Patriotic War—the level of destruction, number of dead, and immense scope of action—is belied by the fact that it was experienced on a very local scale. Soldiers built a front that spanned a continent, but everything they did was determined by local conditions—the commanding heights, little river, or easily visible landmarks had more sway over where and how soldiers dug their foxholes than any projections in Berlin or Moscow. Soldiers using a standard, anonymous, and easily portable object created urban spaces tailored to their bodies and needs, encompassing all aspects of life, including death. People from all over the Soviet Union bore witness to very local tragedies that amounted to cities of rubble. They testified before mass graves and rushed to save their fellow citizens from extermination. Like newcomers to Soviet construction sites a decade before, they built a new environment.

The echoes of Soviet urbanity in the trenches were striking. Production had been key in determining the layout of many Soviet cities, which were often built around one factory. At the front, combat replaced production, playing an even greater role in shaping the space soldiers inhabited. In Soviet cities, projections often ran afoul of reality, forcing urbanites to improvise under often primitive conditions, a phenomenon all the more intensely replayed at the front. Ironically, the green space that was often sacrificed in Soviet cities was preserved at the front as camouflage, not as a space of leisure but as a zone devoid of signs of life. Finally, whenever possible, the transformative institutions of the bathhouse, library, and club were excavated in the city of earth. If previously many had been transformed from peasants into workers in factories, barracks, communal apartments, libraries, and other hallmarks of the Soviet city, at the front civilians were transformed into Red Army soldiers. Through the specific labors of being a soldier, they created a new sense of community in a space that was by necessity practically invisible. This was backbreaking work. The main periodical for army agitators stated in 1943: "war consists mostly of labor. The spade and axe are now in as much demand as the submachine gun."[187] While the spade was necessary to survive, it was also insufficient. A soldier's job is deadly by its nature. He or she not only avoids getting killed but also kills.

CHAPTER 5

"A Weapon Is Your Honor and Conscience"

Killing in the Red Army

> Every soldier knows perfectly well the caprices of their own weapon.
>
> —Ian Palkavneks

Mansur Abdulin begins his memoirs with a reflection on the first two shots he took at the front: "War, the front is shooting. From mortars, machine guns, submachine guns, artillery pieces . . . I took my first shot in combat on November 6, 1942 on the South-Western Front." His unit, cobbled together from men who had not seen combat, had just arrived at the front. He describes in detail how he stalked enemy soldiers, then, shaking and sweating uncontrollably, pulled the trigger and *missed*. A total loss of composure immediately followed, sending him to the bottom of his foxhole on the verge of tears and filling him with self-loathing. He feared that he would die uselessly, "without killing at least one of them," and that his comrades would see him quivering at the bottom of the trench. He gathered himself together and coolly took aim, pulled the trigger, and saw an enemy soldier crumple like a rag doll. The commissar of his unit arrived shortly thereafter, awarding Abdulin a signed notebook, with the inscription: "Abdulin, Mansur Gizatulovich, was the first to open his *boevoi schët* [combat tally], destroying a Hitlerite in honor of the twenty-fifth anniversary of [the] Great October [Revolution]." Abdulin was immediately chosen as the head of the Komsomol organization of his regiment and wrote to his father with pride.[1]

Between 1941 and 1945, millions of other soldiers would be faced with the same moment of truth. Killing was a duty in the Red Army: the inability

FIGURE 5.1 Red Army soldier receiving his weapon (m.1944 Carbine). Mark Markov-Grinberg, *The Oath of War* © 1943 Museum of Fine Arts, Boston.

to kill was scorned, and soldiers were encouraged to keep competitive tallies of their kills. The government provided soldiers with uniforms, rations, and the means to create shelter in order to use their weapons to defend the country and destroy the enemy. Yet many soldiers were not up to the task. A few months before Abdulin took his first shot, a battalion commander complained to the war correspondent Vasilii Grossman: "Sixty percent of our soldiers have not taken a single shot since the war began. The war goes on because of heavy machine guns, battalion mortars, and the bravery of a few individuals."[2] The Red Army was forced to take undertrained soldiers like Abdulin and turn them into efficient killers, often instructing them directly at the front. Soldiers needed to learn to use, trust, and love their weapons.

In significant ways, the Soviet project was a dialogue between humans and machinery.[3] The call to master tractors, lathes, and other industrial means of production articulated during the First Five-Year Plan were echoed in regards to weapons during the war. Just as the First Five-Year Plan made industrial workers out of peasants, the army made soldiers out of yesterday's civilians. The Red Army would use its political skills to stimulate soldiers to fight.[4]

In the previous chapter, we have seen how these soldiers avoided death. In this chapter we explore how they used the arsenal of the Red Army to become the agents of death. This chapter is divided into four sections. The first examines both the universal and specific problems that the Red Army faced in motivating its soldiers to pull the trigger and coordinating weapons systems, concluding with solutions to these problems and reflections on the experience of combat. The second offers a detailed inventory of the Red Army's weapons and a discussion of how soldiers used them. The third explores the social worlds that these weapons fostered, and the last considers the dramatic improvements in both the arms and skills of soldiers.

Pulling the Trigger

Difficulties

Guards Colonel Momysh-uly reflected on the importance of the human factor in getting soldiers to pull the trigger, stating, "The self-preservation instinct forces you to kill another person, and this is much harder than to die yourself." For him, what separated soldier from civilian was the ability to kill.[5] Around the same time, S. L. A. Marshall, who interviewed combat soldiers in the US Army, found that the vast majority of soldiers claimed not to have been able to pull the trigger in combat, even when faced with life-threatening situations. Although Marshall's findings have been challenged, and it appears likely that he overstated his case, they do fit with descriptions from the beginning of the war from the Red Army. Effective killing required training that would make the process of pulling the trigger mechanical and the inspiration of hatred that would overpower both fear and the taboo against killing. This is a process that requires considerable time and resources, both of which were in short supply.[6]

The summer of 1941 was an unmitigated disaster for the Red Army. The historian Anna Krylova dubbed the loss of tanks, planes, artillery, and automatic weapons the "demechanization" of the Red Army.[7] Alongside this demechanization, the army went through a period of what I would call "deprofessionalization": the regular (*kadrovaia*) army that began the war effectively ceased to exist. As a result, the war was waged overwhelmingly by new draftees and reservists, most of whom had little or no experience. These two processes fundamentally shaped the experience of service in the Red Army and contributed to the lopsided losses suffered by the Red Army early in the war.

To create the new army, accelerated guidelines were issued in 1941, providing only a month of training between induction and leaving for the front.[8] Lack of professional cadres was something that the Bolsheviks had been dealing with since they came to power, and in many ways the experience of soldiers replayed those of newly minted factory workers during the First Five-Year Plan, as they learned on the job and worked much less efficiently than trained cadres.[9] The training system devised in 1941 assumed that soldiers would be given time to acclimate and finish their preparation at the front, serving in special training units before being sent into combat. This did not always go according to plan. An order from March 16, 1942, complained that soldiers arrived inadequately trained and once at the front were often used as "faceless marching replacements" sent directly into combat instead of being integrated into units in the rear. This resulted in "a greater number of needless losses, making these reinforcements worthless." The order called for the reduction in new units formed and a new system of rotating units off the line to receive and assimilate replacements before being bled white. In practice, however, soldiers continued to be sent into combat as "faceless marching replacements."[10]

Soldiers often failed to master the basic skills necessary to survive. Reports from the first months of the war noted that soldiers "have not been trained to have the necessary faith in the power and potency of their arms."[11] Even during the largely successful winter offensive of early 1942, Oleg Reutov lamented that a small, well-trained German force could hold back several times its numbers.[12] This situation did not improve as the winter turned to spring. In a report on failed spring 1942 operations, General Samsonov noted that the infantry was inactive, refusing to attack, allowing Soviet tanks to be destroyed and artillery to do most of the fighting. As a result, the technical branches suffered staggering losses. This was due to lack of communication and lack of training and discipline among infantrymen. As Samsonov noted: "A significant portion of the reinforcements did not know how to use their weapons, particularly grenades. It is impossible to say that the five days of training in the division's rear was adequate. This is one of the reasons that the median number of bullets used by one active rifle in March was equal to 2.5 a day." Lacking experienced commanders, there was no coordination and soldiers attacked "as a crowd, at full height, without firing, as small . . . noticeable groups, creating a convenient target for the enemy to open fire on."[13] These suicidal frontal assaults would become the cornerstone of the racist image of a mindlessly advancing horde popularized by Nazis and later largely accepted by Americans during the Cold War.[14] These problems haunted the army through 1942, as a general at Stalingrad noted during the November counteroffensive that infantry often didn't know how to shoot or run properly and "continue to hope that all tasks will be solved by the artillery."[15]

Disorganization would also continue to be an issue, but never as severely as in 1941 and 1942, when Chekalov, an artillery officer, recorded in his diary: "We aren't so much suffering from enemy fire as from our own disorganization."[16] Many soldiers and commanders were still civilians in uniform who had not mastered the basics of soldiering, let alone the more complicated coordination of weapon systems to accomplish their goals.

Lack of skill was compounded by a shortage of arms, machinery, and ammunition that would have been crippling even for well-trained troops. The Red Army lost not only millions of soldiers in 1941 and 1942 but also weapons. Of forty divisions being formed in October 1941, only ten had their full complement of weapons.[17] Despite orders to send replacements to the front with a full complement of arms and equipment, many received weapons only at the front, and many units lacked automatic weapons.[18]

Even more dramatic were the losses in planes, tanks, and artillery.[19] This meant that many specialists served in the infantry, despite valuable training.[20] Soviet tactics and strategy relied on coordinated infantry-tank assault with air and artillery support, not unlike the *Blitzkrieg* tactics of the Nazis. In the absence of these technical means of waging war, soldiers were unable to function in the ways demanded of them. Often the tanks, artillery, machine guns, and even rifles in the hands of skilled soldiers could not be used to full effect due to shortages of ammunition. One infantry commander noted in his diary in August 1941: "At war ammunition is a soldier's life."[21] Artillerists and army commanders frequently complained about their inability to fulfill missions and advance due to a lack of shells.[22]

The situation in the army was desperate in the first two years of the war, and indeed it seemed constantly on the verge of collapse. Yet as one veteran mused in an autobiographical novel: "From experience he had long ago come to know the simple truth: if you were to add up all the defects and shortages, it becomes clear that it is impossible to wage war in this situation. Though wage war they did."[23]

Solutions

The Red Army was forced to deal with the universal issues of motivating men to fight in a specific situation of severe shortage. Intensive training, observation by superiors in combat, valorization of successful kills, hate propaganda, and shame are common strategies to get soldiers to pull the trigger.[24] The Red Army used all of these methods in ways that reflected Soviet culture. Red Army soldiers were subjected to a draconian disciplinary regime that employed threats of terror against them and their families. Soldiers were expected to turn their arms on those among their comrades who failed to fulfill their duties. Yet terror was no substitute for training.[25]

Certain types of weapons and tactics lent themselves well to undertrained soldiers, while others forced the unwilling to take part in fighting. The submachine gun forced a soldier to be aggressive in its use and relied on relatively simple tactics. Weapons crews were more likely to fire due to the pressure exerted by the group and dissipation of culpability for killing among its members.[26] Finally, the Red Army reinstituted an ancient tactic—the volley. A volley is when all soldiers fire simultaneously. This made failure to fire conspicuous, slightly offset the lack of automatic weapons, and was seen as a way to stave off panic and give soldiers confidence in their weapons.[27] Observation by superiors and peers was also utilized as a way to pressure soldiers into doing their part. As Abdulin put it, "at war, to be 'like everyone else'—that is, not worse than others—is sort of like confirming your own value as a person."[28]

Propaganda told soldiers of their individual responsibility to kill, the power of their arms, and the vulnerability and contemptibility of the enemy. Of this last, the Nazis provided ample material. Soviet propaganda focused on what Nazi victory would mean, Nazi occupation had wrought, and the personal responsibility of the soldier to stop the Nazis. These materials graphically depicted rape, murder, pillaging, and destruction committed by the Nazis, implying or explicitly stating that Red Army soldiers were complicit in these crimes by retreating and leaving these populations vulnerable.[29] The only way to redemption was through killing. As the very popular and evocative poem by Konstantin Simonov "If You Value Your Home (Kill Him!)"

graphically described, "She, whom you were too bashful to kiss" would be "taken by force—in agony, in hate, in blood" by three Germans. The author stated clearly that it was each individual's responsibility to kill:

> If the German was killed by your brother,
> If the German was killed by your neighbor,
> It's your brother and neighbor who are taking revenge
> And you can find no justification
> To hide behind another,
> Another's rifle cannot take revenge for you.

Simonov described the war as a zero-sum game that required killing: either the German's wife would become a widow and his house burn, or yours would (he uses the intimate *ty* throughout the poem). The poem reaches a crescendo with a call to violence:

> So kill at least one now.
> So kill him faster!
> As many times as you see him,
> So many times kill him![30]

The incantation of strong emotions reaching catharsis in the act of killing was a well-developed theme in wartime propaganda. In the short story "Tin-Tinych," a mild-mannered schoolteacher becomes consumed by rage: "He bayonetted that vile creature with terrifying strength. For everything. For Private Danilov [a fallen comrade], for his students, for flowers, for the fishermen in Astrakhan, for the steelmakers in the Urals, for the teachers in Saratov, for Donbas miners, for all Soviet people, for life on his native soil, onto which crawled this loathsome fascist beast!"[31] Killing reified membership in the Soviet community: an act of violence can be dedicated not only to fallen comrades but also to students and even flowers. As soldiers came to bear witness to the deaths of their comrades and see the real crimes that the Germans had committed, destroying the enemy became an obsession.[32] Even practices that had been frowned on earlier as relics of the past, such as blood brotherhood and blood vengeance, were encouraged, embedding the call to kill Germans into ancient customs.[33]

In the labor of mastering their weapons and learning to kill, soldiers bound themselves in the greater community of working people throughout the Soviet Union not only as avengers but also as skilled tradesmen. Tulegen Tokhtarov, a Kazakh miner who would later become a (posthumous) Hero of the Soviet Union, is reported to have declared: "A machine gun in battle is like a drill or a jackhammer in the mine. In the mines, extracting lead, I knew that it would be used against the enemy. Now I have been called on to directly guide this lead to its target. Without knowing my weapon perfectly, without becoming a master marksman, can I say that I value the labor of miners, the work that I did only yesterday?"[34] Respect for others' hard work was expressed through the adept use of weapons. Not only did fighting connect soldiers to those laboring on the home front, it was, in its own way, simply another form of work, in which enemy dead were the measure of production.

"I have killed a German, and you?" was a common propaganda trope in 1942.[35] The tremendous social pressure to kill was cemented by adapting shock work to wartime aims. Shock work made competition imperative as teams and individuals attempted to outdo each other in overfulfilling production quotas. At the front this was adapted to the *schët*, *boevoi schët*, or *schët mesti*, which could be translated as "body count/score," "battle tally," or "balance of vengeance." The schët was a body count of the number of enemy soldiers and enemy machinery that a soldier, tank, or weapon crew had destroyed. The government adapted the shock-work tactic of material awards of several hundred rubles for the destruction of tanks. Artillery pieces displayed their schët in the form of a black tank with the number of destroyed tanks written in white on a cannon.[36] Units or individual soldiers could compete for a higher schët, and the opening of one's schët was considered to be the moment when one became a real soldier. Soldiers were sometimes provided with space for their schët in propaganda materials.[37] Without a schët, which translated hatred of the enemy into productive action, one was not a complete person.[38]

Red Army propaganda seldom used euphemism to discuss dying or killing, and the fixation on destroying the enemy could overtake goals such as gaining territory.[39] Momysh-uly wrote to his mistress in the war's final spring that "the results of our actions are measured not by kilometers, but by the numbers we destroy . . . while I am in awe of those who force the enemy to flee, it is better still to dictate your will—to force him to fight, not to let him run but to destroy him, so that he can never wage war again."[40] He enjoyed combat and had become a leading expert on soldiers' psychology.

Combat: Thoughts in Battle

Soldiers in combat often struggled with fear, which could break a person or lead to heroism. According to Momysh-uly: "Overcoming the feeling of fear, a warrior feels himself at ease (relatively, of course) among a multitude of dangers, believes in the strength of his weapon, and rationally, coldbloodedly, prudently acts on the battlefield. Sometimes blistering with the feeling of boiling hatred for the enemy which he has suffered, forgetting even

FIGURE 5.2 The sniper Aleksandra Shliakhova, boasting of her schët, 1943. RGAKFD 0-286776 (V. P. Grebnev).

self-preservation, he throws himself into danger and overcomes it."[41] Soldiers in combat were ideally supposed to be calm and collected with occasional explosions of righteous anger leading them to heroic feats. An iconic wartime song, "Holy War" (Sviashchennaia voina) had as its chorus "Let righteous rage boil over like a wave."[42] Other observers of their own sensations in combat described a variety of feelings from an intense lust for life under fire to a fatalism tempered by excitement or that they were simply too busy to feel fear.[43] Boris Marchenko wrote to his wife: "It has been my lot to live through quite a bit. But you somehow numb your nerves, ignore almost everything around you, lose your feeling of danger and fear."[44] The heroism that soldiers were supposed to show could manifest itself as indifference or a cocktail of emotions.

Tanker Lev Slëzkin described the mix of anxiety and pleasure he experienced while fighting in a letter home to his mother:

> Sometimes I wonder what I am doing. People jump out from a foxhole in front of my machine and run, run, and I cold-bloodedly take aim as if it were training, cut them down with a machine gun, and when they fall I am happy. Or they set up a long antitank gun, a cannon—and you have one thought: faster, faster, and if the black smoke of an explosion takes its place—I am happy. My cause is just, noble, but regardless of

> the cause and how long I am doing this, I have become accustomed to it—this risky business, as opposed to the down time between battles, where time passes without purpose.[45]

This is a professional, someone who had been fighting for years, describing the joys and underlying truth of combat—the necessity to kill the enemy before he kills you. That a member of an intelligentsia family could speak so casually about violence to his mother far from the front attests to its centrality and acceptability. Soldiers described their time out of combat as times of boredom, while moments before battle or when they were under fire but unable to do anything could be terrifying.[46] Battle brought a certain clarity of purpose to seasoned soldiers. They were active, armed agents able to impact both their own fate and that of the country.

Nonetheless, under conditions of demechanization and deprofessionalization, tanks and planes could cause panic even in small numbers. Propaganda materials into 1944 noted that green troops reacted very strongly to planes, tanks, and artillery.[47] A special term, "tankophobia" (*tankoboiazn'*), was coined to denote the panic tanks elicited. Soldiers received special training to overcome this fear—a tank ran directly over their foxhole, after which they threw grenades at extremely close range.[48]

An experienced soldier ultimately knew that these monstrous machines were operated by humans who could be killed and standing one's ground was the only chance one had, even if it was grimly slim.[49] To advance or hold on to territory, soldiers needed to have confidence in themselves and their weapons. This is why a unit's first kill, such as Abdulin's at the beginning of this chapter, was treated with such fanfare: it proved that the enemy was vulnerable. Simply shooting could be good for morale: a veteran commander declared that "if a person doesn't shoot, then he is already demoralized."[50]

Waging war had a strong psychological component intimately tied to the physical tools of the soldier.[51] Overcoming fear and the taboo of killing was a process that every soldier had to pass through individually, and not everyone managed to pull the trigger. Some soldiers could pull the trigger on the war's first day; others would prove unable to in May 1945. Battle required soldiers to overcome common psychological boundaries and extreme stress in order to use their weapons effectively and survive.

Frontline Learning

The battlefield was treated as a classroom. Soldiers and commanders were constantly reminded to expand their skills and knowledge at the front through reading, informal sharing of experiences, and seminars. Commanders were supposed to hold a meeting before a battle in which they inspected all weapons and equipment, demonstrating the power of Soviet arms. They then assembled their soldiers on the eve of an attack to "remind the soldiers of their oath, of the great liberating mission of the Red Army, show our dominance over the defensive enemy, give the fighters practical advice on how to conduct themselves in battle, remember the heroes of prior battles, call them to new feats and vengeance in the upcoming battle with the hated foe."[52] During battle, soldiers were encouraged to observe and learn. After combat, seminars were held to discuss what had worked and what had failed, praise heroic soldiers, and shame those who had failed to do their duty.[53] Survivors who had made mistakes or shown cowardice were publicly humiliated, while the deaths of those who had shown fear or incompetence were used to teach the living.[54] These meetings helped offset the lack of training that soldiers received and created a greater sense of community based on killing and surviving. The worth of individuals to the collective was based on their effective use of arms.

Mastering weapons was everyone's responsibility. Only via the skillful use of arms could the country be saved, and greater mastery would mean less loss of life. Stalin, in his May 1, 1942, address to the people of the Soviet Union, exhorted soldiers of all branches of service: "Study your weapon to perfection, become experts in your work, strike the German fascist invaders until their complete annihilation."[55] Each weapon served a specific purpose and granted a soldier a particular status.

The Tools of the Trade

The Red Army's arsenal consisted overwhelmingly of weapons that were effective, easy to use, and simple to maintain.[56] Soldiers were, as Anna Krylova noted, "partners in violence" with their weapons. Each weapon facilitated a specific type of killing, and different types of combat gave prominence to different weapons; urban environments gave primacy to grenades and compact arms, while open fields favored tanks and artillery.[57] In all environments the careful coordination of specialists wielding a variety of weapons was key to victory.

While constituting an integrated system, the major branches of service all played different roles and used different tactics on the battlefield. Soldiers in signals were tasked with the establishment and maintenance of communication, mostly through telephone but also via radio. Sappers sowed minefields and cleared enemy mines, built and destroyed roads, bridges, and concrete bunkers. Cavalry conducted reconnaissance and deep raids into enemy

FIGURE 5.3 Red Army Soldiers under fire in trenches, the North Caucasus, 1942. The soldier in the front carries a PPSh-41, the one farthest away a Maxim machine gun, the rest Mosin rifles. Their knapsacks, rolled overcoats, pots, and thing-bags are clearly visible. RGAKFD 0-156811 (M. V. Al′pert).

territory, made assaults in pursuit of a fleeing enemy, and was sometimes used to compensate for a lack of armor. Artillery was tasked with destroying enemy tanks and softening enemy positions to allow the infantry and armor to take territory and destroy the enemy. Tanks were used to assault enemy positions, both providing transport and a screen for the infantry. Last, the infantry was tasked with taking and holding territory and was the largest, most maneuverable branch of service.

Soldiers' weapons fell into two major categories, personal and crew-served, which encompassed very different ways of fighting and social organization. Personal weapons (rifles, submachine guns, and grenades) were used by soldiers in coordinated units but were ultimately operated by individuals. Crew-served weapons (machine guns, mortars, artillery, and tanks) could not be effectively operated by one person and required significantly more training and coordination to use. While all soldiers lived and killed collectively, crew-served weapons dictated a different set of relationships. Soldiers developed significant emotional attachments to both types of arms and defined their status and culture relative to their weapons.

Relationships with Weapons

The personal weapons a soldier carried came to feel like an extension of the body. As Loginov mused, "Our hands seem alien without a rifle, submachine gun, or grenade in our grasp."[58] These weapons were imbued with meaning by both the government and soldier from the moment a rifle or submachine gun was placed in a soldier's hands or a crew received their tank or artillery piece. The issuing of weapons was accompanied by a ceremony including the Military Oath, and the weapon was the object that embodied this oath. While arms were mass-produced, they carried with them stories that made them more than simply assemblages of metal and wood. Soldiers often received "named weapons" with long histories that had been used by a soldier in the same unit, accompanied by solemn ceremonies. The script for one of these rituals from a propaganda pamphlet recounted how a wounded veteran gave his submachine gun to a new ("non-Russian") arrival, reciting the names and deeds of previous owners—Paramonov, who killed 114, and Savushkin 121—and ending with the pledge: "I vow not to release this weapon from my hands until the complete victory over the enemy. If I am wounded, I will give this submachine gun, with the permission of my commander, to trusted hands that are capable of maintaining the honor of our Guards weapon."[59] By 1944 a similar ritual had made its way into the regulations of armor formations. Every time a tank crew received a new machine or an old machine received a new crew, a parade was held in which the commander of the tank read aloud the serial number of the tank and names of crew members, followed by a celebratory march, sometimes in the presence of representatives of the factory that had built it.[60] Both rituals celebrate the spiritual weight of a weapon and the obligations it embodied. Other propaganda cited the serial number of the weapon either as a form of remembrance (e.g., "This rifle N. 1591-VB remains as a memory of Gazarov") or a way of shaming irresponsible soldiers (e.g., "Rifle N. 61823," a short story about a rusty, abandoned rifle).[61]

Soldiers' fates were tied to their weapons bureaucratically as well as rhetorically. The serial number of whatever weapons were issued to a soldier were recorded in the Red Army booklet, and loss of a weapon was tantamount to treason. Soldiers could be executed for losing their weapons, but this was often unnecessary—without a weapon a soldier became a "defenseless target."[62] Even wounded soldiers kept their weapons and actively participated in combat until evacuated, and medics were under strict orders to carry both the soldier and his or her weapon from the battlefield.[63] Every day in the army was supposed to begin and end with an inspection. Special attention was paid to the condition of a soldier's weapon, with punishments and public humiliation meted out to those with dirty, rusty, or malfunctioning arms.[64] Every weapon had its own "caprices," and veterans felt uncomfortable with a weapon they had yet to fire.[65]

The successful prosecution of the war and the survival of the soldier hinged on soldiers coming to know and love their weapons. These prescribed bonds of affection found resonance among veterans. Weapons could serve as tokens of affection, exchanged with close friends.[66] Soldiers frequently spoke of their weapons as living beings. Some soldiers dedicated parts of their memoirs to their weapons, sometimes mourning them on the page.[67] Vasilii Grossman recorded that "a cannon after battle is like a living, wounded person. The rubber on its tires is torn apart, parts crumpled and shot through with shrapnel."[68] Many soldiers discussed their weapons in interviews, diaries, or letters home. One soldier wrote in his diary that his submachine gun was "bored from idleness."[69] Some soldiers wrote home about their arms, using terms of friendship or even romantic love. One soldier wrote home that he had met "a pretty special someone"[70]—his submachine gun, while another recorded a ditty: "A wife gets love and affection, And a rifle gets cleaned and oiled."[71]

Soldiers could have close connections with any weapon from a pistol to a howitzer. Aleksandr Kosmodem'ianskii, a tanker, wrote home to his mother that although his crew had been scattered he was happy to receive "my own old fighting machine, tested in battle, all wounded and shot through, diligently patched together in field workshops. Not for nothing has she fallen cleanly into my hands, and now she won't get away from me, she's going to Berlin."[72] Despite being mass-produced and ubiquitous, these weapons felt very personal to those who wielded them. One artillerist described the odyssey of his howitzer to the Mints Commission immediately after the war:

> My cannon is a 120mm howitzer, model 1938. We received her on September 3, 1941. My cannon took part in all of the battles for the duration of the division's offensive. Until October 1942 I was a gunner, and then I commanded the gun. The crew changed after almost every action. I was wounded twice but didn't go to the hospital. Only once for seven days was I in the hospital . . . True, the cannon caught it many a time, four or five times she was repaired at the armorer's workshop, but they never had to do an overhaul . . . My cannon took 10,620 shots, traveled 4,413 or 4,613 kilometers.[73]

This soldier could not separate his experience from that of his weapon, the cannon overtaking his biography and overshadowing his comrades. Such

Figure 5.4 Portrait of decorated Sergeant I. N. Kokurin with his 76mm m.1942 ZIS-3 cannon, First Baltic Front, 1944. RGAKFD 0-82677 (I. Pikman).

strong emotions probably stemmed from the fact that soldiers' very lives, and the fate of the country, depended on their arms.

Individual Weapons

All weapons needed to be used in coordinated efforts to be effective. The most basic building block of the army was the infantry squad of eight to twelve soldiers. Infantry tactics required coordinated fire as soldiers covered the advance of their comrades to destroy the enemy. The formations used by Red Army were open (six- to eight-step intervals between soldiers), and soldiers trained to quickly go from a "snake" (column used to advance, hiding numbers from the enemy) into a "chain" (a line of battle).[74] The squad centered around the machine gun, with most soldiers carrying rifles.

Rifles

Prior to the war a semiautomatic rifle, the SVT-40 (Samozariadnaia vintovka Tokareva obr. 1940 g.) was developed (the weapon with which Abdulin made his first kill) and nearly became the primary weapon of the Red Army.[75] However, soldiers complained that it was "inadequate for combat conditions due to the complexity of its construction, unreliability, and inaccuracy."[76] This weapon probably proved a particular challenge to less mechanically inclined peasants and was unusually flimsy for a piece of Soviet equipment. The SVT-40 violated the simplicity of design that was the hallmark of Soviet arms and was eventually taken out of production.

In 1941 posters called men and women to arms in the defense of the Motherland. In most of these images, the weapon being shouldered is the Mosin-Nagant rifle M.1891/1930. Between 1941 and 1945 the Soviet Union produced roughly twelve million Mosin-Nagant rifles and carbines, which were carried by the vast majority of soldiers serving in the Red Army.[77] A bolt-action rifle 166 cm tall, weighing 4.5 kilograms unloaded with bayonet attached, with an effective range of 800 meters and a five-round magazine, the M.1891/1930 fired a 9.5 gram, 7.62mm caliber bullet at 865 meters per second.[78] A soldier was expected to take ten aimed shots a minute and carried between 100 and 170 rounds of ammunition.[79] This rifle was the updated version of the model 1891 rifle (the bayonet and site had been modified) and reflected what had become the world standard for infantry weapons during World War I. The Mosin-Nagant was effective against enemy soldiers, but those carrying them felt outgunned by the tanks, planes, and artillery that came to shape the battlefield.

Soviet propaganda praised the rifle: "our Russian rifle occupies the most honored place. It is as if she carried the glory of our Russian arms with her."[80] Soldiers were encouraged to think of themselves as part of a proud tradition, carrying a weapon that had been tested in battle over decades and in every condition. They were also taught to think of the rifle as a weapon that, in the hands of a skilled soldier, outclassed more modern arms.

No one proved how deadly the simple rifle could be more than snipers, who formed a "movement" during the war and were declared "Stakhanovites of the front."[81] Armed with a telescopic sight on a regular rifle, these soldiers developed the riflemen's skills to an extreme degree, becoming masters of camouflage and expert marksmen. Stalin exhorted all riflemen to imitate the sniper.[82] These men and women saw their enemy very clearly, silently stalking them, sometimes sitting motionless for hours at a time. Exemplary of the voluntarism and lack of professionalism in the Red Army, snipers often became specialists only at the front and at their own initiative or were "volunteered" by their political officers.[83] They received a special status, with significantly higher pay and a raise in ranks that recognized skill without giving them additional responsibilities.[84] With a tally numbering in the hundreds, successful snipers proved that with training and initiative, soldiers with the Mosin could become predators, "hunting" enemy officers and soldiers and able to kill better armed men.[85]

Snipers were socially important: they demonstrated that anyone could become an effective killer by mastering the most ordinary weapon, regardless of nationality and gender. Poems such as Dem'ian Bednyi's "Semën Nomokonov" celebrated how native peoples of Siberia, hunters in civilian life, became excellent snipers. Bednyi praised the "wild Tungus" sniper Nomokonov, friend of the Russian people, who had a stick with over 320 tallies in his boot—one for each German he had killed. His skills were attributed to his childhood spent "in the harsh taiga" and his heritage.[86] Soldiers drawn from the ranks of prewar hunters utilized specialized skills from their civilian lives. Snipers were often expected to teach their skills to other soldiers, and the passing of skills from one sniper to another often involved interethnic friendship.[87] Sniping gave both "non-Russians" from some of the most remote regions of the Soviet Union and women a chance to prove themselves through killing.

Propaganda and the army used the sniper in a variety of ways. Flirtation with a female sniper often took the form of a challenge to beat her tally, adapting the general tenet that only those who distinguished themselves in battle were worthy of affection.[88] Snipers demonstrated the full potential of a soldier armed with the most basic weapon. The fact that many of them were both autodidacts and teachers who fostered several students made them all the more ideal. Through snipers, more than any other type of soldier, the tally found its realization and was utilized for socialist competitions to kill more enemy soldiers, just like workers at construction sites overfulfilling norms during the Five-Year Plans.[89] Snipers were capable of covering a retreat or destroying a decisive target in combat.[90] The sniper was fully conscious, as demonstrated by proficiency at killing.

The Bayonet

Red Army tactics were very aggressive, and the ultimate symbol of aggression and fighting spirit was the bayonet.[91] It was said that "the bullet clears the way for the bayonet" and that a rifle wouldn't shoot straight without the bayonet attached.[92] The use of the bayonet was closely connected to soldierly psychology, as soldiers were trained to maintain eye contact with an enemy they intended to bayonet. Any sign of weak will, even turning slightly, would lead to death.[93] The bayonet was seen as an integral part of the rifle, to be carried affixed to the rifle at all times. (Other armies provided a scabbard for bayonets.) *Combat Regulations of the Infantry* instructed soldiers "to close with the enemy, attack him, and destroy him in hand-to-hand combat or take him prisoner."[94] Planes, tanks, and artillery all became much less useful when soldiers were engaged in melee. However, bayonets were inconvenient to carry, their form (a cruciform needle) made them useless as everyday tools, and many soldiers threw them away.[95] When the war correspondent Grigorii Pomerants was asked to wage an "ideological battle" with soldiers discarding their bayonets, he recalled that only once had his division used the bayonet. His comrades threw away bayonets as extra weight, as "riflemen are walking dead." Soldiers in most branches of service had to carry all of their equipment on long marches, so items such as bayonets, helmets, and gasmasks—all of which cost the state money and all of which could save one's life—were often discarded for comfort's sake. This became such a problem that a new carbine was issued in 1944 with a swiveling bayonet that could not be removed. The inability to police soldiers' behavior led to the redesign of the army's primary weapon![96]

Regardless, the bayonet continued to have iconographic and mythological importance, even after soldiers discarded them. The central monument to victory on Moscow's Poklonnaia Gora would take the form of a massive bayonet over 141.8 meters tall—ten centimeters for every day of the war. Bayonets also continued to be featured prominently in posters.

Submachine Guns

On the opposite end of the spectrum from the bayonet were submachine guns, which encouraged aggression based on firepower rather than hand-to-hand combat and were immensely popular among soldiers.[97] These were small, hand-held automatic weapons that used pistol ammunition (significantly shortening their range) with either a 71 round disc or a 35 round "horn."[98] The Red Army, at the insistence of Marshal Voroshilov, was initially reluctant to adopt these weapons due to their short range. However, during the 1939 Winter War Finnish troops with submachine guns often proved highly effective at tying down larger Soviet formations and the Red Army soon began mass production.[99]

The three major submachine guns used by the Red Army were the PPD-40, PPSh-41, and PPS-43. The PPSh-41 and PPS-43 were the most heavily employed and represented significant achievements in design. Using stamped rather than machined parts, they could be produced in massive numbers—over six million were made in 1941–1945. They were very easy to clean and maintain, with innovative chromed barrels. The PPS-43 was a specially adapted submachine gun for troops in confined spaces (tank crews, scouts, etc.) and adapted a German-style folding stock and even greater economy of production, being developed in blockaded Leningrad.[100]

The army introduced a submachine gun company into every infantry battalion in October 1941.[101] Submachine gunners were used as an initial force to scout and soften the enemy, take and hold important objects or launch counterattacks. They would advance in either a circular or "T" formation, leaving their own flanks open and attempting to break through enemy lines. They were also used to create the appearance of an attack to distract the enemy. These soldiers engaged at close range, 200–300 meters, and were considered indispensable for taking trenches.[102] Submachine guns allowed troops with less training and underfed soldiers to level the playing field via technology. The PPSh-41 ultimately became the iconic weapon of the war, appearing on posters, monuments, and films.

Grenades

Alongside submachine guns *the* key tools in urban and trench warfare were grenades, popularly called "pocket artillery."[103] A wide variety of grenades—"defensive," meaning that you couldn't throw it farther than it exploded and needed some cover to use it; "offensive," meaning it could be thrown farther than its blast radius; and antitank grenades—were found in soldiers' pockets, pouches, and hanging off their belts.[104]

Soldiers were supposed to carry between two and six grenades on their belts, although many carried several times that. During the Battle of Stalingrad, grenades became the key weapon, according to many observers. Chuikov, commander of the Sixty-second Army, told the Mints Commission: "In these battles our soldiers came to love 'Fenia'; that's what they called the [F-1] grenade. In these street battles hand grenades, submachine guns, the bayonet, knife, and spade are employed. The Germans can't take it."[105] Grenades allowed one well-entrenched soldier to hold off a much stronger foe or to destroy an enemy tank.

"Burning bottles" (Molotov cocktails) were used alongside specially designed grenades to fight tanks.[106] The former came in two forms: one with an ampule, which ignited the liquid as the bottle broke against the target; the other filled with liquid "KS," a napalm-like substance that reeked of rotten eggs and ignited on contact with the air. These were touted as excellent antitank weapons—burning liquid would seep into equipment and telescopes, igniting the engine, ordnance, and crew, while smoke would blind the enemy. However, these weapons were dangerous to use. They were so combustible that they had to be carried in special crates filled with earth.[107] "Burning bottles" exemplified the improvisational, barebones ethos of the army, where glass and fuel crudely combined were used to destroy precision machinery, all at great risk to the soldier using them.

Soldiers were encouraged to sacrifice themselves, trading their lives for the destruction of enemy soldiers and machinery. Trading a person for a tank was considered to be a life well spent. The story of the Red Army sailor Panikakha provides a vivid example. His burning bottle was struck by a bullet, turning him into a human torch. Rather than attempt to extinguish the flames, he ran at an enemy tank, setting it ablaze and killing the crew. This event was immortalized in a poem by Dem'ian Bednyi and later by a monument:

> A burning torch, the avenging warrior . . .
> He burned the enemy with his own fire!
> They will write legends about him,
> Our immortal Red Sailor![108]

Like much of wartime propaganda, this poem promised immortality in return for self-sacrifice and posited that the most powerful weapon in the Red Army was the superhuman will of the soldiers.

To reduce German advantages in planes, armor, artillery, and automatic weapons, soldiers were encouraged to get in close with grenades, submachine guns, and bayonets. An unsuccessful attack would leave the stranded and wounded at the mercy of the enemy, prompting many soldiers to carry a spare grenade for themselves in order to avoid the ignominy and torture of capture.[109] Bullets, grenades, and shells were all essential; they could mark the difference between life and death. They were not, however, something a soldier would form an attachment with, nor an object that helped build a sense of community.

Crew-Served Weapons

The other end of the spectrum was occupied by crew-served weapons, everything from machine guns and mortars to artillery and tanks, which were objects built to last operated by collectives. Infantry soldiers dug their own, individual positions first and slept in bunkers that could be down the line from their fighting positions, weapons crews dug a multiperson fighting position with their weapon at the center and excavated their living space next to or even around it.

Crew members were in continuous contact, knew each other intimately, and functioned as a single entity. A crew could demonstratively distance newcomers, regardless of rank, in the ways that they ate together or slept.[110] Crew members were inseparable and came to rely completely on each other in an exchange of competencies and favors, such as a less literate soldier asking for help writing home.[111] Those lacking a common language could come together around a machine, communicating through gestures and an unspoken understanding of their tasks.[112]

Machine Guns

Machine guns were the nucleus of a squad and important secondary weapons for tanks. The Red Army used a variety of machine guns, but the two most prominent were the DP-28 (Ruchnoi pulemёt Degtiarёva obr. 1928g.) and various iterations of the Model 1910 Maxim machine gun. The DP-28 was a handheld automatic rifle with a 47-round disc. One soldier could operate it; a second was assigned to carry ammunition.[113] The Maxim, while old and heavy, held popular associations with the heroes of the Civil War, particularly Chapaev. The crew of a Maxim was at least two and as many as five soldiers.

The machine gun was a key tool that fulfilled a number of tactical roles, not only killing but also limiting the movement of enemy soldiers by pinning them down. Machine gunners, like snipers, were specialists with higher rank and compensation than regular soldiers.[114] With these privileges came extra work and responsibilities. Each machine gun required its own fire card and reserve positions because they were a primary target of the enemy. At night the vulnerable flanks of a unit were covered by its machine guns.[115] The *Combat Regulations of the Infantry* stated: "A well-functioning machine gun is unassailable for enemy infantry. Therefore machine gunners fight to the end, under any conditions, even in encirclement, sacrificing themselves."[116] They also played an important psychological role, as one commander recalled: "when a machine gun rattled, everyone's mood was elevated."[117] The machine gun made numbers less significant than nerve and skillful use of terrain.

Artillery and Mortars

Equally key to the successful prosecution of modern warfare were artillery and mortars. Red Army defensive tactics called for luring the enemy as close as possible and striking a crushing blow at point-blank range. In offensive operations, artillery provided a shield of fire for infantry to advance behind, suppressing the enemy and destroying vital targets such as tanks, machine guns, and enemy artillery. All of this required coordination by a team of soldiers who aimed, loaded, and serviced the weapon.

Artillery was both an antipersonnel and an antimachinery weapon. Standard artillery calibers ranged from the 25mm anti-aircraft shells to the most common 45mm gun assigned to infantry regiments to massive howitzers firing shells of over 305mm.[118] Artillery shells included both armor-piercing shells and high-explosive shrapnel that could render the human form unrecognizable. Artillerists needed to make advanced algebraic calculations, exploiting the landscape to fire accurately, as the number of shots they could make from one position before the enemy zeroed in on them was limited. The smallest artillery piece, the 50mm mortar, had only a two-man crew that lived like infantrymen, while most artillery was serviced by five or more soldiers. The crew of a larger mortar (82mm or 120mm) or cannon was commanded by a sergeant who was responsible for the actions of all crew members, with a gunner who aimed the piece, a loader, and at least one ammunition carrier. In addition, every unit was supposed to have several soldiers available to replace losses.[119]

Specialized tank-destroying artillery units engaged enemy tanks in a deadly game of peek-a-boo, allowing enemy tanks to get as close as possible in order to destroy them. Soldiers serving in antitank artillery formations received a number of privileges, including a special patch on their sleeves.[120] Their tactics required strong nerves. The artillery was largely immobile on the battlefield, at most having a few reserve positions to maneuver among. Artillerists were forbidden from retreating without their arms. One artillerist explained: "We take heavy casualties. The Germans call us 'dead men' [*smertniki*]. We can't leave the battlefield. Of course, no one can leave, but people run away. It's easier for an infantryman. We can't leave in any event, so they send us to the more dangerous places."[121] This image of artillerists holding their ground against enemy tanks, defending to the last man, emerged during the war and became a major trope in postwar popular culture. Artillery was generally horse-drawn, but increasingly as the war continued became motorized either with trucks or as self-propelled guns (*samokhodki*)—artillery pieces mounted on a tank chassis.[122]

Tanks

Soviet tankers were reminded that their steel behemoths were originally devised as a solution to the problem of the machine gun during World War I. Tanks could destroy strong enemy positions and provide a screen against small arms fire for infantry. By World War II, tanks had developed into rapid-moving, heavily armored machines. Tanks fundamentally changed the battlefield in World War II, deepening the space of the front and allowing for rapid movement that would have been physically impossible in the previous war. The successful use of tanks required trained personnel to crew them and precise coordination to prevent their isolation and destruction by antitank weapons and enemy tanks.[123]

Tanks were divided into light, heavy, and medium and were crewed by two to five soldiers. The Red Army used three chassis for all of its armored vehicles, greatly simplifying manufacture, maintenance, and repair. We will concentrate on the most iconic tank of the war, the T-34, which is often described as the best tank of the war due to the economy of its manufacture and combat effectiveness.[124] A medium tank, the T-34 weighed 28.5 tons, could travel up to 50 km per hour, carried over 700 liters of diesel fuel, a crow bar, a saw, an ax, and two spades. The tank also carried a radio, fire extinguisher, and medical kit. Armed with a 76mm cannon, 71 shells, three DP machineguns with 1,890 rounds of ammunition, 20 F-1 grenades, and the soldiers' own personal weapons, the tank was a veritable mobile fortress.[125]

A tank crew consisted of a commander, turret gunner, loader (*komandir orudiia, bashner, zariazhaiushchii*), mechanic/driver, radio operator, and (depending on model) assistant gunners and mechanics. The commander was responsible for controlling the tank and keeping it provisioned and functional, with specific tasks assigned to each member of the crew: the turret

Figure 5.5 T-34. Cover of *Tank T-34 v boiu.*

gunner kept track of shells, loaded, aimed, and fired the cannon; the radio operator kept the radio in order, communicated with other tanks, and operated a machine gun; and the mechanic/driver drove the tank and maintained the engine, air filter, and tracks.[126] Keeping these machines running was a full-time job, and it has been estimated that about a quarter of tank losses were due to mechanical failure.[127] To operate this behemoth, the crew had to function like an organism, each soldier answering for one part of the system. Using a tank was difficult business. Closed hatches in combat meant that visibility was extremely limited. Gunners often had to aim while moving rapidly over uneven terrain, solving trigonometric equations under fire. They were encouraged to use machine-gun fire or several shots to zero in on their targets, with the mechanic/driver observing where the shot or shell landed (much the way a sniper team worked).[128] The mechanic/driver had to constantly keep in mind the type of ground they drove on, because swamps could swallow tanks and muddy ground immobilize them.[129] Tanks could become giant battering rams, whether to ram through buildings or crush enemy weapons and soldiers.[130] Tank crews needed to keep in constant communication with other tanks and surrounding units, using radio, flares, flags, and hand signals.[131]

Successfully using a tank required the ability to read maps and terrain, make judgments about soil consistency, and coordinate efforts with engineers when crossing swamps, rivers, or lakes.[132] Tanks required a lot of prior planning and information about the enemy to work effectively, and tankers used not only fire maps but also tank maps that recorded information about the landscape, enemy, and points where the tanks were to regroup.[133]

Tankers frequently had to leave the confines of their tank in order to reconnoiter—climbing trees, crawling forward, and gathering as much information as possible before returning to the confines of the tank, where only a few periscopes and slits allowed the crew to see the outside. Vision was so limited that every crew member was responsible for constant observation of a certain portion of the 360 degrees around the tank.[134] When used together, a platoon of tanks functioned exactly like an infantry squad of steel giants, maintaining distance to ensure maneuverability, covering each other, and deploying from columns into battle formations. The turret even moved back and forth while on the march like a giant head.[135] Tanks also provided transportation for soldiers, being exactly the right size to take a squad—they even had special handles for riders.[136] These soldiers provided cover and maneuverability but also required crews to move carefully so as not to crush them and to provide support when these soldiers got pinned down.[137] As soon as the tank stopped anywhere, the crew took stock, replenished supplies, made repairs, and then began a series of new tasks, including the gathering of information, digging a position for sleeping underneath the tank, and setting up a forward position to guard the tank. A tanker's work was never done.

During offensives, tanks destroyed enemy machine guns and artillery, then concentrated on killing soldiers. They were to avoid combat with enemy tanks unless they had a clear advantage.[138] On the defensive, tanks could be used as entrenched artillery, a reserve for counterattacks, or a distraction to confuse the enemy.[139] Tanks used without proper reconnaissance were quickly destroyed.[140] As late as the autumn of 1942, it was found that many crews had yet to master these skills and armored unit commanders often just controlled their own tanks rather than commanding their formations.[141] Measures were soon taken to improve cadres and reorganize the armored forces.[142]

The fact that tanks were so expensive and difficult to operate and required reliable, talented crews led to extreme vetting of armor cadres. In October 1942 a special order to fill the ranks of tank training schools declared that prospective tankers had to have distinguished themselves in battle and have at least seven grades of education, with exceptions to the latter condition made only for those who had attained the rank of sergeant and won combat medals. A special commission screened all candidates. A few months later a directive insured that tankers were used only as specialists, and not sent as replacements to other branches of service.[143] A November 1944 order repeated these requirements, adding that from previously occupied territories only Communists and Komsomol members could be considered to serve as tankers. Soldiers from Western Ukraine, Western Byelorussia, and Moldova—all annexed on the eve of the war—were ineligible.[144]

Tank crews had to be educated enough to master a complicated piece of machinery and dedicated enough to be trusted to function autonomously. An infantry or artillery soldier could be forced to fulfill his duty under the watchful eye of commanders, but a tank crew was often alone and on enemy territory, where the will and ingenuity of its members alone could keep a precious piece of equipment from falling into enemy hands.[145] There was also no way to evacuate wounded soldiers from a tank, and the 1944 *Combat Regulations for Armored and Mechanized Troops* stated, "Every soldier in the crew, if wounded, should exert all of their strength and continue fighting."[146] Tanks were a key weapon of the Red Army in a war that Stalin referred to as a "war of motors,"[147] and the men who crewed them had not only to be dedicated to the cause to the point of being ready to burn alive for it, but also capable of turning a monstrous steel ensemble into a smoothly functioning entity. No weapon better demonstrated the Red Army adage "machinery without people is dead."[148] The connection of soldier with machine could become so great that in one extreme case a soldier used the blood of a fallen comrade to put out a fire in his tank.[149] Tankers without their tanks were particularly pitiable, as one veteran described: "without their tanks, without boots, barefoot, in torn clothing they have a tortured look—filthy, bloody, and burnt."[150] But burning was part of being a tanker.

Corporate Cultures of Different Arms

Soldiers were encouraged to identify with their weapon and branch of service. There were widely accepted nicknames for the various branches, often reproduced in propaganda: "They call the infantry 'the Queen of the Fields,' artillery—'God of War,' sappers—'War's Workers,' and the signal corps—'War's Nerves.'"[151] All of these branches relied on each other to successfully fight, but antagonism was rife as soldiers created more or less exclusive corporate communities. Ivan Iakushin, a cavalryman, recalled an old saying from the tsarist era that could still be heard in the Red Army: "A dandy serves in the cavalry, an idler in the artillery, a drunkard in the navy, an idiot in the infantry."[152] We will look closely here at three branches of service that formed distinct communities and exemplified modern warfare: the infantry, artillery, and armor (tanks).

The infantry were at once the biggest and most maligned branch of service. Virtually all soldiers underwent some basic infantry training regardless of their specialization, because the digging, movement, and use of arms

FIGURE 5.6 Merging of man and machine. Tanker Senior Sergeant E. P. Fëdorov eats in his tank, 1942. RGAKFD 0-57505.

emphasized in the infantry were necessary for everyone. As the only branch of service that could independently take and hold territory, the infantry was the largest and in many ways the most important branch. Despite this, and because it tended to get the least-educated soldiers, many looked down on infantrymen. Momysh-uly proclaimed the infantry to be the "fundamental and universal branch of service":

> No other branch of service requires such toughness and bravery, almost inhuman effort, exertion, as from the infantryman . . . Everything he conquers he takes with his own hands, himself, personally, not thanks to "steeds" or "motor." No one is as deprived of excuses and opportunities to avoid battle, citing a broken machine or lack of preparation . . . In no one else does the natural beauty and that natural ugliness of a person show itself so clearly . . . are man and beast so harmonized, as in the infantryman.[153]

Momysh-uly positions the most humble soldiers as not only those most necessary to get things done but also as the most authentic humans. The infantry suffered the heaviest losses and had the largest circulation of cadres, which meant that cohesion and the establishment of a community was a particularly difficult problem in its ranks. As Lev Kopelev noted: "People there were more mixed, losses [were] heavier, and the soldiers and commanders changed more often, without having time to firm up a real, rooted comradeship."[154]

In sharp contrast to the infantry stood the artillery, where Kopelev noted that young officers called each other nicknames, "played chess and battleship, argued about films, soccer, Maiakovskii, love . . . And at the same time skillfully and heatedly directed artillery fire."[155] The artillery was a highly technical branch, requiring mathematical and mechanical skills that were less crucial to the infantry. Also, in the period of deprofessionalization and demechanization artillery was often the most effective branch, as tanks were too few and infantry undertrained. This led to a sense of superiority among many artillerists that reversed Momysh-uly's lionization of the "Queen of the Fields." Artillery Commander Ivan Bykov told the Mints Commission during the battle of Stalingrad that he would never leave the artillery and contrasted the habits of infantrymen with those of artillerists: "Artillery is a more organized, compact branch of service, and so are its members. You see, we and the infantry get reinforcements at the same time. It seems as if these reinforcements are the same, but they are a different breed . . . Everywhere you see the difference, from their abilities in battle to the way they act . . . Artillery is more culturally developed, its fighting abilities and all other qualities." He saw this as stemming from both training and the material culture of the two branches. He explained that in the infantry: "A person feels less accountability for their actions, and he works on himself less. An artillerist has a rifle, saber, horse, saddle, accoutrements, and so on, and an infantryman just has a rifle and a backpack, maybe a spade and that's it. And he even does a poor job taking care of his rifle."[156] Something similar to class consciousness overtook soldiers, with those from the more technical branches speaking of the infantry in terms that the middle and upper classes often reserved for the poor.[157] The infantrymen were blamed for a poor education and a lack of responsibility tied to the poverty of their material culture, regarded as a mass organization with a lower "cultural level." Artillerists in particular were encouraged to think of themselves as a cut above the rest, as artillery was the "God of War."

Artillerists held a special place in Soviet tactics, being the primary antitank force, particularly in the war's first years. Stalin had a special affection for the artillery, and they alone among ground forces received their own holiday during the war: Artillerists' Day, first celebrated on November 19, 1944.[158] An article from 1944 celebrated Stalin's genius, care for the army, and special relationship to the artillery: "Despite the huge importance of tanks and aviation, Comrade Stalin did not exaggerate their role. The Germans wagered on tanks and planes, underestimating artillery, and have been paying dearly for this. Comrade Stalin called artillery 'the God of War,' and his valuation has been entirely supported by a whole range of events. He devised the doctrine of modern artillery and artillery offensive."[159] The artillery was both being rewarded for the key role it had played in the war and used to gloss over the disastrous first years of the war. Rather than admit that the importance of artillery was in part the result of shortages, the lack of tanks and planes early in the war was presented as part of the wise strategy of the Generalissimo. This was a fitting choice, given that artillerists were likely to be among the few in a division that had long memories, as enough of them usually survived to become part of the *kostiak*, or "backbone," around which units were rebuilt.[160]

Somewhere between the infantry and artillery rode tankers. The elite status of tankers meant that every member of the crew was supposed to be at least a sergeant, and the commander of a medium or heavy tank was either a lieutenant or a senior lieutenant.[161] The rank of a tank commander was the equivalent of that of a commander who could lead dozens of soldiers in the infantry. Yet tanks were surprisingly vulnerable. To be on a tank crew meant to burn, literally. Tanks suffered massive casualties, and escaping from a burning tank was part of the job. One veteran casually wrote his mother that two of his tanks had burned, while another told the Mints Commission of a comrade who had escaped from twelve tanks before being killed outside his thirteenth.[162] Early in the war, it was determined that tanks should be sent to existing units rather than be used to form new units, as a unit would lose 70–80 percent of its tanks in two weeks of combat.[163] Germans investigating destroyed tanks that they captured found that the vast majority were hit within six months of manufacture.[164]

These facts ran counter to an official image of tanks as virtually indestructible. The hit "Tanker's March," made popular by the film *Traktoristy* (dir. Ivan Pyr'ev, Mosfilm, 1939) on the eve of the war, spoke only of the remarkable qualities of tanks, not their vulnerability. It boasted: "Our armor is strong and our tanks fast / And our people filled with bravery / Roaring with fire, shining with the sparkle of steel / The machines go into a furious campaign." The song continued to promise any enemy waiting in ambush that Soviet tankers would fire first and destroy them.[165] Much of the folklore produced by tankers themselves focused on their feelings of vulnerability and spoke to a sense of fatalism that belied official culture, reflecting what they witnessed inside their tanks. The T-34 was often referred to as "a coffin for four brothers."[166] A popular frontline song with many variations spoke of a proud fatalism:

> Motors flare with flame
> licking the turret with fiery tongues
> I accept the call of fate
> with a handshake . . .
> They take us out from the debris
> Carry the carcass out
> And volleys from the turret's cannon
> Escort us on our last voyage . . .
> Farewell Marusia, my darling,
> And you, KV [tank], my little brother,
> You I will see no more,
> I lie with a shattered skull.[167]

Tanks were fearsome machines cared for by their users but often became final resting places. The extreme responsibility that tankers had for their machines, which tended to have short lives in combat, led to a great deal of identification with them. Many had a long list of comrades whose deaths soldiers witnessed in the tight confines of their tanks.

Each tank was individually identified. Tanks themselves often had a name and a story even before arriving in a unit. It was not uncommon for a collective farm, factory, region, or even notable individual to donate money to pay for a tank or column of tanks during the war, often honored by the name of the tank or unit (e.g., Kirovets from the famous Kirov Factory in Leningrad or Battle Girlfriend—a tank paid for by a war widow who later became its driver/mechanic).[168] Many were also given the name of a historical personage such as Chapaev or Suvorov. Naming tanks gave them an affective connection with the rear or reified Russian heroic past. Even unnamed tanks were given a number by the unit commander that was displayed on all sides to make tanks identifiable. Soldiers painted their tanks camouflage colors to break up their outline in the summer and whitewashed tanks in the winter. The potential for personalization was quite high, and soldiers loved their machines, even if they were frequently knocked out of commission and repaired.[169]

Tankers, artillerists, and infantrymen lived very differently. Who you were in the army was determined by what weapon you used and how well you used it. The way that your time would be organized, the activities that you engaged in, how well you were compensated, and a great deal of social capital were an extension of the weapon you wielded. As the war continued and soldiers became more and more competent, these differences continued even as the branches began interacting with each other more and more intimately.

The Emergence of a Professional Army

The experience of veterans, purchased at a steep cost in human lives, was immensely valuable. Just as cadres of workers learned on the job during industrialization, so did soldiers at the front. As the crises of 1941–1942 passed, units were increasingly composed of soldiers who had survived their initial trial by fire and of soldiers who had been given time to more adequately train before being sent into combat. Soviet industry was able to produce arms and armaments on a massive scale, providing soldiers with more of what they needed to kill. The Nazis had also provided ample reasons for most soldiers to pull the trigger. As men and women bore witness to destroyed villages and murdered civilians and POWs, the propaganda proscription to allow "noble anger" to overtake oneself and take vengeance became a reality. By 1943 an officer could tell Vasilii Grossman: "The infantry has become accustomed to shooting. Automation has developed."[170] Beyond simply pulling the trigger, soldiers were increasingly showing mastery of their weapons, as the rising number of Guards units demonstrated. By February 1943 over sixty divisions, corps, and other large units had earned Guards status.[171]

By early 1943 General Chuikov could boast that "Stalingrad is the glory of the Russian infantry. The infantry defeated the entire arsenal of German technology."[172] He presented the clear turning point of the war as the triumph of the most basic, motley, and overwhelmingly peasant branch of the army over the meticulously designed machinery of the Wehrmacht. On Red Army Day (February 23) 1943, Stalin declared that the Red Army had become "a professional [*kadrovaia*] army."[173] The proportion of automatic weapons in the army rose dramatically. Technical formations came to make up a larger portion of the army as armor and artillery were privileged for replacements, often leading to chronically understrength infantry formations. Now expertise and firepower increasingly replaced manpower.[174] The rising competence of both soldiers and commanders led to lighter casualties.[175] The success at Stalingrad provided new tactics and showcased the efficacy of specialized troops such as snipers and submachine gunners. As the army turned westward, it was both remechanized and reprofessionalized. Aggressive tactics that had often been suicidal in 1941–1942 became effective.

Coordination between the branches of service during operations improved dramatically.[176] Beginning in Stalingrad and expanding thereafter, new storm units (*shturmovye otriady, shturmovye gruppy*) played an increasingly important role in combat.[177] Mixed groups of the various arms brought together for a specific task, such as storming part of a city or capturing a hill, provided greater coordination. The storm units were the polar opposite of the poorly trained and barely coordinated formations of the first phase of the war. These groups relied on a small number of highly professional cadres to succeed.[178] This could affect a soldier's understanding of his or her place in the army and foster greater intimacy between soldiers and their commanders. Hero of the Soviet Union General Pëtr Koshevoi described the difference in how units functioned in one of the last operations of the war to the Mints Commission: "every soldier knew who you were . . . We broke the whole existing organization down. We built a new one, prepared specifically for the mission, which was assigned to the troops. Got all people—sappers, artillerists, signals, staffers, tankers, self-propelled gunners—together, in one unit, so that everyone knew everyone else."[179] These new tactics placed greater importance on the individual soldier, as commanders demanded not only more of him or her but also took seriously the rhetoric of a commander being intimate with their subordinates and provided a stable group of well-supplied professionals.

Soldiers had greater access to ammunition and fuel. General Antipenko, head provisioning officer for the First Byelorussian Front, stated that if you took the Battle of Stalingrad as a base level, then by the summer of 1943 the amount of ammunition used daily had increased more than threefold and fuel quadrupled, while by the Battle of Berlin ammunition was being expended at nearly five times the rate of 1942 and fuel over five times.[180] Increasing numbers of Guards units meant that a significant number of infantry formations would have a chance to build camaraderie, because their wounded returned to the same unit.[181] In addition to greater infrastructure and organization, soldiers seemed more willing to kill.

Many observed that soldiers had become fixated on killing and the tools of the trade. One sniper confided to Vasilii Grossman, "I have become a brutal [*zverskii*] person—I kill, I hate them, as if that is how my life should be."[182] A. Luzhbin recorded that one of his soldiers told him, "I have already forgotten how to boil kasha, but I know mines very well."[183] During Stalingrad the desire to kill seems to have reached the level of obsession, as Ivan Vasil'ev, head of the Political Department of the Sixty-second Army at Stalingrad, revealed: "I should say that I didn't see or hear or get any intelligence about Red Army men under any battle conditions showing any sort of pity for the Germans. What is more, even if there wasn't anyone to stick with a bayonet, they would stab the dead."[184] Shortly after the war Boris Slutskii described hatred toward the enemy as "not contempt, not spite, but a disgusted hatred, an attitude on the level of regard for frogs and salamanders."[185] For some in the army, hatred went from passion to a sort of disgusted ennui once Red Army soldiers attained overwhelming superiority in tanks, planes, and artillery.

Becoming a professional soldier could overshadow all other identities, as Loginov recalled: "Sometimes it seems as if we've always been soldiers, as if we never had a family or childhood."[186] Millions of people had been taught to kill and plied this trade for years. Soldiers came to terms with the killing they had done without any government program to reintegrate them into society. Having weapons training and seeing the results of their use was the norm for a huge swath of the population by 1945. After the war, some veterans spent as much of their free time as possible with their comrades, while others avoided other veterans and talk of the war.[187] Some described feeling uncomfortable without a weapon, and firearms were ubiquitous in the chaotic years after the war.[188]

Weapons themselves acquired a new status after the war, as the AK-47 and nuclear warheads supplanted wartime arms and the rifles, machine guns, cannons, and tanks used during the war began to take their places in museums. Often tanks and artillery pieces that had played a part in the liberation of a city became monuments. There was little room for the discussion of the gritty details of what these weapons had done or the difficult process of learning to use them. Instead, what soldiers and their arms had accomplished together, the defeat of Fascism, was emphasized. They had learned a new trade, one that was particularly grim and risky, but which they had mastered. As nomadic soldiers waged an unprecedented war of attrition, they and their government strove to make sense of their wanderings and anchor soldiers in a larger community.

Part Three

Possessions

Figure P.3 Red Army Troops in Schneidemühl, Third Reich, 1945. Note the destroyed bank they are marching past and guitar one of the soldiers is carrying. Thing-bags of various volumes hang from their backs. RGAKFD 0-174769.

Chapter 6

The Thing-Bag

A Public-Private Place

How much can a soldier take with him? A can of conserves. As they say, in battle even a needle is heavy.

—General Grigorii Maliukov

Most Red Army soldiers carried a lot of stuff. According to an official estimate from 1939, the average burden of a soldier was 24 kilograms (52.8 lbs.), while for those servicing machine guns, mortars, and other heavy weapons the weight could rise to 48–64 kilograms (105.6–140.8 lbs.), as weapons, ammunition, and tools to maintain them added to a soldier's burden.[1] Soldiers carried an equipment belt with grenade and ammunition pouches, an entrenching tool, a water bottle, sometimes a pistol or a bread bag. They were also supposed to carry gas masks, which were often discarded and the attendant bag used to carry ammunition and other necessities. Over the left shoulder they wore a rolled overcoat or *plashch-palatka*, depending on the season. On the backs of Red Army soldiers hung the "thing-bag" (*veshchmeshok* or long form *veshchevoi meshok*), a shapeless sack that was nearly monopolized by army-mandated equipment. However, this pack, along with soldiers' pockets, was the only space in which to carry the few personal belongings they might have.

The thing-bag was the closest thing soldiers had to a private place. But like everything else in the army, the soldier's pack was subject to random inspections. Superiors could look into thing-bags if they so desired, and they were encouraged to do so.[2] There was usually little soldiers could fit into their packs aside from regulation items, but the few objects they carried were invested with great meaning, whether they were items from the soldier's home, anonymous gifts, or things crafted at the front. These objects gave some consistency to the nomadic life of Red Army personnel moving from bunker to bunker, peasant hut to peasant hut.

The only nonmilitary items that soldiers were encouraged to carry in their packs were ephemera. Books, magazines, and newspapers were published to raise soldiers' morale, allowing them to expand their knowledge and distract themselves from the miseries of life at the front. In an army learning on the job, they also provided useful information about how to fight and transmitted the latest tactical innovations. Finally, this was the medium where soldiers could find explanations of the meaning of the war and their place in it. Writers from throughout the Soviet Union were mobilized into the ranks of the Red Army as war correspondents, and newspapers were created at the level of the division (thirteen thousand to seventeen thousand soldiers). These newspapers—alongside small pamphlets, agitation sheets, magazines, and the much rarer books—made up most of a soldier's reading. Despite a severe paper shortage, the government created a vast print network of newspapers organically attached to Red Army units. This effort both transmitted information from central publications and generated texts specifically addressing concerns within a unit. The Soviet government was deeply invested in circulating the right kinds of information.

Of even greater importance were letters, which reified connections with home and allowed soldiers to participate in the lives of distant relatives. The army made significant efforts to provide soldiers with opportunities to read and write but also shaped the contours of this communication. The government provided free postal service to and from the army, as soldiers wrote and folded distinctive "triangle" letters without an envelope or postcards with patriotic imagery. In return, the state inserted itself into these correspondences at multiple levels. Commanders were encouraged to write home to their subordinates' families in order to praise or shame them. It was common practice for soldiers to read their letters aloud to comrades, something extolled by official publications and the party apparatus. Finally, the government attempted to censor all letters in all languages. What was potentially a soldier's most personal, intimate communication was made very public.

This was equally true of a soldier's knapsack: the closest space to something private was in fact very public and subject to constant inspection and intervention by government representatives. In this chapter, we too will peer into the soldier's thing-bag and examine this public-private place that was key to helping both state and soldier construct meaning. The few personal items and papers in the thing-bag allowed soldiers to recreate a sense of self while moving from place to place, understand their position in the events of the war, and maintain connections with loved ones. The first section provides an overview of the knapsack soldiers wore, what they were obliged to carry by regulations, and the nonregulation knickknacks that soldiers might carry. From there we shift to the paper goods so important to soldier's morale and the government's mission and examine the production of and responses to print propaganda and literature produced during the war, before turning to a discussion of letters to and from the front, in particular how they were written and read. Soldiers' letters, with their distinctive triangular shape, and the bindle-like thing-bag are two iconic objects from the war, easily recognizable to this day and both born of necessity and economy.

The Thing-Bag

The thing-bag could be seen as a distillation of the ethos of the Red Army. It was crude, utilitarian, cheap and required a little special knowhow to use. It was also, like most of those who would wear it, of peasant origin. The thing-bag was a trademark of wandering peasants as recorded by *Peredvizhniki* (Wanderer) artists of the mid-nineteenth century.[3] It was, in essence, a sack that had a long shoulder strap used to tie it shut, turning it into a back pack. The Red Army's version militarized this peasant accessory by making it out of a green, water-resistant cloth and adding padding to the shoulder straps. The thing-bag was not the first choice of those equipping the army, just as peasants were not seen as the ideal cadres to serve in its ranks. A combination of necessity and utility placed it on the shoulders of Red Army soldiers.

Before, during, and between the world wars, the Soviet Union, Germany, and the United States developed systems of equipment that were designed for specific purposes. The Germans created a number of specialized knapsacks and frames for carrying equipment as compactly as possible. The US Army's standard-issue haversack was designed only to carry a soldier's rations. It was a series of flaps, not a bag, and could not be used to carry anything else, although soldiers could hang a variety of equipment off it. The US Army would eventually develop a more useful "general purpose bag," but this would be issued only to specialist troops such as paratroopers, mechanized infantry, and engineers.

The Red Army had developed its own modern systems of equipment, including several types of knapsack. The models used before the war were the m.1936 *ranets*, a rigid, square construction that resembled a German pack; and the m.1939 and m.1941 packs, which looked more like a classic rucksack, including padded straps and back pads, specialized exterior pockets, and systems to help support the weight of equipment belts. These knapsacks were designed with precise spaces for specific items, and some were packed in layers in such a way that a soldier would have to completely unpack it to get at items at the bottom. These packs were determined to be inconvenient, and the army was headed toward simplification.[4] The thing-bag was officially reintroduced in 1940 as a stopgap measure. Reviews conducted after the Finnish campaign showed that it was difficult to retrieve necessary items, such as weapon repair kits, from the m.1936 pack, while the expansion of the army favored a simplification of equipment across the board.[5] Once hostilities began, this process accelerated as more complicated knapsacks virtually disappeared—those wearing them fell in battle or marched into captivity in 1941. The thing-bag, easy and quick to manufacture, became universal.

The thing-bag was simply a shapeless sack that soldiers tied shut using a special method. According to regulations, this space was to be highly organized and occupied by government-issue items. Soldiers were to carry their plashch-palatkas, a change of underwear and *portianki*, a gun cleaning kit, spare parts, up to five days worth of rations, soap, a razor, a toothbrush, and tooth powder. Soldiers were also encouraged to keep basic tools on hand to repair their uniforms. All of this was packed in a regulation order, which both allowed anyone to stick his or her hand into the pack and fish out the needed objects and formed an amorphous bag into a comfortable shape with nothing poking into the wearer's back. There was very little space for anything else, particularly when people were ordered to carry a change of shoes or extra ammunition. Yet soldiers did find a way to carry a variety of objects.[6]

The government was keen to provide soldiers with the means to entertain themselves, referring to various objects sent to the front to entertain soldiers as "cultural enlightenment objects."[7] Some soldiers carried soccer or volley balls, cards, chess, and checkers.[8] Others carried musical instruments, varying in size from harmonicas and mouth harps to accordions and guitars. The army even tried to send special national musical instruments to units with a large number of soldiers from the national republics.[9] The squeezebox was considered a fine accompaniment to patriotic songs.[10] Some soldiers had access to record players and radios, and Vasilii Grossman claimed that "there [was] no more musical place" than Stalingrad at the height of battle.[11]

Soldiers also carried practical objects from home, such as knives. Boris Marchenko, a school principal from Odessa whose family had been evacuated to Siberia, thanked his wife profusely for a knife she forced him to take: "Remember when I was leaving, I didn't want to take the kitchen knife, but you stood your ground and I took it. It's a good thing that you did—it's been very useful, not just for me, but a whole bunch of comrades use it. I safeguard it and will try to bring it back with the comb." Marchenko wrote of his comb: "I piously preserve the comb that you have ordered me to return and am making all efforts to fulfill your mandate exactly."[12] Combs and knives were among the more practical things that soldiers would bring from home. Some commanders went so far as to bring sets of silverware and china. Momysh-uly wrote home to his mistress: "I am very glad that you forced me to take them [dishes from her home]. Three times a day, while I eat, I feel myself at home—so great are the pleasant memories these 'domestic little nothings' invoke."[13] These objects had great utility. Moreover, they carried significant emotional weight as a familiar object from home that embodied both a connection and the hope of return. Anything a soldier carried would need to have immediate use or great emotional importance, because even a few extra ounces could be felt on long marches.

Religious items, such as Bibles, crosses, and prayers wrapped in leather, were often found among soldiers' belongings. The government was ambiguous about its relationship to religion during the war, as Mikhail Kalinin instructed agitators in 1943: "There can be found among our soldiers, particularly among the older ones, believers who wear crosses and pray. The youth mock them. We must remember that no one has forbidden religion, and we don't persecute it . . . Of course, if someone among the youth makes fun, that's no big deal. But it is important that mockery doesn't turn into humiliation."[14] This quotation hints at the uneasy détente with religion that arose during the war. Religion was identified with older people who could probably never become fully Soviet, but propagandists mobilized much of traditional Russian and "non-Russian" culture to motivate soldiers. The government was forced to reckon with the fact that most in its ranks probably believed in supernatural forces. Pëtr Liubarov, who worked for a frontline newspaper, noted in his diary in August 1943, that crosses and semipublic expressions of faith had become common during the war, with newspapers "writing about religion in a positive light."[15] Many soldiers mention talismans, some of which took the form of *oberegi* (amulets, often prayers wrapped in leather) or crosses, but which also could simply be "lucky" objects with no religious connection, such as a broken cigarette holder.[16]

A key genre of personal items carried by soldiers that could serve as both talismans and utilitarian ends was tobacco paraphernalia. Tobacco was central to the culture of the Red Army and products related to its usage were ubiquitous. When matches became a rarity, soldiers "lived as in the times of [Hans Christian] Andersen," improvising "devilish contraptions" of flint, broken file, and wick.[17] Lighters, taken from the enemy or crafted by soldiers themselves, were very common. Vasilii Glotov recorded how veterans spent their free time "crafting quaint cigarette holders and cases from whatever they can find—wood and trophy plastic."[18] Shell casings were also used to make cigarette holders, which were often inscribed with the names of comrades and places of manufacture, serving as reminders of close friendship in what could be a highly mobile and impersonal institution.[19] *Kisety* or tobacco pouches made of cloth, leather, or even rubber were very popular. Often embroidered with images and patriotic slogans, these pouches were highly valued by soldiers. Vasilii Tërkin spent an entire poem mourning the loss of his tobacco pouch, something inspired by Aleksandr Tvardovskii's interaction with a real soldier in 1942:

> A person with a ruined life (dekulakized, a refugee, etc.).
> He lost his *kiset* with tobacco.
> —Can you believe it, I lost my family, I lost my home, and now I've lost my *kiset*.[20]

That the loss of a tobacco pouch could be put alongside such serious losses as home and family (even in jest) speaks volumes to the importance of this object. The tobacco pouch was an item associated with the key relaxation process of smoking, which brought soldiers together in moments of socialization. It was also an object that had come from elsewhere and been given as a gift, whether passed from soldier to soldier or sewn and sent from civilians in the rear.[21] Irina Dunaevskaia noted: "Former convicts are a strikingly sentimental group: the scout Pasha, nicknamed Bedouin, has a tobacco pouch on which is embroidered: 'Tired after battle, our warrior, sit, and have a smoke.' He often does just that and is very proud of this present, sent to the front without any particular address to 'whoever fancies it,' and without the name of the sender, who Bedouin really wanted to uncover."[22] Tobacco pouches were ubiquitous yet treasured as something deeply personal. They were usually anonymous presents that made material the horizontal connection of all Soviet citizens that the state was keen to mobilize.[23] The tobacco pouch was a constant companion, something that made any place seem more like home for nomadic soldiers as they rolled and often shared a smoke in trains, bunkers, peasant huts, and alongside the road. While sitting enjoying a smoke, soldiers would often be passed by other units, behind which might follow a small cart.

Soldiers could occasionally place their packs on the *povozka*, a word for military conveyances of all types, but usually a two-wheeled peasant cart designated to carry extra equipment and heavy weapons for a unit. However, the ponderous and defenseless carts were among the first things to disappear in combat.[24] In battle or during a retreat, particularly in the summer, soldiers often discarded their packs, much to the chagrin of those in charge of supply. In an attack, soldiers could be ordered to leave their thing-bags at the jumping-off point, which often led to the abandonment or pilfering of their belongings.[25] In general, soldiers lived through cycles of acquisition and impoverishment.

While not in motion, as soldiers crafted their cities of earth, there was a tendency to acquire and manufacture all manner of objects, which would pile up and need to be disposed of as the army became mobile again. This was particularly true of commanders. Marchenko described this process to his wife in the same letter home where he praised her knife and comb: "Thanks in advance for the package, but I ask not to send anything . . . I have reduced my property to a minimum—gave away my books, and even just threw away some things." Marchenko lamented that even so, "You end up burdened down like a mule."[26] An angry letter to Stalin, calling for a purge of rear area services, was less generous, complaining: "Units have acquired a lot of baggage. Someone observing our movement might think that it is a march of gypsies."[27] Commanders often weighed down their supply train dramatically, while soldiers were left to carry all of their necessities on their backs as they marched hundreds of kilometers. Soldiers had minimal carrying capacity and would take with them only what was useful or of emotional importance, something that they were constantly revising as the weight dug into their shoulders over many kilometers. Under these conditions, the government was interested in providing soldiers with lightweight objects that would give meaning to their experiences.

Print

Political officers feared that bored soldiers, overtaken with melancholy, could become a liability. As Shcherbakov, head of the Main Political Directorate, warned: "If we don't keep him busy, he'll start thinking about how things are at home, how his family is doing, etc. . . . It is of the utmost importance that the Red Army man in the foxhole and the bunker has something to read."[28] Print served both to distract soldiers and to educate them about a wide variety of subjects from everyday skills necessary at the front to a proper Soviet way of interpreting the global war they waged. During the war, with little time to train soldiers and a wider swath of people coming into the army than ever before, print spread information about matters both practical and existential.

The army developed an extensive system of military newspapers, with each division, army, front, and military district publishing at least one newspaper. In all, over 1,300 newspapers and journals were produced for the army, with a total circulation of over 4.38 million. There were nearly five thousand full-time journalists and writers in the army and hundreds of thousands of soldiers who wrote occasional pieces and letters.[29] Some soldiers received the army-wide *Krasnaia zvezda* and all-union papers such as *Pravda* and *Izvestiia*. *Krasnaia zvezda* was intended for commanders, reproducing prewar patterns of providing periodicals first and foremost for elite cadres.[30] There was also an army-wide illustrated magazine, *Frontovaia illiustratsiia*, and an army-wide entertainment magazine called *Krasnoarmeets*, similar to *Stars and Stripes* in the US Army. The larger newspapers were printed in major cities, while divisional and army newspapers were produced in mobile print shops that followed their units directly and were sometimes captured or destroyed by the enemy. The army used six hundred tons of paper per month by 1943 and faced constant paper shortages.[31]

Many of the Soviet Union's top writers were mobilized into this massive system as war correspondents. Stars such as Ilya Ehrenburg, Vasilii Grossman, Konstantin Simonov, and Aleksandr Tvardovskii roamed the front collecting material that would be printed in central gazettes and later reprinted in local papers. Lesser-known authors, such as Aleksandr Lesin and Vasilii Subbotin, accelerated their careers as army war correspondents. Some war correspondents chafed at the specifics of this genre, where they would publish profusely under various pseudonyms. One war correspondent complained to his girlfriend serving elsewhere that "the specifics of newspaper work . . . very often run counter to literature."[32]

For some, such as Lesin, service as a war correspondent took them out of the dangers of everyday life on the front lines. However, the life of a war correspondent was far from safe, and many died in service (e.g., Musa Dzhalil' and Evgenii Petrov). Of the 943 professional writers inducted into the army, 225 died during the war.[33] The act of documentation and explanation was considered to be nearly as important as the act of waging war itself and was the object of scrutiny by both the government and soldiers.

Military print served a variety of purposes. The newspaper was a source of practical information needed to survive at the front. Propaganda served more existential purposes as well, providing an explanation of "the meaning of our holy war against Fascism, mobilizing the personnel of a unit for the total defeat of the beastly Fascist horde."[34] The press connected soldier's actions at the front with the efforts of those in the rear. It was to serve as a showcase for acts of heroism to be emulated, providing detailed descriptions of soldier's exploits and lists of those who had earned medals. Newspapers featured articles on how to destroy enemy tanks, dig trenches in different settings, stay warm in the winter, and do many other useful things. This was crucially important in an army that was learning on the job while fighting a war marked by rapid technological development. Newspapers proclaimed that fulfilling Stalin's order "to tirelessly perfect the mastery of arms" was their main goal by 1943.[35] Yet only fully conscious soldiers could truly master their weapons.

Newspapers strove to translate Bolshevik goals and values into language that soldiers could understand. Authors were encouraged to use forceful, lifelike prose, avoid clichés, and appeal to audiences at multiple levels of literacy.[36] Sometimes this translation was literal. The army produced newspapers in all the major languages of the Soviet Union. By 1944 "non-Russian"-language newspapers were to be distributed at the rate of one per every five to ten speakers. In all, during the war 110 newspapers in "non-Russian" languages were produced, with a general circulation of 275,000. Unfortunately, given the constant shifts in the demographics of a unit, these papers were often produced in quantities independent of the number of soldiers of a given nationality serving in a unit.[37] Before the army established a print network in "non-Russian" languages, many political officers tried to secure subscriptions to regional newspapers such as *Krasnaia Tatariia*.[38] Some units took the initiative to print small dual-language dictionaries of key terms (e.g., equipment or landscape features).[39] As the army went abroad, small glossaries of Romanian, Hungarian, and German terms also appeared in military print.

Like prewar periodicals, the wartime press spread a Soviet interpretation of events. Important resolutions were reprinted, and much of the material flowed from the center for reprint in local newspapers.[40] Information about the Allies, the German home front, and the history and culture of places in the path of a military unit all found a place in Red Army newspapers. Newspapers were supposed to provide a window onto what was happening throughout the Soviet Union, paying particular attention to regions that were well represented within a unit.[41] Life within a unit was also documented on the pages of its newspaper. War crimes witnessed by soldiers or their relatives were a major theme.[42] Heroes from a unit were celebrated and made famous, either simply listed next to their medals or photographed with their weapons above stories of their bravery.[43] Soldiers who were undisciplined and chefs who failed to provide tasty food were publicly shamed. Those whose cowardice had cost them their lives were identified and presented as cautionary examples.[44] The everyday life of soldiers, the actions of party and Komsomol organizations, and even poetry were an everyday part of newspapers. As one political officer declared: "The frontline press plays a significant role in educating soldiers. On its pages you can read remarks and correspondences about yourself, your own unit, and well-known heroes. Such materials, if they get to the soldiers in a timely manner, inflame the hunger for acts of bravery [and] help [them] comprehend the importance of the part of the front on which they fight the German occupiers."[45]

Military print was supposed to make sense of the incomprehensibly large-scale events of the war and situate soldiers within those events in a way that showed the significance of their every action.[46] By juxtaposing a wide variety of events, drawn from all the corners of the front, Soviet Union, and world at large, the newspaper created a ritual in which potentially millions of people sat and read the same articles and could visualize both these events in relation to each other and their fellow readers as an "imagined community." Wartime newspapers did not so much create as reify and explain the soldier's privileged position within this community.[47] Military print had a double mission of underlining the larger Soviet community and fostering the smaller community of the specific military unit that the paper served.

Newspapers were to be tightly connected to the life of soldiers within a unit and integrated into their everyday lives. Letters by *voenkory*—soldiers who wrote to the editorial board of a paper—were supposed to receive top billing.[48] Close relations with the party and Komsomol organizations were key, and the newspaper was tasked with providing support for initiatives from below to document and mobilize soldiers, printing small projects such as propaganda sheets and even small books.

These initiatives often took the form of *listki* or *listovki*. Listovki were strips or sheets of paper, sometimes printed, sometimes mimeographed or even handwritten. Not subjected to military censorship, they were produced at the discretion of political officers.[49] Sometimes referred to as "trench papers," listovki reprinted the soldiers' oath in a variety of languages and often with an image, selections from military regulations, and information on local atrocities, particularly the execution of POWs or civilians.[50] Many told the story of an act of bravery, accolades a unit had won, or a declaration to capture a city. According to some, this form of print was very popular and served the functions of military print most effectively. "The merit of the listki," one political officer explained, "was their immediacy, speed, and purposefulness. They showed the best soldiers, explained the immediate goals, translated the experience of heroes, told about unfolding operations, and sometimes gave biting commentary about those who panicked or only cared about their own skins."[51] Often read out loud, these listki were passed from soldier to soldier, sometimes rolled and stored in the used shell of a bullet, as one exemplary propagandist wrote: "It often happened that after reading a listovka, a Guardsman would add his own words straight from the heart." The same political officer gave a special listok to wounded soldiers to encourage them to maintain ties with their unit.[52] Some soldiers even wrote listki during battle, so that if they were killed, others would know of their heroism.[53] These often informal pieces of propaganda encouraged soldiers both to see themselves as historical actors with serious responsibilities and to editorialize their part in the war effort.

In addition to newspapers and journals, the army printed a wide variety of books, from classics such as Tolstoy's *War and Peace* to socialist realist literature to newly published work about the war and Russian history. As David Brandenberger has shown, historical themes and classics were particularly popular with soldiers, who were apt to engage in dialogues with their favorite authors.[54] Diaries, letters, and interviews record soldiers reading Soviet authors; Russian classics by writers such as Chekhov, Lermontov, Tolstoi, Pushkin, and Dostoevskii; and world classics the likes of London, Cervantes, and Maupassant.[55] Many of the classic works of the war were first read in serialization in the military press and agitators were instructed to utilize fiction and poetry in their proselytizing. This continued older Bolshevik practices of blurring the lines between fiction and reality and demanding that entertainment also be instructional.[56] Russian literature was vaunted as providing models of patriotism for soldiers. Works such as *How the Steel Was Tempered* and poets from Pushkin to Maiakovskii were mobilized to show that love of country is what makes one great and that Russians loved their country most of all. A tank commander even claimed that poems were the key to his political work.[57] Partial or full reprints of classics were common, as were print runs of new works.[58] *The Great Patriotic War of the Soviet Union*, consisting of major holiday speeches by Stalin in which he gave a clear definition of the changing meaning of the war and goals of the Red Army, was touted as the single most important book for soldiers.[59] Efforts to provide ample reading in Russian were replicated on a smaller scale in other languages, there even being a short-lived plan to reprint ancient heroic epics alongside translations of key propaganda texts and Russian classics. Every unit and every political worker was supposed to maintain a library that included books, magazines, and small brochures. Each new addition to a unit's library was to be reviewed and announced in the divisional newspaper.[60]

In general, the government's efforts met with mixed success, and soldiers often complained that they had little or nothing to read or that what they were reading was dated. One soldier lamented in 1945 that the newspapers he received were old and that "We don't see any journals, we are sick of it."[61] As a result, soldiers often read whatever they could get their hands on, and a particular culture of reading aloud came about.

Political workers spent much of their time reading newspaper articles out loud and leading discussions of the press, while soldiers informally read papers and books aloud to share them with their comrades.[62] Rank-and-file soldiers usually heard the news, which was either read out loud by a comrade or interpreted by those who had a chance to read. This fact, alongside the transmission of information from center to periphery, helps to explain why much of military print was repetitive and formulaic, as a soldier might have access to only one issue of a paper for weeks at a time. Nonetheless, papers and journals broadcast the official line as it developed throughout the war. Military print was the primary means of transmitting messages to the network of political workers who were supposed to translate a Soviet interpretation of events to soldiers. Even if many soldiers seldom saw a newspaper, much of the discussion at political meetings was drawn from the press.

Given the massive paper shortage in the Soviet Union, newspapers, books, and listovki were scarce resources.[63] The words printed on them had an air of sanctity, explaining the war and recording the deeds of heroes. Yet soldiers often put them to more profane uses. In addition to being reading material, soldiers used newspapers and even books as rolling papers for cigarettes. Khafizov recorded in his diary in December 1942: "My darling! We have a lot of rolling papers and newspapers here. But I will smoke yours, because you sent them and your care—that's what I need! It is not the papers but your care which is dear."[64] That Khafizov, a political officer, saw nothing wrong with smoking print materials, speaks to how standard this practice was. Indeed, since the tsarist period soldiers had often used newsprint to roll cigarettes.[65] This standard practice was not a criticism of the regime but serves to highlight that even semisacred texts could become utilitarian materials.

Soldiers had a range of reactions to the print materials supplied to them. Some soldiers, particularly those who garnered a degree of fame, adopted the clichés of military journalism.[66] Momysh-uly railed against propaganda aimed at demonstrating the connections between the rear and the front as an overworn, bloated, and hackneyed genre. He also noted that few journalists understood battle, using meaningless phrases to describe the war.[67] Assessments in party publications throughout the war mirror these criticisms, calling for accessible, grounded language and serious engagement with soldiers' actual experience.[68]

Others expressed disgust at the distance between their own experiences and what they read in the papers.[69] A joke told by veterans after the war highlighted the press' tendency to exaggerate enemy losses: "As everyone knows, Sovinformbiuro 'destroyed' the German army four times over during the war."[70] Nikolai Nikulin was pointedly acerbic in his criticism of war correspondents after the war: "During the war they made their capital out of corpses, ate carrion. They sat in the rear, had no responsibility for anything, and wrote their articles—slogans written with rose-colored glasses."[71] Nikulin went on to blame journalists for creating and maintaining the myths of the war, ignoring failures, and making it impossible to learn necessary lessons. However, such harsh and direct criticism was rare and dangerous to voice during the war. Some authors—such as Simonov, Ehrenburg, and Grossman—were clearly well liked, and the press was a key means by which the government tried to reach soldiers with its message. However, even the very popular Tvardovskii lamented that he traveled "the wayside of the war"—making short trips to the front, never really getting used to its dangers. He was haunted by "the unforgettable looks" of soldiers as he made his way from the front to the safety of the rear.[72]

Newspapers and other forms of military print explained and documented the war in ways that would motivate soldiers. They were a way of influencing how soldiers thought, embedding them in a narrative of the war and connecting them with far corners of the Soviet Union. Some soldiers carried newspaper clippings about their hometown or a key battle as relics.[73] Reading was considered key to soldier's morale. A major achievement of Soviet rule and difference from soldiers in the tsarist army fighting the previous world war was that Red Army soldiers and their family members at home were overwhelmingly literate, even if the list of languages they spoke and read had a Tower of Babel quality.[74] Soldiers fighting the Great Patriotic War could correspond with their loved ones on Soviet-held territory, whether they were in Leningrad or on a kolkhoz in Abkhazia.

Letters

The majority of letters sent during the war were sent without an envelope, on whatever paper was available, folded into triangles.[75] These random pieces of paper were, as Aleksandr Lesin mused in his diary, "so dear to us—that piece of peaceful life."[76] Another soldier put it more bluntly, using broken Russian peppered with profanity in a letter to a friend: "Write letters. At the front without letter fucking sucks [*bez pis'mo khuëvo* (sic)]."[77] For Tvardovskii, the folksy poet who roamed the front, letters were a distillation of a wife's faithful love "that could be stronger than death."[78] It was a common custom to make soldiers dance before giving them their mail, though many danced out of joy without being asked.[79] The mail occupied a central place in soldier's thoughts. The state was keenly aware of this, tasking political officers with organizing correspondence between front and rear, which it deemed an "outstanding form of political education."[80] A good political officer made sure his soldiers were getting mail.[81] The political department was apt to use letters to motivate soldiers, and the army was deeply concerned that soldiers both wrote home and received mail. Collective reading of letters from home was a common pastime, with commanders and soldiers encouraged to involve themselves in the private lives of their comrades.[82] Letters were a break from the hardships of the front and a means of maintaining and establishing connections. They could be intimate but were also public. They could be deeply personal or generic, and many correspondents didn't see a difference between those categories. The government provided the apparatus and infrastructure for soldiers to correspond free of charge. In return it inserted itself at every level from commanders, who would often read soldiers' letters and write to their subordinates' loved ones without solicitation, to censors, who were tasked with reading every letter sent.

Mail to and from the front ran through a massive system of Field Post (*Polevaia pochta*). It was estimated that soldiers received up to seventy million letters from the home front every month.[83] To get an idea of the scale of letters coming from the army, on July 5 and 6, 1943, in just the Thirteenth Army (approximate strength of 120,000 soldiers), 55,315 letters were processed (6,914 in languages other than Russian).[84] If we use this as a rough baseline (over 20 percent of soldiers sent a letter daily) and apply it to just the active army—that is, those potentially in combat conditions, far from ideal for writing and sending letters (about 5.5 million soldiers at any given time)—we get a minimum of 1.1 million pieces of mail coming from the front every day.[85] A soldier's address was not tied to geography but rather to his or her unit. Initially the number included a soldier's regiment, company, and platoon.[86] This made it too easy for enemy agents to gather information on the army, and a system of numbers and the soldier's name came into place in the fall of 1942. Under the new system an absolute minimum number of people could connect a unit with its Field Post number.[87] If a soldier was wounded or transferred, it would be impossible for their loved ones to write them.[88] Political officers made sure that soldiers knew how to write their address and their family's address.[89] Indeed, soldiers were instructed to check and write their return address on every letter they wrote and sometimes told their correspondents not to write during periods when they lacked a stable address.[90] These letters were both intimate and prescribed, as censorship was an integral part of the Field Post system. Censored letters fell into two basic categories—"A" for "Authorized" (*avtorizovano*) and "K" for "Confiscated" (*konfiskovano*). Letters marked "A" might have some text redacted but would be passed on.[91]

There was a wide spectrum of information that soldiers were forbidden to write about. They were not allowed to reveal their rank, specialization, unit's name, or geographical location, all of which were considered military secrets. Soldiers were also not supposed to write about epidemics, losses, accidents, battles, plans, or the names of high-ranking commanders.[92] They could write about individual friends who were killed or wounded and where they were buried but not about percentages or actual numbers of losses.[93] Censors were encouraged to look for codes in letters that made little sense, underlined words in a suspicious manner, or went into excessive detail about the weather. Drawings and even the location of stamps were to be examined as possible codes.[94] The letters were supposed to undergo chemical analysis to reveal invisible ink and were held up to the light to reveal any hidden layers of text.[95] Anything "anti-Soviet" or "libelous" was marked "K," confiscated, and sent to the NKVD to investigate. This could lead to prosecution, as in the famous case of Aleksandr Solzhenitsyn.[96]

Many soldiers clearly did not know what they were supposed to omit from their letters. Aleksandr Lesin overheard a commander listing what a soldier was forbidden to write and realized that he had just sent a letter that included all of that information.[97] Mikhail Loginov recalled how a soldier read a letter out loud that included a wealth of military secrets. He castigated his soldier: "Cross out the word 'Bagramian.' By naming the commander, you show which front you are sending the letter from. Everyone knows where Bagramian is in command. You can leave 'I saw Ivan Zakharovich, he is alive and well,' but you have to cross out where he is serving. You can't write about there being tanks here. That's a military secret."[98] Instructions to military censors anticipated this ignorance, stating, "In the case of letters containing text which is not to be disclosed in mail or telegraph correspondence but not of an ill-intentioned or hostile character, these letters can be forwarded to their destination, but with mandatory blacking out or cutting out of the forbidden text."[99]

Soldiers were aware that their letters were being read. One soldier appealed to the military censor to allow his diary through.[100] Another wrote to his wife, "I don't know if anything gets crossed out in my letters, nothing is in yours."[101] The few statistics we have about censorship point to there being very few letters sent by soldiers with major complaints, and not very many revealing military secrets. For example, of the over fifty-five thousand letters read by censors in the Thirteenth Army over two days in July 1943 only twenty-one registered complaints, and those were only about lack of food and tobacco.[102] Civilians, however, seemed less aware of what could and could not be written about: Loginov's young soldier complained about half of a letter his mother sent him being crossed out.[103]

What should be censored and what could happen to those who transgressed censorship were not always clear. Soldiers' families' grievances about hunger and poor treatment by local administration were suppressed if they complained too bitterly. However, those letters became part of a regular system of assistance for soldiers' families, in which the soldier's unit automatically provided necessary documents to secure benefits without the soldier ever knowing.[104] Letters from soldiers attacking family members on liberated territory for associating with traitors and describing the destruction that Germans had wrought on occupied territory were sometimes suppressed but ultimately lauded as a patriotic form of correspondence. Anonymous letters accusing the family members of Red Army soldiers of aiding the enemy were to be confiscated.[105] Any mention of arrest or repression of a family member in the army or on the home front was to be purged from correspondences.[106]

Soldiers often found creative means around censorship. Many came up with nicknames for lice, about which it was forbidden to write (e.g., "tankets" or "machine gunners").[107] Others found ways to reveal their geographic location, such as Vladimir Gel'fand, who started several sentences of one correspondence to spell out "I am in Poland near Warsaw."[108] Some simply wrote about where they were, and the censor ignored their transgressions.[109] Late in the war some restrictions were relaxed, as soldiers wrote from the western Soviet Union or abroad in ways that easily identified where they were, and many wrote home about their weapons in ways that revealed their specializations.[110] The army eventually condoned soldiers writing about which front they served on, their participation in important operations, and information about promotion and medals.[111]

Censorship did not always function as smoothly as it was supposed to, as anecdotal evidence reveals. The most educated and reliable cadres had been mobilized into the army, leaving censorship in less able hands. Throughout the war there were complaints about censors who were lazy, incompetent, or overwhelmed. Some complaints spoke of censors who used nails, knives, or bones instead of steam to open letters. This resulted in torn letters, sometimes with unreadable addresses and damaged photographs.[112] Censors could cut information from texts in such a manner as to render the letter unreadable, destroying harmless text on the other side of the page or smearing it with ink. Some censored content for no apparent reason, "so distorting the text that its contents becomes even less desirable" than when text was blacked out in ways that left the intent of the author clear, another common problem. Some even allowed "anti-Soviet" letters to pass, merely crossing out the offending passages, rather than confiscating them and sending them to the NKVD. Some of this seems to have stemmed from a lack of infrastructure, as a major corrective stated that censors "should be fully supplied with all of the necessary equipment for high-quality removal of texts, including ink of different colors, pencils (regular, chemical, and colored, India ink, rubber bands, etc.)."[113] Some letters simply passed without censorship.[114] A report from Tatarstan in October 1942 noted the censoring of information being sent to NKVD Director Lavrentii Beria, putting letters in the wrong envelope (98 cases), failing to read letters (175 letters), and failing to cross out text (290 letters) out of 49,338 letters checked.[115] As late as 1944, many of those in charge of censorship were unqualified "to independently and properly decide what to remove" and many censors didn't understand "basic political issues." Some couldn't match Soviet leaders with their posts, other thought that the Soviet Union was at war with Japan. Many censor offices didn't have access to newspapers.[116] Proper censorship required an immense amount of work and competent cadres.

A major complicating factor in all of this was the wide variety of languages in the Soviet Union. Momysh-uly complained at the end of the war that a letter he had written in Kazakh, in the old Arabic script, was never delivered, presumably due to a lack of censors who could read it. Writing in "non-Russian" languages could dramatically slow down the mail, as censors fluent in the relevant languages needed to be found wherever the front lay.[117] Some "non-Russian" soldiers transitioned to Russian in correspondence with their relatives and friends. This may have reflected the fact that their day-to-day language was now Russian, and possibly a desire to show both educational achievement and their close identification with their status as Red Army soldiers.[118] More practically, some requested that their family members also switch to Russian so that their letters would pass more quickly through the censors.[119]

Censorship was a tool to monitor and control the moods of soldiers and their families. The military collected information about soldiers' and civilians' moods and the character of their complaints, including quotations from their letters in their reports. The main purpose of censorship and the postal service more generally was to boost soldiers' morale by connecting them with those at home. That connection was supposed to be free of unpleasant facts that could depress soldiers, such as relatives neglected by the local government. However, horror stories that could stoke the flames of hatred for the enemy, such as accounts of murdered friends and relatives, had a privileged place.[120]

Letters were key to soldiers' morale and major effort was put into organizing the field post system.[121] Initially some units lacked dedicated mail carriers, giving letters to whoever happened to be headed to this or that unit.[122] Later the army established a system of dedicated letter carriers, usually older men or young women.[123] Letter carriers were instructed to deliver mail once a day to a unit regardless of conditions, even bringing mail to the front line under fire.[124] Despite these instructions, when the army was on the move or a unit far away from rail hubs, mail could take weeks or months to reach soldiers. Hundreds of kilometers from a rail line, Boris Suris complained in his diary on December 17, 1942, that he hadn't received mail for a month.[125] Nikolai Inozemtsev, who had been in the army since before the war, received his first letter from home in November 1941.[126]

Both delivering the mail and writing letters presented a number of challenges. Soldiers often had a hard time finding paper. During the Finnish campaign in 1939–1940, Vladimir Zenzinov, a Russian émigré who wrote a book based on letters recovered from the battlefields of the Winter War, noted that the paper soldiers wrote on was of consistently poor quality and overwhelmingly torn from notebooks produced for schoolchildren. Some made crude envelopes, while many folded their letters into a peculiar triangle shape.[127] During the Great Patriotic War the situation became much worse, and the folded triangle would become an iconic object reflecting the paper shortage. Paper designated for soldiers was often pilfered by commanders and supply troops.[128] Throughout the conflict high-level Soviet organs complained about a lack of paper available to soldiers, forcing many to write on "newspapers, listovki, the wrapping of ammunition, the cardboard from concentrated rations, etc."[129] One article at the end of the war describes soldiers' letters written "on sheets torn from a notebook, trophy postcards, or on the blank spaces of a read-over newspaper."[130] Trophy paper and postcards presented a special problem because they could contain enemy propaganda, and many censors simply confiscated all correspondences with foreign language text. Instructions later prohibited swastikas, Fascist slogans, or images of German soldiers and their equipment on the paper and postcards soldiers used. Furthermore, all foreign text required translation before being forwarded.[131]

Once paper was found, writing could still prove difficult. Soldiers had to find a place to write, often on their helmet or mess tin.[132] Then they had to find time to write. Tat'iana Atabek responded to her mother's anger at not writing as follows: "Nowhere do you think so much about family and home as in the army . . . Sometimes you can't write a single letter for weeks, but there isn't a day when I go to sleep without thinking of you."[133] Viktor Nekrasov wrote home complaining, "It always happens that when you have

FIGURE 6.1 Soldiers write a letter to their kolkhoz in faraway Oirotiia, Western Front, 1942. RGAKFD 0-57152.

something to write about there is no time, and when there is time there is nothing to write about."[134] Soldiers in encirclement had no ability to send letters out; Kharis Iakupov recalled that four months passed in 1941 before he could send a letter home.[135]

To send letters home, soldiers had to know where to write. Given that millions of civilians had been evacuated, finding loved ones could be a difficult task. It took Mikhail Vail eight months and the help of a radio show that read soldiers' letters to find his wife.[136] Many soldiers had to lean on their political officers, draft boards, or local government to help them find their loved ones' addresses. "Non-Russian" soldiers often did not know how to write home, and propaganda from 1942 and 1943 emphasized the importance of establishing correspondence with their relatives in the rear and teaching everyone how to write addresses in Russian.[137] Letters were particularly important for "non-Russians," and their sense of isolation and trepidation about their families' fate was frequently compounded by a lack of knowledge of such vital information as where the front lay. One article for political workers presented the example of a Kazakh soldier, unable to write home because he could not write his family's address in Russian. He was in a state of panic in the face of his family's silence and feared Kazakhstan (far in the Soviet hinterland) was under German occupation.[138] Millions of people remained on actual occupied territory, leaving many soldiers with no one to write.

Political officers were tasked with organizing pen pals for those whose families were living under occupation or had been murdered by the Nazis. One exemplary article, "Patriotic Letters," proclaimed: "Any soldier can find thousands of friends on the Soviet home front. Our responsibility is to help him find these friends." Some received thousands of letters or found pen pals in this manner.[139] Many soldiers wrote to their family, friends, or factories asking them to find female correspondents for "suitors" in their units, while one political officer described using the letters sent by women from a local factory at "critical moments" to motivate his soldiers.[140] Some soldiers wrote to exemplary workers who had been featured in the military press. Charles Shaw has posited the possible emergence of a "unified romantic community" in the Soviet Union as a result of similar exchanges.[141]

Alongside these personal correspondences was the genre of the *nakaz*—mass letters, often published in the press, in which collectives from a unit at the front, factory, city, or republic described their accomplishments and vowed to do even more.[142] For example, soldiers might write to the factory that had made their cannon with a report of the enemy artillery and tanks they had destroyed and a vow that "you at your factory and we at the front will double our strengths, abilities, and knowledge" in order to fulfill Stalin's order to destroy the enemy.[143] Group letters such as these were part of a concerted effort to help soldiers imagine themselves as part of a larger Soviet community united in the war effort. Membership in this community was based on what individuals accomplished to bring victory closer, whether working tirelessly in a factory or risking their lives to destroy the enemy.

In Soviet culture, it was exceedingly difficult to separate the personal from the political. Many soldiers adopted stock phrases and repeated propaganda clichés in their personal letters.[144] Commanders would write home to shame or praise their soldiers, leaning on families in the same way a teacher might in the case of an errant pupil. The case of one Kiselëv near Stalingrad was exemplary:

> Narovishnik, deputy political commander of the company, decided to write to his parents: "That's how outrageously your son behaves; perhaps you can help." That letter was read to the company. The company knew that the letter had been sent to his parents as did he himself. The very fact of sending the letter to the parents got him thinking. On arrival at the front the parents' answer was received: "Why are you bringing shame on our gray heads? We can't look our neighbors in the eye. Have you forgotten what we were saying when we were seeing you off—be worthy!"

Conversely, soldiers who received a medal could have a letter of praise sent home by their commander, sometimes with a photograph. Indeed, the commanders of the above Kiselëv "had to write another letter about how he had made things right and already distinguished himself and that he had received a decoration from the government."[145] The army's head political officer, Aleksandr Shcherbakov, imagined the tremendous impact that letters about decorations sent to family members, kolkhozes, and factories could have on the home front, imagining a fictional Vasia who destroyed an enemy tank:

> First the old folks will cry, then they will go around the whole kolkhoz with this letter, to young guys about to go into the army. These guys will think, if Vasia can do that, why can't I become like him? This means that the German isn't so frightening. Then, you can be certain, most kolkhozes will have a meeting, the old man will go up and read this letter out loud, and there will be people who will say, "Your Vasia distinguished himself, and we will take it on ourselves to hurry up with the harvest and deliver grain to the government . . ." There will be a good mutual correspondence and a wonderful political response.[146]

Shcherbakov, certain of the massive dividends of these letters, encouraged the adoption of this practice as standard policy throughout the army. Whether to shame or praise, the government called on loved ones at home to apply pressure to soldiers at the front.

Aside from inserting themselves directly into correspondences, commanders also arranged, monitored, and interpreted them. Andrei Orlov, commissar of the Eleventh Guards Rifle Division, told the Mints Commission in June 1942:

> Now our goal was to show that it is a poor political worker who isn't approached by Red Army soldiers with their letters. Why doesn't the *politruk* [political officer] know that Red Army man Petrov has received a letter and not answered it? Why doesn't he know that Private Petrov's family isn't writing? What is he doing to fix this? Is he in touch with his family, with local organizations? Here people are experienced, battered, but when they get a letter from home some valve starts to work better in their bodies. If a soldier can't write, the *politruk* should come and help him, connect him with his family. If the family isn't well appointed, then the regimental, divisional, or corps commissar needs to help out in any way possible.[147]

It was the job of political officers to aid soldiers in securing assistance for their families, whether that be by arranging for tax breaks, sending part of their salary home, or obtaining a cow for family members evacuated to faraway kolkhozes.[148] Instructions to political officers also encouraged the use of letters to and from soldiers as material for political meetings.[149]

Whether formally at political meetings or informally among comrades, it was common practice to share letters and read them out loud. In May 1944 Viktor Nekrasov wrote to his mother: "Letters are our greatest joy. First everyone reads his own letters; then we start reading them to each other."[150] Pëtr Veselov wrote to his wife in July 1942, "the guys brought me your letter in the trench, and we always read your letters together in the squad, because in our correspondence there is no lack of self-control."[151] Whether this was a warning not to write anything saucy or a stamp of approval is difficult to tell, but it is clear that the arrival of mail and the contents of letters were something of public interest. Soldiers discussed these letters, judging their authors and guarding the interests of their friends. David Samoilov recorded in his diary how in late November 1943 soldiers (*slaviane*) gathered around to listen to letters one had received from "a chick" (*ot babënki*): "We'll see whether or not she is worthy of becoming the wife of a sergeant of the Red Army." They then continued to judge various letters, one of which was

FIGURE 6.2 Guards Red Army man V. Podrezov shares news from home with letter carrier M. Shcherbakov, Third Ukrainian Front, 1944. RGAKFD 0-323035 (O. A. Lander).

"sweet naïve babble, in which there was much freshness, purity and unselfishness," while another was a "letter written to fulfill the plan . . . without a word about love."[152] This culture of open reading—in which a comrade or superior could feel entitled to read a soldier's letter before delivering it to its addressee—allowed some soldiers to hide bad news, such as "dear John" letters, from their friends or wait to break the news to soften the blow.[153] These discussions brought soldiers together, sometimes as a collective defined in part by contradiction to the civilians whose letters they discussed. Nonetheless, Grigorii Baklanov reflected that a soldier who received a letter when his friend did not was "embarrassed at his good fortune [*schast'e*]. In a war such as this you don't boast about your good fortune, the same way that you don't boast about bread in front of a starving person. But you can share bread. How do you share good fortune if it has smiled on only one?"[154] Even so, the practices of letter reading were directed toward soldiers sharing their emotions as they shared their bread.

The contents of soldiers' letters to their loved ones and comrades were, like their everyday experience, a mix of the mundane and the epic.[155] Soldiers' letters, given that they were headed away from the front, were much more likely to survive than letters addressed to soldiers.[156] Lev Slëzkin lamented in his memoirs that despite the fact that his mother wrote to him nearly every day, none of her letters survived, deeming their preservation impossible during wartime.[157] However, given that letters were written in dialogue, we have a good idea about what family and friends wrote to their loved ones in the active army. The observant reader will note that letters have served as a major source for the previous five chapters of this work. Soldiers wrote extensively about how they were eating and what they were wearing, either asking for material assistance or reassuring their families that they were well provided for. They wrote about their accomplishments: medals received, number of enemy soldiers killed, promotions, or joining the party—something that was strongly encouraged by commanders and political officers. Many repeated material, fearing that not all of their letters were making it to their addresses.[158] Letters also had an important ritualistic function: they served as proof of life. As Martyn Lyons has noted of French soldiers in World War I, the contents of letters were not so much important as the mere fact that a letter served as physical evidence that a loved one was still, or at least recently, alive.[159] However, soldiers could have meaningful communications with their loved ones. Some wrote friends to brag about romantic conquests.[160] Many admitted that they would be able to tell the truth of their experience only when and if they returned home: "If I live, then I will tell you everything, only it is hard to believe that one can remain alive in such a hell, that is just a miracle."[161] Some documented what they had seen and done in a way that did little to shield their loved ones from the dangers they faced (e.g., Slëzkin's letter to his mother quoted in chapter 5).

Some resented the gulf of understanding between their experiences and those of their loved ones at home. Boris Marchenko wrote his wife an angry letter in November 1943 in response to her complaints while in evacuation from Odessa in Siberia: "You've had enough of Siberia. Surely. But I would give a lot, to spend even a week in your 'awful' conditions . . . And in terms of my own circumstance I can only say that yesterday, on the holiday night of November 7–8, I lay in water until six in the morning."[162] This became a running theme in his letters, as he warned his wife that the war had changed him dramatically, writing in 1944: "Maybe I write harshly, excuse me, in our circumstances your character becomes coarse and you become meaner. It's too bad that I wasn't this way in Siberia. In general, if we even meet again, you won't recognize me by either my external or internal form."[163]

Like millions of other men mobilized and taken far from their families, Marchenko turned to letters as a way to maintain a role in his family's life from afar.[164] Soldiers were by the nature of their enterprise displaced, while family members were often also on the move; for example, Marchenko had no idea where to write to his wife in June 1944.[165] People were forced to manage complicated family relationships—adolescent children, younger siblings having difficulty studying, and so on—from afar. Marchenko had to write to convince his son to return to his stepmother.[166] Rafgat Akhtiamov frequently expressed concerns about his little brother's progress at school, warning that "without knowledge life loses all meaning."[167] Aleksandra Shliakhova pressured her younger sister to become a proper Komsomol member and chided schoolchildren for their poor studies.[168] Wives sent word of children born since their husbands had left, some inventing fictional children to buck up the morale of their husbands, others sending string to show how big real children had grown in their absence, while some husbands fretted about the health of their children.[169] Soldiers were deeply concerned with the financial lives of their family members, making sure they had enough to eat, often encouraging them to sell their possessions to provide for their family's comfort. Others, such as Akhtiamov, asked family members to guard prized possessions, writing several times about how he missed his accordion, telling his parents to preserve it in his memory and urging his father to play it on occasion for his mother.[170] Finally, letters home were a chance to gather information about friends and family who had been scattered by the war. A typical letter home from Akhtiamov, showing concern for the financial well-being of his family and a desire to know who was where, began as follows: "My most heartfelt, ardent greetings to all of you. How are you? How was the potato crop this year? The grains? What news from the front? Do you get letters from Uncle? What is Auntie doing? How is Grandpa? Has Aunt Mukhtariama arrived yet? Any letters from [my] brother-in-law? Who is Roza living with? Do you get letters from Uncle Gabdulkhai, Uncle Khalim? Is Askhat going to school in Cheremshan? Father, how are you doing?"[171] Soldiers like Akhtiamov and Marchenko—whether sons separated from parents or fathers and husbands separated from children and wives—strove to insert themselves in the everyday lives of their loved ones and maintain the web of connections with friends and family that make life meaningful under conditions in which death was a likely outcome. As Akhtiamov wrote home, "In spirit I am still in the village, with you."[172]

Silence on the part of loved ones was cause for alarm and reason to be upset. Soldiers frequently reminded their correspondents of how long they had waited for letters, blaming either them or the mail service. Akhtiamov habitually noted how many months he had waited for a letter. Momysh-uly and his lover fought bitterly in 1945 over a lack of correspondence.[173] Marchenko wrote his wife in June 1943: "I have decided not to write anyone except you. And I will only write one letter for each I receive. There is no need to write anyone—if people still don't understand that for us a letter is our only comfort and a tiny bridge connecting us to another life."[174] Any break in correspondence could lead to fears of infidelity.[175] Fears of being forgotten by loved ones were a common theme in letters. Conversely, custom prevented some soldiers from more patriarchal corners of the Soviet Union from writing to loved ones. Genatulin discussed an Uzbek soldier who could not write his fiancée due to a taboo against correspondence between women and men to whom they were not related.[176] This soldier perished in combat having never written or received a letter from his beloved.

Letters for and about the dead deserve special attention. Mail addressed to soldiers who had perished was marked "impossible to deliver" and returned to the sender.[177] Soldiers would sometimes read mail found on the dead. Nikolai Inozemtsev recorded in his diary how a love letter "filled with life's elixir" he took from the pocket of a dead tanker "does not at all harmonize with the picture of death, of nonexistence."[178] Commanders were responsible for sending death notices to the families of their soldiers, and comrades often wrote personal letters attesting to their love for their fallen comrade, promising vengeance. These letters usually described the battle in which soldiers had fallen and its significance. This was done to soften the blow of loss, highlight what the sacrifice of a loved one had given the larger community, and promise revenge. Nikolai Chekhovich's commander wrote his bereaved mother that her son was beloved among all in the unit, but more importantly, "Your son fulfilled his combat mission with honor and gave his young life for the Motherland, to liberate the city of Lenin" and "His platoon savagely avenged the Fascist fiends for their commander."[179] Aleksandra Shliakhova's commander wrote her parents a similar letter, closing with a declaration: "Let the memory of our heroine inspire us to new feats of bravery. Do not despair [*ne skuchaite*]; you have many good friends who will avenge Sasha and never leave you in need."[180] Letters constituted communities of both the living and the dead, as the bereaved communicated via post and came to terms with their losses through the rhetoric of vengeance and sacrifice for the greater good. Official propaganda encouraged correspondences between surviving members of a unit and the next of kin of the fallen.[181] Some soldiers promised to show their comrade's family members where friends had been buried after the war.[182] These letters, prized possessions left behind alongside a few medals or belongings mailed home by comrades, would be the only physical connection to the departed buried in a distant, often anonymous, grave.

The practices of letter writing that connected Red Army soldiers with their loved ones could be seen as fundamentally shaped by a totalitarian regime inserting itself into the private lives of its citizens. Although the Soviet state did go to impressive lengths to censor and shape what soldiers wrote, it is worth noting that several of the potentially "totalitarian" aspects of soldiers' letters can be seen as a combination of the particular place of literacy in Soviet culture and of traits common to soldiers mobilized to wage mechanized war. Martyn Lyons's work on everyday writings in Italy, France, and Spain shows that many of the characteristics of soldiers' letters that could be ascribed to a cowed populace who knew their letters were being read might just have easily stemmed from the fact that the majority of these letter writers were peasants, who were not used to regularly expressing themselves in written form, uncomfortable with the genre, and loath to frighten and burden loved ones with details of an extreme situation that they often simply did not have the words to express.[183] Under a much gentler system of censorship, one lacking a massive repressive infrastructure such as the NKVD, French and Italian soldiers in World War I demonstrated a similar lack of knowledge about what might be considered a military secret, resorted to many of the same tricks, such as coded letters to hint at where they were, and used generic phrases, sometimes adopting official language.[184] They also often showed the same fatalism, faced with similar conditions of mechanized killing while attempting to stay connected and not lose authority within families from which they had been separated.[185] Last, and perhaps most surprising, was the expectation that letters would be read aloud and have a wider audience, being more akin to a newsletter than a private, intimate correspondence.[186] This would be less true of more urban and educated soldiers, but nonetheless, the assumption of multiple readers was implicit to most wartime correspondences, with the exception of love letters, missives between intimates discussing issues of sex, or in cases where the author explicitly forbade the addressee from sharing it (e.g., a sister telling her sibling not to show their parents a letter).[187]

Despite all of this, for both the soldiers of the Great War and the Great Patriotic War, letters were, as Marchenko asserted, a soldier's "only comfort and a tiny bridge connecting us to another life." They were a refuge that allowed soldiers to preserve a part of their identity from the disorienting effects of witnessing and participating in mass carnage. War led to an explosion of writing as people who otherwise would have had no reason to write became narrators of the events in which they participated.[188]

Last, a few words should be said about another major form of writing that soldiers engaged in—diaries. Those who kept diaries generally served in capacities where keeping some sort of notes was a necessity. War correspondents, party workers, translators, or commanders (particularly artillerists) were more likely to have the ability to write than ordinary soldiers. It would have been virtually impossible to keep a diary without comrades and superiors being aware of it, and soldiers knew they were forbidden from recording anything, whether a diary or simply personal notes, that could be of use to enemy intelligence.[189] Not everyone understood these orders to mean that writing diaries was forbidden, and for some their diary served the role of close friend with whom to share secrets.[190] In at least one case a political officer dictated to a soldier what was worth recording, leading to a diary that was indistinguishable from the official press.[191] Aleksandr Lesin's commander admitted after the war that he could have had him prosecuted for keeping a diary, but that he liked its contents and hoped Lesin might turn it into a history of their regiment.[192] Some soldiers were prosecuted on the basis of "anti-Soviet" musings in their diaries, and there was a superstition among soldiers that those who kept diaries were fated to perish.[193] The observant reader will note that diaries, like letters, have served as a major source throughout this book and by now can judge for themselves how freely and sincerely many soldiers expressed themselves in their notebooks. All diarists understood that they were experiencing something extraordinary and strove to record for family or for projected literary works their impressions of the people, occurrences, and personal experiences of an event of unfathomable scale. Many diarists were or would become writers. Whether writing diaries, articles for the military press, or letters home, writing helped to anchor a sense of self, record changes taking place in body and mind, and place their authors in both a network of loved ones and the scheme of massive war they were waging.

Baggage

Like most of the soldiers who carried them, thing-bags were of peasant origin and barely indistinguishable to outside observers. Yet every pack was different and contained individual objects that carried immense emotional weight and could serve as connections to the past and a hoped-for future. More often than not, this meaning was only truly legible to the owner, even if others might steal or commanders might inspect a pack's contents. The thing-bag's primary purpose was to carry the necessities of war that did not fit on a soldier's belt. Rations, rain cape, a change of underwear, and extra ammunition were all key to the soldier's physical survival and comfort. These objects also weighed him or her down and left only marginal space for anything else. That space was filled by things that emotionally anchored these nomadic soldiers whose lives were in constant danger. Whether a lighter, knife, tobacco pouch, book, newspaper clipping, diary, or letters from home, they helped soldiers create and maintain meaning in the face of dislocation and mass violence. The thing-bag serves as a fitting analogy for processes taking place in the soldiers' minds. Both had little space for anything not related to waging war and were marked by the intermixing of the personal and public, the increasingly successful attempt by the government to identify its goals with those of soldiers, and the constant threat of encroachment by superiors, which could have dire consequences. Both this convergence and the soldiers' sparse material culture took on new meaning as the war entered its final phase and soldiers went abroad.

Chapter 7

Trophies of War

Red Army Soldiers Confront an Alien World of Goods

> They stole and extorted enough from our country; now we will steal and extort.
>
> —Boris Suris

On March 29, 1945, Guards Colonel Strukov, Head of the Political Department of Field Operations, First Ukrainian Front, described the range of reactions by Red Army soldiers to the Third Reich for correspondents from the Mints Commission. His account focused almost entirely on the soldiers' interactions with *things* and outlined the four major responses to alien goods in the course of the war: a sense of inferiority, a sense of superiority, covetousness, and rage, as well as the government's suppression, cultivation, or tolerance of these attitudes. As he told his interviewer:

> A certain category of people, seeing the external everyday culture, its livability, became enraptured by it. One major said outright that the living conditions in a German village were higher than those of our kolkhozes, and that, therefore, our system lagged behind their culture. The major was, of course, expelled from the party, but they should have explained the roots of this prosperity. First, Germany was living off the theft of all of Europe; second, this is prosperity of the second order. Comfort, cozy things, pictures—this is domestic comfort, but Germans don't have any real aesthetic feeling. They have a lot of pictures, but they are lithographs in rich frames; there are no good canvases. *Meshchanskii* [Bourgeois philistine] comfort, very close-minded.

Strukov continued, decrying the poor quality of German goods and explaining that historical circumstances had spared Germany the terrible fate of Russia. In part, the latter explained the reaction of many soldiers:

> Another group of people, a large one, was overcome by the desire to take vengeance on the Germans, to beat, to destroy. One warrior, for example, burned nine houses down. And when he was reprimanded, answered: "Do what you want. I will burn ninety-one more. I promised myself for all their villainy to burn a hundred houses." Of course, you can't punish him, but you need to rein him in a bit. They destroy furniture in the same fashion, break and wreck everything.
>
> Finally, there is another, although small group, that has morally decayed. Soldiers and officers, seeing nice things, were overtaken by an acquisitive instinct. To take as much stuff as possible, put it all in sacks. We had to shoot one military commandant in front of the soldiers. In a month he managed to rape twelve women, and they found forty gold watches on him.[1]

Between 1941 and 1945 soldiers of the Red Army, whose modest material culture has occupied the last six chapters of this book, were confronted with an enemy who was often better dressed, wealthier, and initially much more effective. First on Soviet territory and then abroad, Red Army soldiers confronted an alien culture. For average citizens this trip abroad was a unique chance to go beyond Soviet borders, one that came at great personal risk and with a clear objective—to destroy Fascism and the Third Reich. What soldiers saw along the way was puzzling. They not only reckoned with material objects and institutions that the Soviet Union had purged but were also left to wonder why people who lived materially so much better than they did had waged a genocidal war against them, marked by systematic rape, pillaging, and wanton destruction.

This chapter discusses how Red Army soldiers confronted an alien material world through the practice of trophy taking. In so doing, it draws on how particular Soviet perceptions of class, criminality and property informed soldiers' understandings and actions.[2] The first section analyzes the term *trofei* and the formal structures and policies developed in the army to control the collection and usage of trophies until 1945. The second section describes the encounter of Red Army soldiers with enemy soldiers, weapons, living spaces, and personal possessions on Soviet territory in the context of Third Reich occupation policies. The third section follows the Red Army into Europe, examining how soldiers interpreted their encounter with the bourgeois world and what they did there. The chapter ends with a discussion of items taken from the Third Reich back to the Soviet Union.

Trophies carried a variety of meanings. They were physical proof of victory over the enemy. The Soviet government staged and filmed ostentatious displays of trophies as proof of victory during the war in ways that harkened back to the age of Caesar. Trophies were not only markers of victory; they could also be useful and included evidence of the enemy's depravity.[3] Last, trophies could be objects of desire, rare consumer goods that were beyond the reach of most Soviet citizens.

Trophies elicited a combination of attraction and revulsion, embodying dynamics of inferiority and superiority that have been a leitmotif of Russian-"Western" interactions for centuries as well as ambivalence about the place of luxury in Soviet society.[4] Anxiety about material objects and meshchanstvo (and its adjectival forms *meshchanskii*, *-oe*, *-aia*), a concept Strukov used in his dismissal of supposed German superiority, had deep roots in Russian culture. Meshchanstvo, a culturally fraught term, was posited as one of the greatest threats to morality and the Soviet project. It could corrupt from within as the desire for comfortable living overtook revolutionary consciousness, and early Soviet culture often defined itself in contradiction to traditional refinement and frivolous things. On the eve of the war many of the accessories of bourgeois culture returned among the Soviet elite, including servants, summer homes, traditional refinement, and conspicuous consumption.[5] Fear of meshchanstvo and praise of *kul'turnost'* (culturedness) existed in an uneasy balance, as the acquisition of certain objects marked one as a cultured person, but desire for these trappings in and of themselves could be seen as bourgeois greed.[6] Moreover, the traditional peasant societies from which most soldiers came were deeply hostile to conspicuous displays of wealth. The affluence of Polish *pany* and German burghers was shocking to Red Army soldiers—it marked their enemies as potentially more cultured than the Soviets—but the obsession of real-life *meshchane* (bourgeois philistines) with their things was disturbing. Meshchane were somehow pathetic, but their material wealth was a cornucopia of most enviable forbidden fruits.[7]

Until 1945 everything left on the battlefield was treated as government property. As Soviet troops moved beyond the borders of the USSR, policy shifted to allow soldiers to send trophies from the Third Reich to their families. This shift from restriction to invitation could be seen as a reiteration of early Bolshevik attempts at leveling through the violent redistribution of property. In 1918, the party called on working people to "loot the looters," confiscating and nationalizing the possessions of class enemies. Collectivization (1928–1933) was accompanied by dekulakization, the process of liquidating a class that had been declared criminal and the expropriation of their possessions to be used by the newly formed collective farms. During the Great Terror of 1937–1938 expropriation of the property of "enemies of the people"—cast as foreign agents undermining the regime—was also common. While in these cases confiscated objects were supposed to become socialist property, pilfering for personal gain was common.[8] In 1945 Red Army soldiers were again invited to loot their enemies' property, this time with the official sanction to keep what they took for themselves. Class and nationality merged in propaganda that treated all Germans as inherently criminal bourgeois philistines.

Beyond extraordinary events such as dekulakization and the Great Terror, the right to take from criminals was engrained in Soviet jurisprudence. The confiscation of property was part of the sentence for a variety of felonies. Soviet law and policy had openly targeted groups before and during the war, leading to the repression of categories of people, with the parallel confiscation of their property.[9] The criminal nature of German occupation led many soldiers to see confiscation of property from *any* German as just. Expropriation in the Soviet context meant that a person had lost his or her rights, and by hinting that soldiers could loot, the government also implied that Germans could be abused without consequences. Red Army soldiers brought particular understandings of property and justice with them to Eastern Europe and the Third Reich, which included government monopoly and the right of confiscation.

Words and Structures

"Trophy" came to modern languages by way of the ancient Greek term *tropaion*, the practice of demarcating the place of one's victory using the weapons of the vanquished.[10] The Russian word *trofei* is used slightly differently from its English equivalent, and the units responsible for collecting trophies, called trophy teams (*trofeinye komandy*), had a wide range of competencies. Trofei, according to a dictionary published on the eve of the war, was defined as "spoils taken during victory over the enemy" and "a mark or symbol of victory."[11] In practice, trofei meant not only things that belonged to the enemy but anything useful left on the battlefield, including damaged Soviet tanks, abandoned weaponry, and even whole cities.[12] *Trofeinyi*—the adjectival form—could also be used colloquially to indicate anything foreign: for example, American cigarettes. Trofei covered everything from scrap metal to weapons to works of art, and the army's expanding understanding of trophies and what to do with them moved in roughly that progression.

Early in the war no one was assigned to collect scrap metal and enemy valuables. At the beginning of the war evacuation of scrap and "trophy goods" was a part of the competency of rear-area services without dedicated personnel.[13] In the first months of the war evacuation of Red Army weaponry, machinery, and supplies was often unrealizable, as units retreated and disintegrated, leaving the Wehrmacht with innumerable trophies. However, the reversal of fortunes in early 1942 and throughout 1943 led to an expansion of the Soviet trophy-collecting apparatus.[14]

The first dedicated teams of soldiers tasked with collecting and evacuating army property, trophies, and scrap metal were organized on September 29, 1941, as a way to prevent resources from going to waste. This order created something like a department of sanitation for the army and called for the collection of everything from the battlefield. Commanders were permitted to recruit local civilians to assist with evacuation and use military transport vehicles returning from the front to carry evacuated goods. A system was created to sort between items for immediate use by the collecting army and those that should be evacuated farther to the rear for repair or investigation. Trophies, scrap metal, and equipment requiring serious repair were evacuated far to the rear. Reusable shell casings, weapons, and ammunition were sent to the collecting army's base for immediate repair and/or reuse.[15] Trophies were categorized as something soldiers should simply collect and seemed like an afterthought in an order primarily concerned with evacuating scrap metal and reusable Soviet equipment, reflecting the precarious state of the army in the fall of 1941. By March 1942 the Red Army confronted a massive number of trophies during the Moscow counteroffensive. On March 22, 1942, the State Committee of Defense created groups of soldiers (three per division, eight to twelve per front and army) tasked with collecting trophies and scrap, with temporary commands of up to two hundred assistants. Soldiers received premiums for collecting, and the army continued to mobilize local civilians.[16] A year later, after Stalingrad, the army won even more trophies, and in March 1943 larger trophy commands were established. Orders highlighted that trophies were government property and that taking trophies was the same as stealing from the state. This order also emphasized regular reporting of trophy evacuation.[17] By April 1943 a separate trophy committee was established, which made the creation of a museum displaying enemy equipment a priority.[18]

In 1944—the year the Red Army left Soviet borders—another reorganization of the trophy-collecting apparatus made the process more professional and efficient. In January 1944 trophy teams became responsible for cataloging and protecting works of art and other cultural valuables, such as the Third

Reich's collection of vintage wines.[19] In April trophy teams were tasked with evacuating everything left behind by soldiers, guarding recaptured objects of national economic interest (e.g., factories), as well as recovering weapons and equipment in the hands of civilians. They were also instructed to report on any new enemy weapons they observed. This order established a new weapons section that was to safely handle and repair captured armaments at the level of the division, army, front, and the entire Red Army. The newly expanded apparatus integrated specialists from different branches of service into the trophy teams that evaluated captured equipment. The trophy teams were given more resources, including fuel, and access to all means of transport, including railroads.[20]

Trophy-collecting units were the army's way of ensuring that all resources left on the battlefield were under control. Service in the trophy teams could be particularly unpleasant for frontline soldiers, and some of their functions, particularly the collection of food and equipment from beleaguered civilians, recalled the food detachments of the Civil War and collectivization.[21] Conversely, trophy teams could be ordered to help sow grain and repair agricultural equipment at recaptured collective and state farms. Trophy teams were often understaffed and could fall far behind advancing units. Frontline units often used enemy equipment and food stores to meet immediate needs, rather than giving them to the state to apportion.[22]

Despite orders clearly marking everything as government property, soldiers often helped themselves to whatever was at hand that could make their lives easier. Red Army personnel were known to tear apart houses for firewood or to construct their bunkers. They also could steal food from civilians both inside and outside the Soviet Union.[23] Commanders were reprimanded for sending packages of trophy goods home to their families in March 1944.[24] During the Battle of Kursk the local secret police noticed that abandoned tanks were not collected by trophy teams and were instead "dekulakized" (i.e., cannibalized) by Red Army soldiers to repair their own tanks.[25] Weapons and other items taken from the enemy were popular among Red Army soldiers, and there was often little that could be done to wrest these objects from the hands of those who took them.

There was another reason that the Soviet leadership wanted to control trophies. Trophies had an aura around them; they were things that had belonged to people now dead or in captivity, had been used against those who took them, or at the very least had been constructed on foreign territory and came from an alien world. Foreignness was dangerous in the Soviet Union. Soldiers who were recruited from occupied territories could be referred to as "trophy soldiers" and regarded with suspicion.[26] That the enemy had penetrated so deeply into Soviet territory was both a physical and an existential threat, and the detritus left behind as the enemy retreated was of particular interest—the objects of the terrifying world of capitalism.

Battlefield Trophies, Living and Dead

Red Army soldiers encountered enemy POWs, corpses, weapons, bunkers, and equipment on battlefields from the Caucasus Mountains to Berlin. They also confronted what the enemy had done on Soviet territory. These encounters took place on or near the battlefield, where enemy soldiers had actively been trying to kill them. What Red Army soldiers saw changed their understanding of who they were fighting, the stakes of the war, and what human beings were capable of. It also set the stage for how Soviet soldiers would act when they arrived on enemy territory.

Foreign objects could infect and pollute. Listovki, small leaflets that attempted to convince soldiers to surrender, were produced by the Germans and their collaborators in a variety of Soviet languages and dropped by the millions on Red Army trenches.[27] They promised safety and often appealed to antisemitism, nationalism, or class antagonisms within the army. Some claimed that the listovki were produced on special paper to make them more attractive to soldiers, who would use them to roll cigarettes. Finding German listovki in a soldier's possession was considered proof of intent to commit treason.[28] Surgeon A. A. Vishnevskii boasted in his diary: "we use them [*listovki*] for another purpose entirely," presumably as toilet paper. However, the head of the Red Army's Political Directorate complained in 1942: "All we do is collect and burn these listovki, but the soldiers read them anyway."[29] Listovki were dangerously seductive to soldiers facing unbearable conditions and could announce inconvenient truths, such as the fact that Stalin's own son had been taken prisoner, or lies to demoralize soldiers, announcing the fall of cities that never surrendered.[30] The war exposed Soviet citizens to alternatives, whether in the form of listovki or under occupation. Although by and large these alternatives were worse than Stalinism, for certain groups and individuals the Third Reich offered an opportunity to forward nationalist dreams or take revenge. It took months of bitter experience to convince many disgruntled Soviet citizens that the Fascists were worse than the Communists.

Although not everyone realized it, the Nazi invasion presented dangers heretofore unknown. Planning in the Third Reich imagined Soviet territory as a space to be cleared of inferior people to make way for Germans, using the settlement of the American West as a model. This led to a barbarization of warfare combined with modern technology that facilitated horrendous violence. Many of the hard-earned gains of Stalin's crash industrialization were rent asunder by the occupying forces. Soviet citizens were deliberately starved, executed in large numbers, enslaved, and deprived of property.[31] The Wehrmacht planned to feed the army at the cost of local civilians. German soldiers were allowed to send home parcels for free, initially with a one kilogram weight limit, later without any weight limits. They were also authorized to take whatever plunder they could carry with them home on leave with no customs fee.[32] Soldiers of the Third Reich functioned in an environment virtually without law, where they were encouraged to take and destroy whatever they wanted, completely ignoring the humanity of the people inhabiting the spaces they controlled.

As the Red Army recaptured towns and cities, even those who had suffered greatly under Stalin realized that a victorious Reich meant extinction or slavery. The war correspondent Vasilii Grossman overheard women in Stalingrad say, "That Hitler is the real Antichrist, and we used to think that Communists were Antichrists."[33] Soviet propagandists had no need to fabricate stories about German atrocities, and soldiers both witnessed the aftermath of atrocities firsthand and read newspapers filled with calls to kill German soldiers as an act of collective vengeance.

Enemy dead were a clear sign of victory, a visceral reminder to soldiers who saw them that they were alive and their enemy was not. Enemy corpses could evoke a sense of pride, as Boris Marchenko wrote home on May 1, 1943: "I used to consider the aphorism 'an enemy's corpse has a sweet smell' savage, but now I understand . . . It is pleasant to look at these dead Aryan faces with their long teeth showing. You see, they stood between you and me and our children . . . So the more corpses we make, the sooner our life together will come and with it our inevitable happiness."[34] Soldiers were often struck by the wealth of things found on the German dead. Nikolai Nikulin recalled: "What surprising hoarders [*barakhol'shchiki*] these Germans are! Some sort of rags, women's underwear, dishes, rugs, even a delftware toilet. And in their pockets—photographs, letters, condoms, and whole collections of pornographic postcards."[35] Soldiers were free to take what they wished from these corpses, who were treated as litter.

Cleaning battlefields of corpses was a major priority, both to prevent disease and to efface the war's toll.[36] The Germans had left many well-groomed cemeteries in their wake, turning sites of cultural patrimony in the Soviet Union into graveyards.[37] These cemeteries would eventually be removed without a trace. A directive from April 1942 called for the "liquidation" of all enemy graves on Soviet territory and the reburial of enemy dead "far from population centers, highways, or the graves of Red Army soldiers and commanders."[38] Boris Suris, in the aftermath of Stalingrad, realized that he and his comrades had "lost any respect for death. We say 'croaked [*dokhlyi*] Fritz'; we laugh at the sight of German cemeteries with their symmetrical rows of crosses. I saw a soldier peeing on a German grave."[39] This lack of respect for the dead was paralleled by objectification of the living.

Prisoners had been a rarity in the first year of the war. With the first major surrender of a German army at Stalingrad, German POWs became trophies. General Chuikov, commanding the Sixty-second Army at Stalingrad, recalled that soldiers would refuse to hand prisoners over to other units unless a commander intervened and gave them a receipt.[40] Displaying prisoners was an important act of vindication. A political officer recalled that German prisoners looked so awful, with their soiled clothing and herds of lice, that they were shown to soldiers as a form of "visual propaganda."[41] Parades of German POWs through major cities began in Leningrad in 1941 and culminated in the convoying of fifty-seven thousand POWs taken during Operation Bagration through Moscow, followed by a demonstrative washing of the street. This parade was filmed and distributed.[42] High-ranking prisoners were particularly prized and often publicly identified in the media. Many German POWs were put to work rebuilding what they had destroyed. In 1944 General Erëmenko recorded with pride that German prisoners were rebuilding Kiev.[43] Today many post-Soviet cities have neighborhoods built by German POWs after the war.[44]

In any war prisoners run the risk of being killed by soldiers still caught up in the ecstasy of battle or because of the danger prisoners present.[45] Senior Sergeant Vladimir Kukel' recalled executing a group of surrendering soldiers: "We ceased fire and later—I can't explain how it happened—we took them all out with our rifle butts and bayonets."[46] On the battlefield captives had to be disarmed, searched, and escorted to the rear, all of which diverted soldiers from combat. The man who obediently awaits his fate one minute could take up arms against his captors if the fortunes of battle change. The prisoner and his comrades were recently trying to kill their captors and may have killed some of their friends. Beyond the microcosm of the local battlefield many soldiers had scores to settle, whether family members killed by the Wehrmacht or the humiliation of the first years of the war.[47] One correspondent from the Kalinin Front wrote to Kliment Voroshilov in 1943 that "troops totally refuse to take prisoners."[48] Collaborators were often shot on sight.[49] An officer in the Latvian Rifle Corps recalled at the end of the war, after taking over two hundred prisoners: "In truth, few of those prisoners got to us. In the process of battle it always happens that when soldiers are exasperated [*ozlobleny*], they take few prisoners."[50] Others understood that taking prisoners could shorten the war, as one soldier told the Mints Commission: "I never shoot prisoners, because if you shoot at one German, the second won't come over, he'll say that Russians shoot prisoners. Quite the opposite, I would treat them well so that they would understand that the Russian soldier was freeing them from under the German yoke. I trained my submachine gun on him and said, 'give me your watch.' If they resisted, I killed them."[51] This soldier's casual statement that he was ready to kill anyone who resisted alongside his unabashed talk about taking what he wanted from prisoners highlights both the danger and opportunity that prisoners presented. To mitigate the dangers, prisoners had all of their buttons removed so that they couldn't run away because their pants would fall down.[52]

The government and the soldier wanted different things from prisoners. The army needed POWs to interrogate, and each division had its own translator.[53] To the soldier, prisoners were a source of danger, pride, and trophies. Orders existed not to rob POWs of their personal belongings, but prisoners ultimately had little recourse if Red Army soldiers began rifling through their pockets.[54]

Watches, which German soldiers had bought or pilfered from all over Europe, became increasingly common after Stalingrad. Watches became a sort of currency and were an ideal trophy, being lightweight and of relatively high value. Galina Golofeevskaia told the Mints Commission how after Stalingrad:

> The Germans had a startling number of watches—both their own and stolen, each had two or three. As a rule, when we would lead a column [of prisoners], the soldiers would take watches before anything else. So every soldier had eight to twelve watches. I had eight watches. On the road I traded these watches. Whatever you took on the road was measured in mess tins: a mess tin full of apples—a watch, a mess tin full of moonshine—a watch, somebody also bought some boots for a watch.[55]

Boris Komskii also acquired watches. He recorded in his diary how he took a watch from a German soldier he had wounded and refused to bandage in heavy fighting near Kursk. Shortly afterward Komskii himself was wounded and traded the watch to a nurse for extra food. He acquired "two bags, two pistols, binoculars, a camera," and three more watches a year later.[56] Trophies needed to be useful for soldiers at the front, even if they were carried for only a few days and exchanged for food or drink.

Like prisoners, material objects were displayed as a sign of victory. Military units recorded the machinery and weapons they destroyed and trophies they took in battlefield reports and tallied them for regimental histories.[57] Local displays of captured or destroyed enemy technologies were common throughout the Soviet Union, and in the summer of 1943 a display of an unprecedented scale was prepared in Moscow.[58] Thirty-four enemy planes, 58 tanks and armored transports, and 128 cannons were displayed in Moscow's Gorky Park. This display was intended as a graphic demonstration that the Red Army was winning while highlighting the high-quality equipment of an enemy who had proven so difficult to defeat.[59] A film was made of the display, emphasizing that the Germans had taken trophies from the nations they had defeated.[60] Over a million people—among them Allied, neutral, and even Japanese diplomats—saw the display. According to a report filed by Kliment Voroshilov: "As a rule" visitors were convinced that "if the Red Army was capable of winning victories against the enemy, of capturing and destroying such powerful machinery, then it is undefeatable and the day of victory cannot be far-off."[61] Seeing a variety of machines that had been designed with the express purpose of killing those people who now viewed them as trophies provided material proof of official proclamations that had often rung hollow in the war's first two years.

There was a wide array of responses to the Wehrmacht's things among soldiers. Grigorii Baklanov recalled how everything in the Wehrmacht was "thought out to the last detail, so a soldier could fight."[62] Many soldiers sought German weapons (submachine guns and pistols were particularly prized), and at the army encouraged the practice of maintaining a stockpile of trophy weapons.[63] Some commanders gave friends captured weapons engraved with touching messages.[64] Soldiers learned to convert German shells for use in Soviet mortars.[65] These encounters, improvisations, and stockpiling of extra weapons were a double-edged sword. Additional weapons in the hands of soldiers expanded their ability to kill, but an affinity for foreign-made weapons and equipment threatened Soviet claims of supremacy, as could visions of the alternative world that German bunkers offered.

German dugouts and bunkers could be both repellent and alluring. Vasilii Grossman wrote in his diary about "the obligatory desire to wash your hands" after touching German things.[66] Lev Kopelev, who worked closely with German POWs, recalled the specific smell of German dugouts and prisoners: "the sour-musty smell of wet wool, dry wax, cold tobacco ash, dirty, sweaty underwear, a polluted water closet."[67] Anatolii Soldatov, the deputy political officer of the Sixty-second Army, also described "an unbelievable stench" near the German headquarters at Stalingrad, where German soldiers had used the corridor as a bathroom, leaving piles of feces "up to your chest."[68] While this was an extreme case, other soldiers described German trenches as full of miasmas, junk, or other corrupting influences. Given the German claims to cultural and racial superiority, finding evidence that the enemy lived in squlaor was great for morale. Some recalled getting lice from German bunkers.[69] Lev Slëzkin expressed a mix of wonder, envy, and rage in describing a German bunker near Novgorod:

> you couldn't help but notice (and how many times!) the difference between our soldierly lifestyle and that of those coming here from far away. Under the table were empty bottles with seductively colorful labels of rum, cognac, French and Spanish wines. In a cardboard box near the door were empty cans and jars of olives, sardines, and jam. On the table was a pile of vibrantly illustrated journals with many pages of beautiful women, tanned bathers, luxurious southern hotels, beautiful landscapes . . . We tell ourselves that those bastards are showing off, but their end is nigh.[70]

The fact that the Germans' comfort came at the price of robbing all of Europe did not make their luxury any less real.

Red Army soldiers were happy to take many of the goodies the Germans left behind, but the Wehrmacht would often poison and booby-trap whatever they could. Germans mined their own dead, knowing that curious soldiers would hunt for trophies.[71] They also booby-trapped foodstuffs and weapons, something Soviet propaganda was keen to highlight.[72] The Wehrmacht purposefully left stores of alcohol behind to slow the Red Army's advance. The wealth that so impressed many who entered German bunkers could be deadly but also beckoned. However, not everything that belonged to enemy culture proved appetizing.

In the pockets of German soldiers and on the walls of German bunkers soldiers found a genre of photography that did not exist in the Soviet Union.[73] Mansur Abdulin described the reaction of his Uzbek comrade to something they found: "To us "illiterates," it was hard to understand what they were doing. But when we finally discerned and understood, we couldn't believe our eyes. This provoked a feeling of abhorrence, and my Uzbek friend even vomited . . . That's how we came to know about pornography."[74] Abdulin's comrade was not alone in being appalled by pornography.[75] Stalinist official culture had no place for explicit sex, so these images would have been shocking. Any pornography found in the mail was to be confiscated and destroyed.[76] However, German observers noticed that Red Army soldiers were particularly impressed with German erotica and that even underwear advertisements were quickly snatched up by curious soldiers.[77]

Another genre of graphic media found on German soldiers could only disgust and perplex. Many German soldiers had cameras that they used to document their experiences in the Soviet Union. This meant that Red Army soldiers frequently found pictures of massed Red Army prisoners, dead Red Army soldiers, and the executions of Soviet partisans and civilians on the persons of German dead and POWs.[78] Occasionally German soldiers were found with trophies taken from dead Red Army soldiers. A particularly vivid article for agitators described finding a drawing of the medic Frida Fel'dman, mutilated, with her medals attached to it, in the map case of a dead German.[79] Photographs, letters, diaries, drawings, and anything else found on prisoners and the German dead were used extensively to incriminate them, as Jochen Hellbeck and Karel Berkhoff have shown.[80]

The desire for revenge was to be paid not only in blood but in *things* as well. In Tvardovskii's *Vasilii Tërkin*, the title character is given shelter by an old peasant couple while retreating in 1941. In 1944 he comes across the same couple again, after they have survived German occupation and had their prized grandfather clock stolen. Tërkin promises to bring two new clocks from Berlin.[81]

The Wehrmacht had proven to be a vicious foe capable of anything. From the early days of the war "learn from the enemy" was a watchword of Red Army propaganda.[82] The Wehrmacht was in many ways the model of a successful army to Red Army soldiers. The soldiers of the Red Army came to Europe with vivid memories of what the Germans and their allies had done on Soviet territory and a Stalinist understanding of the world: the war had rent asunder Proletarian Internationalism by the time the Red Army entered Europe.[83]

The Alien World: The Red Army Goes Abroad

In the Other World

Many Soviet soldiers felt as if they were in a foreign land before technically leaving the borders of the Soviet Union. Fighting their way through the Baltics, Western Ukraine, or Bessarabia—all territories annexed by the Soviet Union on the eve of the war—Red Army men understood that they were on alien territory where the locals could be openly hostile and nationalist guerillas were active. The wealth of the land and houses could leave quite an impression on those who saw them because, as Kovalevskii confided in his diary, in Western Ukraine apartments resembled "palace museums" or "the stage of good theaters."[84] Aside from the dramatic difference in living standards, the customs and languages of these territories were often confusing. Nonetheless, this territory was still at least formally part of the Soviet Union, and soldiers were not expected to change their behavior significantly, as they performed before a domestic audience.

As the army moved beyond Soviet borders, new policies emerged. Orders for security forces aimed to separate soldiers from locals emphasized the necessity of making a good impression while trusting no one. The army was also deeply concerned with how soldiers would react to foreign ways and social structures, particularly religion. The fact that the army would have to function in places where Russian was not the lingua franca created great consternation about how to document and control the locals. Finally (and ironically), units were encouraged to lighten their baggage in order to move more quickly while abroad.[85]

Setting foot abroad led to immediate changes in the way soldiers acted. There was an increased emphasis placed on soldiers' appearance and behavior, reminiscent of prewar attempts to manage appearances in front of foreigners.[86] Guards Lieutenant-Colonel Iakov Babenko, who had previously worn regular soldiers' clothing, recorded in his diary that in Germany "we came to demand that officers and soldiers [wear] their uniforms strictly according to regulations."[87] His notes reflected a general trend in the Red Army's attempt to manage its image. An order from July 1944 noted, "It is common for our officers to walk around in dirty clothing, without *pogony* and in such a state fill the shops, taverns, and restaurants of the cities of Romania and Moldavia, appearing on the streets drunk and conducting themselves in a contemptible manner."[88] Boris Komskii, echoing this order, complained to his diary that being abroad "forces us to remake ourselves to a large degree. We need to demonstrate our culture, but swearing thrives."[89] The emphasis on appearance and behavior was tied to the need to craft a certain image of the Red Army, highlighting both pride and insecurity before foreigners.[90]

While the Political Department was concerned with how the outside world perceived its soldiers, it was also influencing how soldiers understood the lands through which they marched. Soldiers were warned to be vigilant against local counterrevolutionaries, White Guards, and female spies.[91] The Red Army produced information about the places it trudged through, including brief guides to the history, politics, and customs of different countries and regions as well as phrasebooks of local languages.[92] Some soldiers felt a sense of disappointment about how much more they knew about the people they were meeting than their foreign interlocutors knew about the Soviet Union.[93]

Soldiers were often quartered in the houses of civilians as they moved west, leading to prolonged interactions. Ivan Turkenich wrote to his family about how exotic Poland was: "Here, where we are, I have encountered much of which I only read about. And here, as you know, they live by their own customs. Here you can meet a lord, their workers. Though it's not a life such as ours, but some sort of mess [*nedorazumenie*]."[94] Turkenich, who was tasked with sowing wheat, was deeply disturbed by the primitive agriculture and attitude toward the world among the Poles: "To talk to the people about these issues was very difficult, because their view of life is totally different, or rather the people have not been raised in our spirit and know only private property and private interests."[95]

The mores and creature comforts of people living outside the Soviet Union could have a demoralizing impact on soldiers, as one translator reported being annoyed by the "meshchanskii coziness" of the "snail's world" of Poland, which made it "harder to fight in Poland than on other fronts."[96] The world outside the Soviet Union was strange and often confusing, a place where soldiers confronted real life meshchane and their things. Many soldiers recorded in their diaries, memoirs, and interviews how impressive these things were. Boris Shumelishskii remarked how the high quality and standardization became depressing and oppressive, despite the obvious comfort and quality.[97] Tat'iana Atabek described in great detail the first German house she encountered: "A wealthy estate! A splendid two-story house. Every room has its purpose. An exhibit of beautiful china and porcelain figurines. How beautiful! . . . Ten types of crystal glasses. A library, bedroom, kitchen, and bathrooms. Halls, the walls of which are hung with hunter's horns, shotguns, pictures, and rugs. They have an organ."[98] Atabek found repulsive things among this beauty as well, including many "tasteless objects" and a giant swastika banner that had been hidden. The wealth of the houses and their former inhabitants was astonishing, and Atabek was by no means alone in reflecting on the incredible abundance found in the Third Reich.

David Samoilov was also deeply impressed with the things he saw in Germany but was highly critical of them and their owners, reflecting in his diary in April 1945:

> Landsberg. A person can become a slave to things. This is an old truth. Perhaps in Russia it was easier to accomplish the Revolution, because "things" were never sovereign there. Never, I think, was there such a minuteness of the everyday [*byt*], such a predominance of things. Here things are not just objects of everyday life. No! Things instruct, things have their own philosophy, things pompously speak the truth. Oh, flat,

> wooden, self-assured philosophy of things! . . . Things are sentimental and self-satisfied. Just like their owners. They were also things in their own houses. And they have been given out for demolition, like their homes, like the most abominable thing in the world—Germany.[99]

Samoilov went on to decry Hitlerism as the natural end point of burghers and meshchanstvo, the ultimate expression of the self-obsession, hatred, and envy that accompanied a self-centered and materialistic view of the world. Samoilov, much like Colonel Strukov quoted at the beginning of the chapter, viewed something dangerous and pathetic in the meshchanstvo of the enemy, and saw the obsession with things and acquisition as part of what had driven the "respectable burghers" to such horrible crimes. Samoilov's understanding—a personal interpretation of the greater Soviet narrative of class enemies and the wartime message of narrow-minded, self-obsessed German thieves—still gives class primacy over ethnicity as an explanatory factor. But these concepts bled into each other; meshchanstvo had been associated with Germans in classic Russian literature. Samoilov was a highly educated Muscovite who spoke German. The depth of his reflection on Germans and their things was far beyond that of many soldiers. Some of these soldiers would agree that destruction was a fitting end for German things, but many simply saw an opportunity to enrich themselves and undo some of the damage the Germans had inflicted on them. The latter group saw no corrupting aura around their trophies.

Parcels and the Right to Take

The Soviet government could be seen as the master trophy taker—it took control over and sometimes exported whole factories, businesses, banks, mines, agricultural concerns, and any institution or business of scale.[100] It did so initially not as a form of reparations but simply because the Red Army was often the only governing body in any given territory. Local government officials often fled before the Red Army's arrival. This helped to contribute to a sense of lawlessness that created an ideal situation for looting. The government continued to maintain a monopoly on large-scale trophies, but it relinquished its control over soldiers' trophy taking in the last months of the war.

In late December 1944 the army dramatically shifted its trophy policy. It was announced that, with their immediate superior's permission, all soldiers in the active army who were "fulfilling their duties well" would be allowed to send parcels home starting on New Year's Day (coincidentally the main gift-giving holiday). Soldiers were allowed to send home five, officers ten, and generals sixteen kilograms once a month. Soldiers sent their packages for free, while higher ranks were charged two rubles per kilogram. Parcels could even be insured. Hundreds of dedicated personnel were added to the Red Army's postal infrastructure to handle the influx of packages.[101] A subsequent order laid out further rules. To send a parcel home, soldiers would have to provide a hard case—either a wooden box or a suitcase, leading some units to set up special workshops.[102] Soldiers were forbidden from sending parts of Red Army uniforms or their rations home—anything that was government issue. Weapons; perishables; liquids; medicine; anything explosive, poisonous or flammable; currency; letters; or any printed material were likewise prohibited.

Soldiers were also barred from using anything printed in any language as packing material.[103] Perhaps most curious at first glance was the total prohibition on any form of anything written—both soldiers' notes and print of any kind. The reason was simple: any print material would have been subject to censorship, overwhelming the parcel system. This final point illustrates the dangerous nature of foreign goods as a potential agent of corruption. The Soviet government could accept the acquisition of foreign goods, but the transmission of foreign ideas was inimical.

The parcel orders never explicitly told soldiers to loot, but it was easy to read between the lines. The order says nothing about where soldiers were to acquire the items that were to fill these parcels, and the only geographic limitations as to where a soldier could send parcels from was "the active army," which was already beyond Soviet borders. This meant that how many trophies a soldier could take depended on local conditions, without an explicit army-wide policy. While in Yugoslavia, Czechoslovakia, and Poland there was good reason not to alienate locals, who were Soviet allies, in the Third Reich and other enemy states, soldiers were given carte blanche to take whatever they wanted. General Erëmenko noted in his diary: "As opposed to Germany, the Czech population cannot be wronged [*nel'zia obizhat'*]."[104] Soviet propaganda would explicate the justness of looting to soldiers, but many were sympathetic to this policy without any explanation.

The government's position on trophies had at long last come into line with that of most soldiers and in significant ways mirrored German practices discussed above. The dedication of resources, including vital railways, to transport trophies to families in the rear was impressive. What had been a crime became policy. Boris Suris, an astute observer of the parcel policy, wrote in his diary: "This is a natural phenomenon. A soldier has broken into a foreign country and wants to feel like a victor . . . The justification for marauding is the introduction of parcels. They stole and extorted enough from our country; now we will steal and extort." He also noted the confusion created by the order, citing a rumor that Mekhlis, one of Stalin's representatives at the front, berated the commander of soldiers held by court martial for looting, shouting: "You don't understand a thing about our policy. And what are parcels for? What do you think, the soldier is going to send his dick home?" After this the soldiers were immediately freed and their trophies returned. The head of the Political Department of Suris's army, realizing during an audit that few soldiers were sending parcels home, declared, "What a lousy bunch of occupiers!" Suris himself was ambivalent about this policy.[105] Expropriation of enemy wealth was official policy, even if never explicitly stated.

Prior to the Red Army's entry into the Third Reich, instances of marauding had occurred but were looked down on by soldiers and could be prosecuted. Lieutenant Kiselëv noted in his diary that in the gray zone of the prewar Polish-Prussian border "the understanding of 'marauding' no longer exists, and if in Ukraine you could end up in a penal battalion for stealing a stinking goat from your landlady, here nobody will say anything no matter what you steal from anyone at all."[106] It became clear to all, particularly after the order concerning parcels, that in the Third Reich the rules were different.

This policy did not favor everyone equally. It clearly privileged officers and generals, doubling and tripling the limits of what they could send vis-à-vis soldiers. Commanders decided who could send parcels home, allowing them to use the parcel as a carrot in their personal dealings with their subordinates and giving them the opportunity to take items as a form of bribery. Soldiers in poorer regions and rural areas often had less to take, and the difference between the possibilities open to combat soldiers and rear-area personnel was dramatic. Boris Suris noted in his diary that a soldier needed to get permission to send parcels, gather booty, make a wooden crate and wrap it in sturdy material, find clean paper to write on, and then go to the field post, all of which were difficult tasks for frontline troops. In the meantime, "Rear-area guys [*tyloviki*] are excited; it's a holiday on their street."[107] Boris Slutskii went so far as to declare that by the spring of 1945 "a group of professional marauders and rapists" formed among those with "relative freedom of movement: reservists, *starshiny*, rear-area personnel."[108] Others noted that frontline soldiers were less likely to pillage anything they couldn't eat or immediately use, but rear-area personnel were known to loot even under fire. Despite the disparities, soldiers often felt that this policy was just, even if pilfering of the parcels themselves was common. Suris recorded one soldier saying, "if it gets stolen, then all the same Russia will be richer for it."[109] Some understood this policy as compensation for all that soldiers had suffered and as a way to help feed families suffering from hunger in the rear. All of the trophies Boris Tartakovskii recalls sending home were traded for food by his family.[110]

Red Army soldiers found items stolen from all over Europe and the Soviet Union, which added to the sense of looting as justice. An officer interviewed by the Mints Commission recalled, "We found our uniforms and civilian clothes with our labels, our stamps, our TeZhe brand soap, our chocolate in almost every home."[111] The Red Army's primary journal, *Krasnoarmeets*, ran a story titled "Samovar" that served as a clear and direct justification of the parcel policy. In this story Mitia Riabchenko—whose parents had all of their possessions, including their prized samovar, stolen by one Otto Kesler of the Wehrmacht—fights his way to Landsberg, where he retrieves his parents' samovar from the deceased German soldier's terrified family.[112] This allegory was a clear message to soldiers that plundering was just. Many seemed to draw this conclusion about the enemy spontaneously. Gabriel Temkin recalled that his comrades, on entering Nazi-allied Hungary with its rich farms, exclaimed, "What a bunch of Kulaks!" Resentment and hatred stemming from their wealth made looting "morally painless, justified, almost an act of social justice."[113] Some political officers interpreted the parcels as a way to motivate soldiers to continue fighting beyond Soviet borders.[114] News about parcels traveled with lightning speed, with some civilians writing their family members at the front with requests and reminders of their sizes as early as January 1945.[115] Flirtatious soldiers even offered to send parcels to girls as they struck up correspondences, while soldiers with families wrote tantalizing inventories of what they were sending to their spouses and children or asked how they were enjoying gifts already sent.[116]

For certain soldiers, however, this policy was confusing or abhorrent. A furious General Erëmenko recorded in April 1945 how most of the vehicles in his column carried "completely useless things that weigh down transport."[117] Some Red Army personnel felt that stealing was tempting fate or betraying their principles. Nikolai Inozemtsev refused to send parcels home because he felt that it was unlucky.[118] In the waning days of the war Junior Lieutenant Rashid Rafikov wrote his father a heartfelt letter concerning his refusal to loot, not wanting to enter the "ranks of the pillagers" and directly citing his father's experience: "I can't raise my hand to take these things, my heart hurts, and before my eyes I see your difficult existence in the Thirties." Hinting at the experience of dekulakization, Rafikov reflected that despite the horrors he had seen and the plunder found in German homes, he couldn't "touch these things and cannot send you a present; I hope that you agree with my sentiments. You always told me that wealth is something to be gathered with your own sweat, and if we survive, we can't avoid wealth."[119] (Rafikov was killed in action a month after penning these lines.) He was not alone in this sentiment. The writer Aleksandr Tvardovskii (also the son of a kulak) wrote a similar letter home to explain why he hadn't sent any parcels: "It's strange and shameful to see our *kultur-traegers* [sic], gathering, 'organizing' underwear, rags, worn shoes, and the like and sending parcels. I don't know how you would take it, with your practicality, if I were to send you children's clothing (a little worn!) or a dress, etc., but I suspect that you would be ashamed of me for it."[120] Tvardovskii found the pilfering of secondhand goods pathetic and shameful. Veniamin Tongur wrote with disgust about the "clothes horses [*shmotochniki*] and ragpickers" who shook with greed before trophies.[121] Others were frustrated that they were looting the poor who had been left behind, as the (despicable) wealthy people had been able to flee.[122]

Some feared that looting would erode morale and morality in the army and society. Lev Kopelev's memoirs are filled with trepidation that the army would become exactly like its enemy, while Suris noted in his diary that "the army is disintegrating" as soldiers became obsessed with looting.[123] Tvardovskii, reflected that "only now has it become clear how the Germans conducted themselves in our lands, when we act this way," deciding that "everything accompanying the occupation is almost inevitable . . . it is naïve to think that our occupation, even justified by the fact that it comes later, as an act of vengeance, could happen differently."[124] Soldiers hardened by war and thirsty for revenge had been given total power to take whatever they pleased, and many of them did just that.

So what were they taking? Trophies varied widely, from the immediately necessary or useful to the aspirational and curious. They could be tiny, like pornographic cards, or as large as motorcycles or furniture; exotic, such as a short-lived parrot (it repeated *Heil Hitler* until shot by soldiers), or utterly mundane, such as the spotlessly white chamber pot hanging off a soldier's knapsack.[125] Much of what soldiers took was what they needed at the moment. Boris Shumelishskii grabbed eyeglasses.[126] Tat'iana Atabek, a medic, took medical instruments.[127] A German woman forced to do the Red Army's laundry noted the creative reuse of an embroidered nightstand cover as a handkerchief.[128] One Uzbek soldier took a hunk of gold, which was fashioned into false teeth after the war.[129] By 1945 almost everyone describes eating mostly trophy food. Enemy warehouses fell into Soviet hands and led to soldiers eating a variety of luxury goods.[130] Abandoned farms provided a wealth of meat and vegetables.[131] Several observers noticed that they or those around them were gaining weight.[132] Some smoked cigars constantly, dragging on them like cigarettes and shocking the bourgeois folk who were used to enjoying them.[133]

Clothing was a particularly common trophy. Female soldiers in particular were glad to be able to find women's wear, including bras and slips, as the army provided only unisex underwear. Iakov Babenko described sharpening his image in March 1945: "they brought me two leather coats, one of which I gave to the division commander Volkovich, and I have to say I took off my wadded jacket and got a proper officer's look, because it is shameful for the commander of a regiment to walk around Germany in common soldiers' clothing."[134] Many soldiers sent or brought back fine European garments. Pregnant women gathered hard-to-find children's clothing.[135] Suits, cloth, and even enemy uniforms became common items in parcels and soldiers' suitcases after the war. A Komsomol organizer recorded taking a tuxedo as a trophy.[136] One report immediately after the war noted that demobilizing female soldiers were not only proud to have "raised their cultural level" during the war but were also "satisfied that they don't have to worry about clothing for now. Most of those going home have a lot of things, 50–100 kilograms [110–220 lbs.], many gifts; most have watches, some bicycles, radios, and other things."[137]

The Soviet Union had extensively censored the public sphere and taken many books out of circulation. This world of print included Russian-language editions of books long forbidden in the Soviet Union and texts that were outside Soviet orthodoxy, ranging from religion to politics to sex.[138] A soldier could bring these books home but not mail them. Many soldiers used trophy paper to write home or keep diaries, or simply burned or trampled them.[139] Books could be raw material or a highly valued commodity. It was all in the eye of the beholder.

The informal visual culture of the Red Army was fundamentally altered by trophies. Before 1944 cameras were a rarity in the hands of Red Army soldiers, being almost exclusively the purview of war correspondents and the occasional commander. In 1943 *Krasnaia zvezda* went so far as to run an article decrying the difficulties facing soldiers who wanted to have their image taken at the front.[140] This was in contrast to the US Army and the Wehrmacht, where many soldiers carried cameras and documented their experiences, also photographing their friends and exchanging snapshots. In the Red Army soldiers wishing to send their image home or trade keepsake images with their friends had to sit for studio photographers or talented soldiers who could sketch their likeness.[141] At the war's end Red Army soldiers began to acquire cameras in large numbers. Soldiers often posed with their trophies using the ruins of conquered enemy cities as a backdrop. Grigorii

Baklanov used such a photograph with his friends (in which he prominently displays a trophy SS dagger) to find them thirty years later via a newspaper.[142]

This sudden wealth lead to a culture of lavish gift giving. Babenko recorded a present he gave his commander: "After discussion, I, the chief of staff, and the division commander had dinner together, and I gave him a beautiful and well-crafted accordion as a sign of my love for him."[143] Immediately after the war he recorded gathering his officers together and using his capacity as a commanding officer to give them such trophy goods as motorcycles and bicycles.[144]

Soldiers appropriated all forms of conveyance, from bicycles to cars. Babenko could only give his men motorcycles, because orders forbade the awarding of automobiles without government approval.[145] Grigorii Baklanov recalled that soldiers young and old in his regiment became obsessed with bicycles, despite the difficulty of riding in their full kit. Eventually several soldiers became victims of nighttime accidents on military roads, and their commander gathered all the regiment's bicycles and ordered tractors to run them over.[146] As soon as trophies became dangerous, it was a commander's prerogative to liquidate them. Risks to the safety of local civilians were a much lower priority, as they too were often treated as trophies.

The sanctioning of parcels gave soldiers carte blanche to rob locals, especially Germans, but not all soldiers were interested in trophies. Aleksandr Lesin confided in his diary shortly after the war:

> The older soldiers are occupied with trophies. And what about the youth? We don't need any old German junk [*barakhlo*]. We find this to be beneath us. You see, we are busy with something else, that we discuss among ourselves but there is no way we would write about in our diaries . . . something in us was muted during the war . . . We waged war and it is entirely understandable, that all energy—physical and otherwise—was given to the cause. And now? The energy of youth roars in us, demands release. You see we ended the war not exhausted, but devilishly hungry.[147]

While soldiers with families were busy collecting parcels to send home, many younger soldiers sought sexual contact. Vladimir Stezhenskii confided to his diary: "Our Ivans continue their rampage, despite the most severe orders and warnings. Now in all of the areas we have occupied you can't find one house that hasn't been pillaged or one women aged fifteen to sixty who hasn't become acquainted with a Russian [*ne izvedavshikh russkogo*—euphemism for sex and given the context, probably rape]. And more than once."[148] For him and many others, sexual conquest and pillage were part of one process. Lesin's account does not imply that his encounters were nonconsensual, while Stezhenskii's is quite ambiguous. Other soldiers were explicit about the sexual violence that they witnessed, committed, or interrupted. Some tried to save unfortunates from their attackers; others saw a sharp distinction between what the Germans had done and they were doing.[149] One soldier explained to his Polish hostess: "With us, in Ukraine, the Germans raped young girls and then shot them; that's what they did. And our fools saw them and learned what to do, but they never shoot them, not even when they are drunk . . . That, you see, is the difference."[150] Grigorii Pomerants recalled that a political officer near Stalingrad consoled himself that while his wife was probably sleeping with a German: "It's no big deal. We'll go to Berlin and show those German girls!" Pomerants was shocked by the desire to imitate the enemy, but would live to see it: "At the end of the war the masses were overtaken by the idea that German women from fifteen to sixty years old were the lawful prize of the victors."[151] At least one victim of rape described herself in these terms: "we're nothing but booty, dirt."[152] As with looting, sexual violence had occurred on Soviet and neutral territory but was generally censured and never reached the same scale as on the territory of the Third Reich.[153]

Rape at the end of the war remains a highly controversial and vexing subject. The scale of rape in the Third Reich was massive, with the historian Mark Edele estimating 0.6 rapes per Red Army soldier in East Prussia, over a million cases of rape.[154] Historians have argued that the epidemic of sexual violence at the end of the war can be traced back to brutalization, indiscipline, the lack of a leave system in the Red Army, vengeance, male bonding among perpetrators, and the desire to both humiliate the male population and claim total domination over the country. There is also debate as to what extent these rapes were managed from above by the party or a visceral reaction from below. Sexual dominance is one of the most widespread and brutal means by which power relations are expressed. The greater the cultural difference between perpetrator and victim, the more brutal these encounters can be. By not punishing Red Army soldiers who raped German women, the government demonstrated its total mastery over the Third Reich, while in taking control of women's bodies, some Red Army soldiers gained a level of mastery over others after years of their own bodies being government property.[155]

A significant factor in these rapes and other crimes at the end of the war was alcohol.[156] Boris Slutskii noted that "many Europeans" came "to the conclusion . . . that a Russian person is good, as long as he is sober."[157] Alcohol became a problem at the end of the war, as soldiers gained near unlimited access to stores throughout East Central Europe and Germany. Nikulin claimed in his memoirs: "It was spring and the whole army was drunk. Alcohol was found everywhere in abundance, and we drank, drank, drank. At no other time in my entire life did I drink so much alcohol as in those two months! It might be that the war ended so quickly because we, besotted by wine, forgot about danger and asked for trouble. Explosions, bombardment, fire—and right there is a squeezebox and a drunken waltz."[158] Others recall acts of drunken rampage and measures taken to deal with the potential for alcohol to derail operations.[159] Nikolai Inozemtsev refused to take drunken soldiers with him, while noting that they had access to as much as they wanted to drink.[160] Babenko smashed bottles to keep his subordinates in line, while Tartakovskii wrote in his memoirs of the need to guard and control reserves of alcohol.[161]

In the war's final months a spirit of uncontrollable vengeance was in the air that Grigorii Pomerants came to associate with feathers: "Down—the sign of a pogrom, the sign of total free will, that makes you dizzy, that rapes, burns . . . Kill the German, and then take a German woman. There it is, the soldier's holiday of victory. And then place a bottle upside down!"[162] It is telling that the destroyed remnants of an object so closely associated with home and meshchanskii comfort became for Pomerants the symbol of his comrades' rage.

Destruction

The objects of bourgeois coziness often became the targets of soldiers' anger. Boris Komskii and his comrades awoke on the morning of January 24, 1945, in an East Prussian town. They had just crossed the border from Poland into the Third Reich the night before. Komskii recorded in his diary:

> In the whole village only two houses were burnt down; all the rest were intact. The houses, like all their property [*khoziaistvo*], are highly cultured, city-like. All their belongings were left behind with the exception of clothing and valuables. Furniture, household goods, grain, livestock, dishes, farm machinery, bicycles were all left in the yard. All of their hatred, all of their thirst for vengeance the fighting men instantly vented on these things. They started breaking windows, mirrors, records; they lit up stoves with beds, tables, and dressers.[163]

This is a participant's view of soldiers acting out their desire to "break and destroy everything" mentioned by Strukov at the beginning of the chapter. The cultured nature of their enemy's homes aroused a visceral response among men and women who had traveled thousands of kilometers of scorched earth to find an enemy whose peasants lived more richly than many Soviet bureaucrats. Lesin confided in his diary that "not a small number of our soldiers have become unbridled avengers . . . Many don't have a home, their home was burned down, their mothers have been shot, their sister or wife or fiancée has been shot. Try and immediately snuff out or even temper the vengeful hatred when a person is overtaken by the demonic hypostasis of an eye for an eye."[164] This desire for vengeance could justify the taking of enemy property, rape, and the destruction of anything belonging to the Germans, so that they might feel some of the emptiness that overtook many Red Army soldiers who had lost homes and families.

The destruction of property was also meaningful. The noblewoman Alexandra Orme, living near Budapest in the winter of 1944–1945, recalled that soldiers had not only burned all of the furniture they could find but also destroyed photo albums. She believed that the burning of furniture was not done out of spite but rather due to ignorance of its use and value, as familiar items were left untouched: "Kitchen stools were not burnt, although they most certainly must have been easier to chop up; but then the Russian knows what a kitchen stool is, he has them in his own home and so he respects them."[165] It did not occur to Orme that there could be a symbolic or political dimensions in the fact that those items not marked by meshchanstvo—that is, used by servants rather than masters—were left intact.

The clash of cultures that she witnessed involved not only destruction but scatological acts. Orme, her servants, and her family scrubbed their house furiously once Red Army soldiers left, but she was surprised to find that the W.C. had been left untouched. Instead soldiers had taken an old peasant wardrobe into the yard to use as a latrine, explaining: "We're a cultured nation, not pigs like you. We don't stool in the house we live in."[166] Aversion to foreignness could also lead to symbolic marking of territory. Lev Kopelev recalled that the men in his unit had agreed in advance to "mark their crossing of the border in the appropriate manner"—by peeing. This proved to be an emotional moment and men embraced each other afterward.[167] These incidents continued to the very heart of Germany. At the end of the war, Nikulin described the two ways in which soldiers marked a major symbol of the Reich:

> Many wrote their names on the Reichstag or thought it their duty to piss on its wall. An ocean was poured around the Reichstag. And the requisite smell . . . The best autograph that I saw was located, if memory serves me right, on the pedestal of the statue of the Great

> Elector. There was a bronze plaque with the genealogy of the great names of Germany: Goethe, Schiller, Moltke, Schliffen, and others. It was thickly crossed out with chalk, and lower was written: "I fucked all of you up! [*Ebal ia vas vsekh!*]—Sidorov." Everyone, from generals to soldiers, was moved, but the chalk was later erased and the priceless autograph lost to history.[168]

Sidorov's autograph—much like the memory of the looting, sexual violence, destruction, and urination on enemy property—had no place in postwar mythmaking and official memory.

Exposure and Mitigation

The policy of trophy taking could serve to erase the humiliating memory of 1941. Lev Kopelev imagined a soldier weighed down with trophies in 1945: "A recruit from Riazan, Orël, or somewhere near Moscow is riding around Germany, as if 1941, German trenches at the gates of Leningrad, and tanks in Khimki never happened, as if Stalingrad and a flag with a swastika on Mount Elbrus never happened."[169] Kopelev found this amnesia, made concrete by the previously hapless soldier now traveling Germany as a conqueror, disturbing. How could so many people forget the humiliating defeats of 1941 and 1942 and begin to conduct themselves in a manner similar to their enemy? However, he seemed to be a rarity among his comrades in expressing this sentiment.

Soviet soldiers' arrival in Europe and their subsequent actions led to three major consequences. The first was a fear that Soviet citizens had become uncontrollable. Several soldiers expressed trepidation about their comrades' lack of restraint. Nikolai Inozemtsev recorded a conversation in late January 1945:

> You know, I don't pity the Germans in the least; let them be shot and do with them what you will. We could never do anything comparable to what they did to us, because that was on the level of the government. But it is offensive that all of these rapes lower the dignity of the army as a whole and every Russian individually. And besides that, this inevitably brings with it the disintegration of discipline and lowers the combat effectiveness of the army. All of these unbridled animal instincts will be very difficult to stamp out.[170]

Immediately after the war measures were taken to reimpose harsher discipline and dramatically limit contact between Red Army soldiers and locals.[171]

The second consequence was that Soviet people had been exposed to the disease of meshchanstvo and all of its glittering objects. As Strukov mentioned, some soldiers saw the wealth of their foreign enemy as proof that the German system was more advanced. These doubts led to an intensification of propaganda proclaiming the preeminence of the Soviet system. Some articles took an economic tack, particularly in regard to Soviet agriculture and the kolkhoz, explaining that the Soviet Union could defeat the Third Reich, which had acquired all of the riches of Europe, only because of its superior, socialist means of organization.[172] This superiority often took on a spiritual dimension, as the great Russian authors of the past and present united in their hatred of meshchanstvo. An article in the army's main political magazine compared German and Russian literature: "It turns out that the ideal of meshchanskii comfort created by German literature attracted a horde of rapists and child killers, becoming ammunition for Hitlerism . . . Russian literature did not praise meshchanstvo, but on the contrary, unmasked and castigated it."[173] In the immediate aftermath of a policy that invited soldiers to covet and take the very objects of meshchanskii comfort, the Communist Party attempted to inculcate the sort of revulsion that Samoilov had felt in German homes. The objects associated with meshchanstvo were the objects of mundane domesticity, markedly feminine in a way that hinted at seduction and even Eve and the Fall.[174]

This fear of contamination had a physical manifestation as well. In 1944 a monthly medical exam was introduced for all personnel, as well as a medical exam before (rare) furloughs, demobilization, and returning to the front. The order warned that Germans had spread venereal disease on Soviet territory and that "on enemy territory currently occupied by the Red Army, prostitution is highly developed." Soldiers were told that German agents had left behind loose women both to infect soldiers and to gather information. The corrupting influence of some of the trophies of war required an immediate response, while others would lead to vigilance after the war.[175] The ability to take was everywhere accompanied by the dangers of contamination.

The government was deeply concerned about whether soldiers would return from Europe with "Athenian pride" or as "Decembrists" and "political westernizers." According to many observers, the wealth and ways of the West repelled as much as they attracted. Boris Slutskii, a perceptive witness to the effect of Europe on Red Army soldiers, noted that the sight of prostitutes at first fascinated him and his comrades, but that "the initial elation at the fact of free love quickly went away. This was the result of not only fear of infection and the price but also disdain for the very possibility of buying a human being." He and his comrades found other customs, even how people washed, quite repulsive. Soldiers exaggerated the pluses of Soviet life to foreigners, underlining the "justness of life in Russia." Ultimately, Slutskii (who traveled through Romania, Bulgaria, Yugoslavia, and into Austria) noted by the end of the war that the "traditional esteem for foreign things" had been undermined.[176]

This led to the third major result of the Red Army going abroad: a realization that Soviet people were fundamentally different from those of the bourgeois world, an epiphany that often played out through things. We have already seen how pornography and other objects could repel soldiers. This repulsion could often unite the very diverse cadres of the Red Army. As Boris Slutskii recalled: "The campaign abroad united all the nations" serving in the Red Army.[177] Uzbeks and Russians could see more in common with each other than with the people of the bourgeois world. To foreigners "Russian" and "Soviet" were interchangeable, and soldiers on both sides noticed that "Russian" and German soldiers even smelled different.[178] Exposure to foreign lifestyles and manners convinced many soldiers that the Soviet system was superior. As Pëtr Pustovoit wrote in his diary in September 1944: "View on Europe. Much is new here, and the more I encounter it, the more I value everything kindred [*rodnoe*] to me, everything Russian."[179] At the end of the war, many Red Army soldiers felt like tourists who were very much ready to go home.[180]

Back in the USSR

Red Army soldiers returned to a country destroyed by war with whatever trophies they could acquire. Some lost their trophies along the way, others traded them at flea markets that had become part of the landscape during the war and filled with trophy goods in the war's aftermath. Many of them returned to their families with previously unattainable objects. But as these soldiers came home, the rhetoric around trophies changed. An article singing the praises of the "Great Russian People" reminded soldiers that "hundreds and hundreds of times Russians have built on ruins—others would not be able to bear this."[181] It went on to highlight the many selfless acts of Stakhanovites and ordinary Soviet people in creating wealth not for themselves but rather for the greater collective and government. These sentiments cast the policies of trophy taking in a shameful, alien light.

Many soldiers stayed for years in occupation forces in Europe, where they witnessed the separation of local civilians from soldiers and the institution of strict controls over soldiers' behavior. These controls included trophies. As the war concluded and soldiers were demobilized, trophies again became the exclusive domain of the government. The military committees of an army or front were permitted to award demobilizing soldiers a variety of items, including cameras, radios, bicycles, and watches. Soldiers could purchase cloth, clothing, dishes, and a variety of other household goods for themselves and their families at fixed prices from military stores. All demobilizing soldiers were given a sizable bonus on leaving the service, with which they could purchase these goods.[182]

Trophies played a central part in crowning the Soviet Union's victory. During the victory parade in June 1945, two hundred captured enemy banners were paraded and flung in a pile at Stalin's feet as he stood on the tribune at Red Square.[183] Shortly after the war a series of Soviet films were made on captured color film stock, including *The Fall of Berlin*, which told the official Stalinist version of the war (without looting). In addition to trophies shot in the Soviet Union, captured trophy films, from Hollywood as well as the Third Reich, were also shown both in the Soviet Union and to troops awaiting demobilization in Europe.[184] Joseph Brodsky called these films "the greatest spoils of war" and claimed that they had a tremendous impact on him as a youth.[185] In the wake of the influx of non-Soviet goods and films, the emerging Cold War reified the danger of all things foreign.

The postwar period saw a rise in xenophobia as the Soviet Union shut its borders more tightly. A campaign against "rootless cosmopolitanism" in favor of Soviet patriotism was launched in the early days of the Cold War, making it dangerous to praise and appreciate all things foreign. However, all of this did not stop what historian Stephen Lovell has referred to as "trophy westernization," the continued obsession with Western goods that led many Soviet citizens to seek out foreign commodities.[186]

People after the war looked very different from those before the war. Postwar fashion was a strange combination of women wearing the trophy fineries of Europe and men who often wore their uniforms for lack of other clothing. Women gained access to European fashion via both the trophies that they or their husbands sent and books and magazines that could now be purchased. This tendency, which began with the annexation of the Baltic states and Western Ukraine (L'vov and Riga became centers of fashion in the year before the war), was cemented by the influx of clothing and material from East Central Europe. Expensive fabrics were briefly attainable, and even the way women used makeup changed.[187] The sartorial impact of the war was unmistakable, as men in uniform escorted women who resembled European fashion plates.

Certain types of trophies, particularly weapons and radios, did not draw much attention during the war but could become a threat afterward. Pistols and knives were ambivalent trophies during the war. Once their owners returned to civilian life, however, weapons became a major problem, as crime rates rose and trophy pistols could land their owners in prison.[188] Nonetheless, many demobilized soldiers felt uncomfortable without weapons and continued to carry them.[189] Radios could be the source of a more existential danger; they could be used to listen to Western broadcasts during the Cold War, offering not only returning veterans but later their children a window onto forbidden worlds.[190]

The very act of trophy collecting was used to discredit some highly ranked veterans, most notably Georgii Zhukov, the marshal most associated with the victory who had commanded the Moscow victory parade. A set of accusations concerning his excessive trophy taking led to the confiscation of his trophies, loss of his post, and near expulsion from the Communist Party. Described as "money-grubbing" and "clearly criminal," Zhukov stood accused of stealing 70 pieces of gold jewelry, 740 pieces of silverware, 30 kilograms of silver, 50 rugs, 320 pelts, 60 paintings, 3,700 meters of silk and other items.[191] Many but not all of these items were later confiscated, highlighting the ambiguity of how much of the spoils should be left to the individual victor and how much to the government.[192] The desire to have the trophies of war could be used against the victors, whether they were highly positioned or of low standing. By the late 1940s these glittering objects could be dangerous, and the story of their acquisition had no place in the mythology of the war.[193]

Every belligerent in World War II targeted property and civilians, each in its own way. The Wehrmacht waged a war of annihilation to steal, enslave, and clear territory for settlement, both bombing enemy civilians and engaging in face-to-face acts of brutality. The British and Americans used advanced technology, distancing themselves from what they wrought. The Royal Air Force and US Air Corps both eventually pursued incendiary bombing of cities as a way to crush the economy and morale of their foes, consciously targeting civilians and their homes.[194] The Red Army brought its Marxist-Leninist-Stalinist worldview to bear on enemy civilians, complete with ideas of class warfare waged through expropriation. This made looting an act of social justice that highlighted the criminality of those who were expropriated. Every belligerent made claims of vengeance, and the Red Army's were clearly valid, even if many felt things went too far. After the war Allied and German means of waging war could be censured as cruel and cowardly, while the erasure of expropriation at the war's end left the Red Army a spotless liberator, sacrificing Soviet lives to free the peoples of Europe.

The control and marshaling of resources had been key to the Soviet victory in the Great Patriotic War, yet trophies existed in a gray zone. Trophies lay at the conjuncture of several ambiguities: attraction and repulsion, symbolic and functional, state and personal; and even when policy concerning trophies was clear, it was often subverted or subject to sudden change. The government remained the master trophy taker, seizing factories, minerals, Nazi scientists, and populations. It also choreographed ostentatious displays of its trophies. But there was space for soldiers to acquire much of what the Soviet economy could not provide to the masses—the fineries of meshchanskii comfort. Soldiers were left to fend for themselves at many points in the war, but in the conflict's final months Soviet leadership looked the other way and even provided an infrastructure for soldiers to help themselves to whatever they desired. The government in turn could take back whatever it wanted, whenever it wanted, as the case of Zhukov shows. Just as soldiers' material culture embodied the relationship between the government and soldier, so too did trophies embody the dynamics of repulsion and attraction to all things Western and "meshchanskii."

Greed was not something that was acceptable under socialism. And it was precisely this aversion to greed, which had roots in both Bolshevik ideology and peasant tradition, that could make the wealth soldiers saw abroad both revolting and attractive. At the end of the war, standing in the ruins of a rich apartment in East Prussia or Berlin, few Red Army soldiers doubted either the moral or material superiority of their ideology and their people. Meshchanstvo had been trampled into the dust, while the objects it had given birth to, whether standards emblazoned with the Swastika or bolts of silk, were on their way to the Soviet Union in parcels carefully packed by the victors.

Conclusion

Subjects and Objects

This has been the story of a diverse collection of individuals coming together around a set of objects, learning to use them and winning a war of epic scale. Our narrative has alternated between the personal and the grandiose, and it may be worthwhile to revisit the socialist framework that undergirded Soviet society. Marxism-Leninism-Stalinism divided the world into objective and subjective phenomena. *History* was seen as a series of massive processes that were objective and inexorable. The transition from capitalism to communism would happen just as the transition from feudalism to capitalism had happened, regardless of what any one individual or group tried to do about it.[1] How these massive changes were experienced by individuals was simply subjective, a mere byproduct of these processes. The war served both to highlight and to undermine this logic of history: Soviet victory had always been touted as inevitable, but what individual soldiers did seemed to matter very much in its outcome. Heroes had been lauded and even retrieved from the dustbin of history to inspire soldiers. Apparently individuals did matter very much in Soviet society, but the enemy had lost any trace of individuality.

The strange behavior of the enemy seemed to support a Marxist viewpoint. Whoever the Germans had been as *subjective* individuals, as a collective they were *objectively* dangerous. The Germans came to be seen as a nation that embodied the dying bourgeois *Geist*. They were obsessed with accumulation to such an extent that they had seen fit to rob and murder their neighbors. As Red Army soldiers came to see the incredible wealth of the Third Reich's citizens, it was impossible to disassociate this wealth from the murder and destruction that Soviet citizens had witnessed. The "objective" process of the violent death throes of capitalism seemed to align very closely with the "subjective" emotions that the experience of war had aroused. Moreover, the Soviet system had produced the items and fostered the citizens that had defeated the Third Reich. Whether they saw the experience of the war in Marxist terms or not, soldiers emerged from the war as dramatically different people.

In place of a Marxist narrative of "objective" and "subjective" experience, this has been a story of *objects* and *subjects*—things and people. Specific subjectivities, or senses of self, formed around the use of specific objects, as soldiers came to identify themselves with the weapons they wielded, invest meaning in rituals both mundane (e.g., eating together) and extraordinary (e.g., "Meetings of Vengeance"), and inhabit the idiosyncratic ranks of the army and its cities of earth.

The calling together of the Soviet people to defend the state succeeded for several reasons. A Nazi victory would have been apocalyptic, as the cities of rubble attested; the physical survival of soldiers and their families was contingent on victory. The Soviet government excelled at crafting objects that were utilitarian, cheap to manufacture, and easy to use. They could clothe, feed, and arm an army of such tremendous scale, despite incredible loss of territory and resources, because they did it on the cheap. Everything, from *portianki*, mess tins, tunics, thing-bags, and shovels to submachine guns and tanks, was made as simply and economically as possible. The only exceptions to this rule of practicality—the introduction of scores of medals in precious materials and the shift to a new uniform in the midst of war—were perceived to have had such an important morale component that they outweighed purely materialist concerns.

Ironically, this victory was achieved in part by mobilizing what had been seen as "backward" to fight a highly mechanized war. The professional army that emerged in 1943 drew inspiration from both a Bolshevik-revolutionary and ancient past, and its material culture reflected that, wiping away visual traces of the shame of 1941. The items a soldier carried were animated more often than not by peasant bodies and seemed to have something of a peasant soul themselves, many items being remarkably simple, humble to look at, born of improvisation and frequently just militarized peasant gear. This was in contradistinction to the soulless, mechanical and malevolent material world of the enemy. Defeating such a vile enemy had elevated the status of both Soviet citizens and the Soviet government.

A September 1944 article for political officers, "Friendship, Sealed in Blood," spoke of the changes that had occurred in Soviet society in the course of the war. Two of the major changes it noted touched on how soldiers saw themselves and how they saw their country. According to the author, himself a Georgian: "A Soviet warrior of any nationality proudly calls himself a Russian soldier. The word 'Russian' has become a symbol of friendship, strength, courage, victory." The uniforms and medals soldiers came to wear had an increasingly Russian flare, slippages between Russian and Soviet became more and more prevalent. Abroad, Red Army soldiers were perceived as "Russian," regardless of nationality. But even before they went abroad, soldiers' senses of space and self were changing: "The war has immensely expanded the conception of the Motherland among Soviet people. Until now, many of those who are now defending the Soviet land with weapon in hand had only heard and read about the vastness, wealth, and diversity of our Fatherland. At war they have seen with their own eyes and measured with their own soldierly tread the unimaginable expanses of the Soviet Motherland."[2] In the decade before the war, many members of the peasant majority would have had few opportunities to leave their collective farms and explore the vast expanses of their country.[3] It is virtually impossible to imagine how else, other than in the army (or perhaps the expansive Gulag system) an Uzbek cotton picker and a Ukrainian swineherd could come to live together, criminals and intelligentsia could become friends, and hundreds of thousands of déclassé people could be ushered into the Soviet fold. It was only in uniform that the journey around the Soviet Union and into foreign territory was sanctioned.

The war transformed people. A Leningrad confectioner passed through the hell of the Nevskii Bridgehead. A Tatar poet went from the Gulag to command a mortar platoon, dying from wounds in East Prussia. A forty-year-old Russian peasant who made *valenki* for a living became a machine gunner. One Jewish student became an artillery observer, another a machine gunner. The young Muscovite grandson of an imperial army general lost an eye as a tanker. A disgraced NKVD officer was released from prison to fight, crippled, but never deemed worthy of rehabilitation or a pension. The son of an Armenian and a Georgian who had been repressed in the Great Terror defended the state that had destroyed his family. Russian teenage girls became snipers. The son of a dekulakized Kazakh herdsman led troops into Berlin, helping to raise the Red Banner over the Reichstag. A young Jewish paratrooper turned sculptor would both get in a shouting match with Nikita Khrushchev and design the premier's tomb.[4]

In the ranks of the Red Army, these diverse people would be turned into a united fighting force that ultimately won the largest war in history. Future celebrities—including the most popular bard of the 1960s (Bulat Okudzhava) and the most famous Soviet clown (Yuri Nikulin)—and a host of anonymous collective farmers, factory workers, and clerks had all worn the same uniform, eaten from the same pot, slept in the same bunkers, fired the same guns, and could take similar trophies. Millions of small-scale changes had a dramatic impact on Soviet society.

Years after the war, David Samoilov reflected on the transformation the war had brought to the Soviet people:

> It was precisely at this moment that the nation experienced a sharp period of its development, when the makeup of the people changed dramatically. We still think very little about the role the war played in accelerating the process that we call urbanization.
>
> The exit from the stage of history of the folk-*muzhik* became the grand finale of the peasant tragedy.[5]

Many of those who exited "the stage of history"—whether they were peasants, workers, or members of the intelligentsia—found their final resting place in the trenches they had dug. Both those who perished and those who survived were subjected to a specific form of urbanization and modernization in the ranks of the Red Army. Making peasants more urban had been part of the Soviet project from the very beginning. The army provided an urban space in which to mold soldiers, exposing them to propaganda, the wider world of the Soviet Union, and even foreign capitals. Many veterans from the countryside would use demobilization and the opportunities presented by postwar construction to move permanently to cities.[6]

The Soviet process of urbanization, was, in practice, the process of yesterday's peasants adapting to the vicissitudes of cities. This meant learning to deal with shortages and improvising means of survival, sometimes adapting peasant ways to a new milieu, sometimes adopting new behaviors. The army repeated these processes under conditions of extreme surveillance and danger. The government claimed total control over its soldiers, yet time and again those soldiers found ways around these claims: whether ignoring censorship, throwing away equipment, deserting, pilfering rations, or otherwise subverting attempts to control them. What they did with government issue items demonstrated a great deal of agency. The state was constantly reacting to the ways in which soldiers and commanders undermined its authority, often forced to compromise and acknowledge the limits of its power. This compromise may even have been built into the system.[7] Yet the constant exposure to official propaganda, which soldiers listened to together and often co-authored, colored soldiers' perceptions and offered an idiom that celebrated them as heroes. Much of what they had seen undermined Soviet claims, but the horrors the enemy had committed, the end result of the war, and the sacrifices Red Army soldiers had made created significant overlaps between how the government and soldiers understood the cataclysm they had weathered. As they returned home, soldiers couldn't help but speak a language that borrowed heavily from official Soviet discourse. The experience of the war could unify and standardize as no prior event.

Like immigrants to a new city, the multitudes of people who fought in the war had rubbed shoulders with people from every walk of Soviet life, often living intimately with people very different from themselves in age, nationality, class, and experience. This contributed to an antielitism that infused the memoirs of those who survived: even encounters with those who could not speak the lingua franca or whose "backward" customs seemed bizarre were often described in terms of fascination rather than condemnation. After all, they were reduced to the same set of objects, their everyday lives made uniform in the most literal sense. For those who fought in the war, the massive mobilization that they had participated in was often described as both their and the Soviet Union's finest hour. Yet for years after the war many soldiers refused to talk about it. As Anatolii Genatulin wrote: "It wouldn't come to anyone's mind to boast that they fought in the war . . . Men talked about anything except the war . . . It was not so much a forbidden topic, but something too fresh. Neither the pain of it had been forgotten, nor the fear of death, the bitterness and the curse."[8] The experience of the war and the transformations it brought about were so ubiquitous as to go unspoken by many, but eventually inspired massive outpourings of text by veterans turned authors, including Genatulin, as veterans tried to come to grips with the massive war they had waged as it faded into the past.

For the first three decades of Soviet power, mobilization had been the ethos that kept the project going, promising a brighter future for sacrifices today. Under Khrushchev, mobilization began to give way to promises of abundance sooner, rather than later.[9] What had been always on the horizon was now supposed to be just around the bend. If those who fought in the war were somewhat fearful of the corrupting nature of foreign goods and prepared to live humbly, but in peace, their children—fascinated by the items veterans had brought back and frustrated that the people in countries that had fought against the Soviet Union (East Germany, Hungary, Romania) seemed richer than Soviet citizens—demanded that the government provide material comfort and plenty. During the Cold War, it became increasingly obvious that the United States did not present the same sort of existential threat as the Third Reich and that the Soviet system was not able to provide as well for its citizens as capitalism, or even neighboring socialist states.[10] Taking these territories into Soviet orbit became dangerous precisely because of this perceived wealth, as Soviet citizens came to covet imported products from the Eastern Bloc. Once the state ceased to promise existential and spiritual change with any credibility, it was left only with *stuff*.[11] The Soviet planned economy could produce for defense but had neither the necessary flexibility to adapt, nor much of an interest in pandering, to the everyday desires that drive a market. The Soviet government could make a great tank, overcoat, or thermos but was less capable of making perfume, producing stylish coats, or providing fresh fruit out of season. Coveting fashion and frivolous comfort seemed dangerously close to meshchanstvo, the source of evil that the Red Army had defeated.[12] In this regard the war could be seen as serving a double role in securing Soviet legitimacy: the privations of the current generation paled in comparison to those suffered during the war, while the obsessive drive to consume on the part of the enemy (whether German or American) was part of what made them so different from—and ultimately morally inferior to—Soviet people.[13]

Pride in the victory and fear of the destructive power of war were key to instilling a sense of prestige and purpose in Soviet citizens, even as the everyday objects that had been the war's substance began to collect dust. Sometimes anonymously, sometimes as named objects connected to a personal story, what had been the tools of war took their place behind glass in homes and museums around the Soviet Union. Others were repurposed, such as a soldier's overcoat reworked into a New Year's decoration.[14] Still later, these items—from medals to tunics to rifles—became the purview of collectors and dealers, transforming into commodities. Battlefields to this day offer up the material of the war, from tanks to spoons to human remains, providing a less well-preserved but more visceral record of the war and those who did not live to see the full transformation of those in the ranks or the collapse of the government they defended.

Notes

Prelude

1. Daniil Granin, "Prekrasnaia Uta," in his *Nash kombat: Sbornik* (Moscow: AST-VZOI, 2004), 218–20.

2. Kharis Iakupov, *Frontovye zarisovki: Zapiski khudozhnika* (Kazan': Tatknigizdat, 1981), 26.

3. Tat'iana Reutova, *Gvardeets. Uchenyi. Diplomat: Frontovye dnevniki akademika Reutova* (Moscow: Zvonnitsa-MG, 2011) (hereafter Reutov), 43, 45, 49 (quotation), 51, 52, 54, 57, 72, 78 again in the spring of 1942, 109–10.

4. Reutov, *Gvardeets*, 53–54.

5. N. N. Inozemtsev, *Frontovoi dnevnik* (Moscow: Nauka, 2005), 64.

6. Reutov, *Gvardeets*, 56.

7. Iakupov, *Frontovye zarisovki*, 63.

8. Inozemtsev, *Frontovoi dnevnik*, 172. Indeed, all of the men cited here had truly exceptional experiences in that they served from 1941 on and survived the war.

9. Yuri Slezkine, *The House of Government: A Saga of the Russian Revolution* (Princeton: Princeton University Press, 2017), 606.

Introduction

1. Nikolai Chekhovich, *Dnevnik ofitsera: Pis'ma leitenanta Nikolaia Chekhovicha k materi i neveste* (Moscow: Molodaia gvardiia, 1945), 5.

2. Ibid., 3, 96.

3. On learning Russian in the Red Army, see Sarsen Amanzholov, *Opyt politiko-vospitatel'noi raboty v deistvuiushchei armii* (Ust'-Kamenogorsk: Reklamnyi daidzhest, 2010), 24–25. For a more complete discussion of the "non-Russians," see Brandon Schechter, "'The People's Instructions': Indigenizing the Great Patriotic War among 'Non-Russians,'" *Ab Imperio*, no. 3 (2012): 109–33.

4. *Posobie dlia boitsa-tankista* (Moscow: Voenizdat, 1941), 39. The system of passwords was standardized so the challenge would be a piece of military equipment, and the answer a geographical location, both starting with the same letter (e.g., "Tank-Tiumen'").

5. N. Rubinshtein, "Zametki ob agitatsii," *Agitator i propagandist Krasnoi Armii*, no. 14 (1942): 43.

6. Boris Suris, *Frontovoi dnevnik: Dnevnik, rasskazy* (Moscow: Tsentrpoligraf, 2010), 215.

7. A note on terminology: Throughout this book, I use the terms "government" and "state" more or less interchangeably, using "government" more often as it implies both institutions and the people who constitute them. This decision stems from idiosyncrasies inherent to the Soviet project. The typical juxtaposition of "state" as a permanent institution and "government" as the people currently in power is not functional for several reasons. First, the Soviet project consciously tried to efface any distinction between state, society, and government. There was no way to imagine a change of government without the fall of the state. The Soviet Union was a one-party dictatorship in which individuals serving in administrative posts can and did change, but all belonged to one overarching political organization. Second, the country was run in such a way that one person could hold a confusing and overlapping set of positions, some within the party, some at various institutions, others in the government or army. For example, Aleksandr Shcherbakov simultaneously ran the Communist Party organization of Moscow, the Main Political Directorate of the Red Army (becoming a general), and the Soviet Information Bureau, while also serving on the Politburo and in the Supreme Soviet. Since the party and the government were formally separate, and power and competencies often doled out in informal ways, I have decided to spare the reader a prolonged discussion of how Soviet leadership functioned and decisions were made. Instead, how those decisions were implemented and perceived by soldiers is at the center of this book. As such, I will be loose with my use of "state" and "government"—for our purposes, there really is no functional difference between them.

8. A process mirrored by the experience of many of their loved ones in the rear, who engaged in similar mixing due to the evacuation of civilians.

9. Citizenship need not imply liberal democracy, as Yanni Kotsonis has argued: "citizenship could be understood as forced participation in the construction of an integrated, scientifically understood society rather than simply the conferral of rights" ("Introduction: A Modern Paradox—Subject and Citizen in Nineteenth- and Twentieth-Century Russia," in *Russian Modernity: Politics, Knowledge and Practices, 1800–1950*, ed. David Hoffmann and Yanni Kotsonis [New York: Palgrave, 2000], 9).

10. Bruno Latour, *Reassembling the Social: An Introduction to Actor-Network Theory* (New York: Oxford University Press, 2005), 68.

11. See, e.g., Leora Auslander, *Cultural Revolutions: Everyday Life and Politics in Britain, North America, and France* (Berkeley: University of California Press, 2009), 20.

12. For a thorough discussion of modernity in Soviet historiography, see Michael David-Fox, *Crossing Borders: Modernity, Ideology, and Culture in Russia and the Soviet Union* (Pittsburgh: Pittsburgh University Press, 2015), 1–71. Another key aspect is unrootedness and adaptability. See Yuri Slezkine, *The Jewish Century* (Princeton: Princeton University Press, 2004).

13. Norbert Elias, *The Civilizing Process: Sociogenetic and Psychogenetic Investigations* (New York: Urizen Books, 1978); Kotsonis, "Introduction," 1–16; Vadim Volkov, "The Concept of *Kul'turnost'*: Notes on the Stalinist Civilizing Process," in *Stalinism: New Directions*, ed. Sheila Fitzpatrick (New York: Routledge, 2000), 216; Stephen Kotkin, *Magnetic Mountain: Stalinism as Civilization* (Berkeley: University of California Press, 1995), 356.

14. Marx famously declared that "being determines consciousness" (*byt'ë opredeliaet soznanie* in Russian). See Karl Marks [Marx], *K kritike politicheskoi ekonomii*

(Moscow: Goslitizdat, 1949), 7. For an excellent account of this process in practice, see Emma Widdis, *Socialist Senses: Film, Feeling, and the Soviet Subject, 1917–1940* (Bloomington: Indiana University Press, 2017).

15. Sheila Fitzpatrick, *Everyday Stalinism: Ordinary Life in Extraordinary Times. Soviet Russia in the 1930s* (New York: Oxford University Press, 1999), 10–13.

16. Volkov, "Concept of *Kul'turnost',*" 217–21.

17. On kulak markers, see Lynne Viola, *Peasant Rebels under Stalin: Collectivization and the Culture of Peasant Resistance* (New York: Oxford University Press, 1996), 35.

18. Volkov, "Concept of *Kul'turnost',*" 212, 217–18, 220–25.

19. Kotkin, *Magnetic Mountain*, 246.

20. For a discussion of the centrality of the peasantry to the Soviet experience, posited as "the rural nexus," see Moshe Lewin, *The Making of the Soviet System: Essays in the Social History of Interwar Russia* (New York: Pantheon, 1985), 11–18. On hostility to the peasantry, see Viola, *Peasant Rebels under Stalin,* especially chapter 1.

21. Laura Engelstein. "Morality and the Wooden Spoon: Russian Doctors View Syphilis, Social Class, and Sexual Behavior, 1890–1905," *Representations* 14 (1986): 169–208.

22. A. T. Tvardovskii, *Vasilii Tërkin: Kniga pro boitsa* (Moscow: Detskaia literatura, 1977).

23. Slezkine, *House of Government*, 957; Amir Weiner, "Robust Revolution to Retiring Revolution: The Life Cycle of the Soviet Revolution, 1945–1968," *Slavonic and East European Review* 86 (2008): 208–31.

24. Amir Weiner, *Making Sense of War: The Second World War and the Fate of the Bolshevik Revolution* (Princeton: Princeton University Press, 2001).

25. *Pis'ma s fronta* (Alma-Ata: KazOGIZ, 1944), 84.

26. Grigorii Pomerants, *Zapiski gadkogo utënka* (Moscow: Tsentr gumanitarnykh initsiativ, 2012), 126.

27. Yuri Slezkine, "The USSR as a Communal Apartment, or How a Socialist State Promoted Ethnic Particularism," *Slavic Review* 53 (1994): 414–52; Terry Martin, *The Affirmative Action Empire: Nations and Nationalism in the Soviet Union, 1923–1939* (Ithaca: Cornell University Press, 2001); Francine Hirsch, *Empire of Nations: Ethnographic Knowledge and the Making of the Soviet Union* (Ithaca: Cornell University Press, 2005).

28. David Brandenberger, *National Bolshevism: Stalinist Mass Culture and the Formation of Modern Russian National Identity, 1931–1956* (Cambridge, MA: Harvard University Press, 2002); David Brandenberger, *Propaganda State in Crisis: Soviet Ideology, Indoctrination, and Terror under Stalin, 1927–1941* (New Haven: Yale University Press, 2011).

29. Kotkin, *Magnetic Mountain*; Fitzpatrick, *Everyday Stalinism*.

30. Peter Holquist, "'Information Is the Alpha and Omega of Our Work': Bolshevik Surveillance in Its Pan-European Context," *Journal of Modern History* 69 (1997): 415–50; Kate Brown, *A Biography of No Place: From Ethnic Borderland to Soviet Heartland* (Cambridge, MA: Harvard University Press, 2004).

31. Jochen Hellbeck, *Revolution on My Mind: Writing a Diary under Stalin* (Cambridge, MA: Harvard University Press, 2006); David Hoffman, *Stalinist Values: The Cultural Norms of Soviet Modernity, 1917–1941* (Ithaca: Cornell University Press, 2003); Anna Krylova, "The Tenacious Liberal Subject in Soviet Studies," *Kritika: Explorations in Russian and Eurasian History* 1 (2000): 119–46.

32. Sheila Fitzpatrick, *Stalin's Peasants: Resistance and Survival in the Russian Village after Collectivization* (New York: Oxford University Press, 1994); S. Chuikina, *Dvorianskaia pamiat': "Byvshie" v sovetskom gorode (Leningrad, 1920–30-e gody)* (St. Petersburg: Izd-vo Evropeiskogo universiteta v Sankt-Peterburge, 2006).

33. Mark von Hagen, *Soldiers in the Proletarian Dictatorship: The Red Army and the Soviet Socialist State, 1917–1930* (Ithaca: Cornell University Press, 1990); Joshua A. Sanborn, *Drafting the Russian Nation: Military Conscription, Total War, and Mass Politics, 1905–1925* (DeKalb: Northern Illinois University Press, 2003).

34. Peter Holquist, *Making War, Forging Revolution: Russia's Continuum of Crisis, 1914–1921* (Cambridge, MA: Harvard University Press, 2002).

35. David M. Glantz, *Stumbling Colossus: The Red Army on the Eve of World War* (Lawrence: University Press of Kansas, 1998); David M. Glantz and Jonathan M. House, *When Titans Clashed: How the Red Army Stopped Hitler* (Lawrence: University Press of Kansas, 1995); David M. Glantz, *Colossus Reborn: The Red Army at War, 1941–1943* (Lawrence: University Press of Kansas, 2005); Roger R. Reese, *Stalin's Reluctant Soldiers: A Social History of the Red Army, 1925–1941* (Lawrence: University of Kansas Press, 1996); Roger R. Reese. *Why Stalin's Soldiers Fought: The Red Army's Military Effectiveness in World War II* (Lawrence: University of Kansas Press, 2011); Walter S. Dunn Jr., *Hitler's Nemesis: The Red Army, 1930–1945* (Mechanicsburg: Stackpole Books, 2009).

36. E. S. Seniavskaia, *1941–1945: Frontovoe pokolenie. Istoriko-psikhologicheskoe issledovanie* (Moscow: IRI RAN, 1995); E. S. Seniavskaia, *Psikhologiia voiny v XX veke: Istoricheskii opyt Rossii* (Moscow: ROSSPEN, 1999); A. E. Larionov, *Frontovaia povsednevnost' Velikoi Otechestvennoi voiny: Upravlenie, organizatsiia i material'nye usloviia zhizni RKKA v 1941–1945 gg.* (Moscow: Zolotoe sechenie, 2015).

37. Weiner, *Making Sense of War*.

38. Mark Edele, *Soviet Veterans of the Second World War* (New York: Oxford University Press, 2008); Mark Edele, "'What Are We Fighting For?': Loyalty in the Soviet War Effort, 1941–1945," *International Labor and Working-Class History* (2013): 84, 248–68; and other works cited later in the text.

39. Filip Slaveski, *The Soviet Occupation of Germany: Hunger, Mass Violence, and the Struggle for Peace, 1945–1947* (New York: Cambridge University Press, 2013); Robert Dale, *Demobilized Veterans in Late Stalinist Leningrad: Soldiers to Civilians* (New York: Bloomsbury Academic, 2015).

40. Anna Krylova, *Soviet Women in Combat: A History of Violence on the Eastern Front* (New York: Cambridge University Press, 2010); Roger D. Markwick and Euridice Charon Cardona, *Soviet Women on the Frontline in the Second World War* (New York: Palgrave, 2012); Amandine Regamey, "Falsehood in the War in Ukraine: The Legend of Women Snipers," *Journal of Power Institutions of Post-Soviet Societies* 17 (2015), https://pipss.revues.org/4222. Roger Reese also dedicated a chapter to this topic in *Why Stalin's Soldiers Fought*. Svetlana Alexievich was among the first authors to deal with women's experience of the war (*The Unwomanly Face of War: An Oral History of Women in World War II*, trans. Richard Pevear and Larissa Volokhonsky [New York: Random House, 2017], from *U voini—ne zhenskoe litso . . .* [Minsk: Mastatskaia literatura, 1985]).

41. Oleg Budnitskii, "The Intelligentsia Meets the Enemy: Educated Officers in Defeated Germany," trans. Susan Rupp, *Kritika: Explorations in Russian and Eurasian History* 10 (2009): 629–82; Oleg Budnitskii, "The Great Patriotic War and Soviet

Society: Defeatism, 1941–42," trans. Jason Morton, *Kritika: Explorations in Russian and Eurasian History* 15 (2014): 767–97; O. V. Budnitskii, "Muzhchiny i zhenshchiny v Krasnoi Armii (1941–1945)," *Cahiers du monde russe* 52 (2011): 405–22. He has also been instrumental in bringing to light and publishing several diaries.

42. Catherine Merridale, *Ivan's War: Life and Death in the Red Army, 1939–1945* (New York: Metropolitan, 2006).

43. Leora Auslander, "Beyond Words," *American Historical Review* 110 (2005): 1015–45; T. H. Breen, *Marketplace of Revolution: How Consumer Politics Shaped American Independence* (New York: Oxford University Press, 2004); Auslander, *Cultural Revolutions*; Lizabeth Cohen, *Making a New Deal* (New York: Cambridge University Press, 2008); Lizabeth Cohen, *A Consumer's Republic* (New York: Knopf, 2003); Benedict Anderson, *Imagined Communities: Reflections on the Origin and Spread of Nationalism* (London: Verso, 2006).

44. Some notable exceptions are Vadim Volkov, cited above; Lewis H. Siegelbaum, *Cars for Comrades: The Life of the Soviet Automobile* (Ithaca: Cornell University Press, 2008); Julie Hessler, *A Social History of Soviet Trade: Trade Policy, Retail Practices, and Consumption, 1917–1953* (Princeton: Princeton University Press, 2004); Jukka Gronow, *Caviar with Champagne: Common Luxury and the Ideals of the Good Life in Stalin's Russia* (New York: Berg, 2003); V. V. Lapin, *Peterburg: Zapakhi i zvuki* (St. Petersburg: Evropeiskii dom, 2007); and Kate Brown, *Plutopia: Nuclear Families, Atomic Cities, and the Great Soviet and American Plutonium Disasters* (New York: Oxford University Press, 2013).

45. Vera S. Dunham, *In Stalin's Time: Middleclass Values in Soviet Fiction*, enl. and updated ed. (Durham: Duke University Press, 1990); Alexei Yurchak, *Everything Was Forever until It Was No More* (Princeton: Princeton University Press, 2006); Olga Shevchenko, *Crisis and Everyday Life in Postsocialist Moscow* (Bloomington: Indiana University Press, 2009); Sergei Oushakine, *Patriotism of Despair: Nation, War, and Loss in Russia* (Ithaca: Cornell University Press, 2008); Widdis, *Socialist Senses*.

46. Yurchak, *Everything Was Forever*, 39–44.

47. For a critique of the Mints Commission, see Oleg Budnitskii, "A Harvard Project in Reverse: Materials of the Commission of the USSR Academy of Sciences on the History of the Great Patriotic War—Publications and Interpretations," *Kritika: Explorations in Russian and Eurasian History* 19 (2018): 175–202. While many of Budnitskii's criticisms about the interviews—in particular the questions of representativeness and editing—are valid, they still represent one of the most remarkable and fruitful sources available.

48. Jochen Hellbeck, "A Galaxy of Black Stars: The Power of Soviet Biography," *American Historical Review* 114 (2009): 620–21.

49. David Samoilov, *Pamiatnye zapiski* (Moscow: Mezhdunarodnye otnosheniia, 1995), 233. On the problem of representativeness, see Roger Reese, "Ten Jewish Red Army Veterans of the Great Patriotic War: In Search of the Mythical Representative Soldier's Story," *Journal of Slavic Military Studies* 27 (2014): 420–29.

50. Slezkine, *Jewish Century*.

51. Anna Krylova, "'Healers of Wounded Souls': The Crisis of Private Life in Soviet Literature, 1944–1946," *Journal of Modern History* 73 (2001): 307–31; Merridale, *Ivan's War*, 17, 268, 284–85, 356, 364, 387–88.

52. G. A. Stefanovskii et al., *Poslednie pis'ma s fronta, 1941* (Moscow: Voenizdat, 1991), 1:336.

1. The Soldier's Body

1. Mikhail Loginov, *Eto bylo na fronte* (Kazan: Tatknigizdat', 1984), 9. Epigraph: Vladimir Stezhenskii, *Soldatskii dnevnik: Voennye stranitsy* (Moscow, Agraf, 2005), 17 (October 5, 1941).

2. Loginov, *Eto bylo na fronte*, 6–7.

3. See Brandon Schechter, "*Khoziaistvo* and *Khoziaeva*: The Properties and Proprietors of the Red Army, 1941–45," *Kritika: Explorations in Russian and Eurasian History* 18 (2017): 487–510.

4. V. V. Gradosel'skii, "Velikaia Otechestvennaia: Voennoe stroitel'stvo. Komplektovanie Krasnoi Armii riadovym i serzhantskim sostavom v gody Velikoi Otechestvennoi voiny," *Voenno-istoricheskii zhurnal*, no. 3 (2002): 6–12.

5. G. F. Krivosheev, ed., *Grif sekretnosti sniat: Poteri Vooruzhennyh sil SSSR v voinakh, boevykh deistviiakh i voennykh konfliktakh. Statisticheskoe izdanie* (Moscow: Voenizdat, 1993), 143.

6. Gradosel'skii, "Velikaia Otechestvennaia," 7.

7. Arkhiv Prezidenta Rossiiskoi Federatsii (AP RF) f. 3, op. 50, d. 270, ll. 146–56, in *Voina: 1941–1945. Vestnik Arkhiva Prezidenta Rossiiskoi Federatsii*, ed. Sergei Kudriashov (Moscow: Arkhiv Prezidenta Rossiiskoi Federatsii, 2010), 216–21.

8. Krivosheev, *Grif sekretnosti sniat*, 144, 139, 153, 154.

9. Nikolai Tikhonov, "O prostykh sovetskikh liudiakh," *Krasnoarmeets*, no. 14 (1945): 1.

10. Grigorii Baklanov, *Zhizn', podarennaia dvazhdy* (Moscow: Vagrius, 1999), 40. Michel Foucault used identical language to describe the modern concept of being able to turn anyone into a soldier (*Discipline and Punish* [New York: Pantheon, 1977], 135).

11. Iu. A. Poliakov, ed., *Vsesoiuznaia perepis' naseleniia 1939 goda: Osnovnye itogi* (Moscow: Nauka, 1992), 20, 28, 50, 57–58, 82,

12. AP RF f. 2, op. 50, d. 266, ll. 27–34, in Kudriashov, *Voina*, 124. Although official planning still attempted to ensure 20–30 percent of all soldiers in units were Komsomol or party members, and 25 percent were veterans returning from hospitals.

13. See, e.g., AP RF f. 3, op. 50, d. 266, ll. 110–11, in Kudriashov, *Voina*, 144. This order ended many deferments.

14. Lev Slëzkin, *Do voiny i na voine* (Moscow: Parad, 2009), 420–21; Samoilov, *Pamiatnye zapiski*, 206–10.

15. Rossiiskii gosudarstvennyi voennyi arkhiv (RGVA) f. 4, op. 12, d. 107, ll. 651–53, in *Prikazy narodnogo komissara oborony SSSR 1943–1945gg.: Dokumenty i materialy*, ed. A. I. Barsukov et al., vol. 13 (2–3) of *Russkii arkhiv: Velikaia Otechestvennaia* (Moscow: TERRA, 1997), 109–11. Those guilty of refusing to testify against others (*nedonositel'stvo)* could serve.

16. AP RF f. 3, op. 50, d. 272, ll. 83–86, 91–95, in Kudriashov, *Voina*, 306.

17. Amir Weiner, "Something to Die For, a Lot to Kill For," in *The Barbarization of Warfare*, ed. George Kassimeris (New York: New York University Press, 2006), 109–10.

18. Krivosheev, *Grif sekretnosti sniat*, 129.

19. AP RF f. 3, op. 50, d. 265, ll. 112–12ob., in Kudriashov, *Voina*, 99–100.

20. Nicholas Ganson, "Food Supply, Rationing, and Living Standards" in *The Soviet Union at War, 1941–1945*, ed. David R. Stone (Barnsley: Pen & Sword, 2010), 75,

Before a shift in policy in February 1942, 2.8 million Soviet POWs died in German captivity, and only 400,000 of those left alive were capable of working. By the end of the war 57.5 percent of all Soviet POWs had died.

21. Gabriel Temkin, *My Just War: The Memoir of a Jewish Red Army Soldier in World War II* (Novato: Presidio, 1998), 87–90; AP RF f. 2, op. 50, d. 266, ll. 27–34, in Kudriashov, *Voina*, 126; AP RF f. 3, op. 50, d. 270, ll. 146–56, in *Voina*, 218; AP RF f. 3, op. 50, d. 272, ll. 83–86, 91–95, in *Voina*, 306. Of over 46,000 soldiers and civilians held by NKVD of the South-West Front between November 20, 1942, and January 20, 1943, filtration "unmasked" 1,946 "spies, traitors, deserters, and marauders." See Tsentral'nyi arkhiv Ministerstva oborony Rossiiskoi Federatsii (TsAMO) f. 232, op. 590, d. 147, ll. 32–39, in *Preliudiia Kurskoi bitvy: Dokumenty i materialy 6 dekabria 1942 g.–25 aprelia 1943 g.*, ed. A. M. Sokolov et al., vol. 15 (4–3) of *Russkii arkhiv: Velikaia Otechestvennaia* (Moscow: TERRA, 1997), 334–38.

22. Viola, *Peasant Rebels under Stalin*, 13–44; Lynne Viola, *The Unknown Gulag: The Lost World of Stalin's Special Settlements* (New York: Oxford University Press, 2007), 32, 170, 205. Viola puts the number of executed at well over 10,000 and the numbers exiled as well over two million (32).

23. See, e.g., Rossiiskii gosudarstvennyi arkhiv sotsial'no-politicheskoi istorii (RGASPI) f. 17, op. 125, d. 188, ll. 22–30, 31, 46–49.

24. Pavel Polian, *Ne po svoei vole . . . Istoriia i geografiia prinuditel'nykh migratsii v SSSR* (Moscow, O. G. I.–Memorial, 2001), www.memo.ru/history/deport/.

25. See, e.g., AP RF f. 2, op. 50, d. 266, ll. 27–34, in Kudriashov, *Voina*, 124; AP RF f. 3, op. 50, d. 272, ll. 83–86, 91–95, in *Voina*, 306.

26. RGASPI f.17, op. 125, d. 85, l. 64.; Nauchnyi arkhiv Instituta Rossiiskoi istorii Akademii nauk Rossiiskoi Federatsii (NA IRI RAN) f. 2, r. I, op. 28, d. 30, l. 10.

27. See, e.g., Boris Slutskii, *O drugikh i o sebe* (Moscow: Vagrius, 2005), 119–20; NA IRI RAN f. 2, r. III, op. 5, d. 36, l. 12; Schechter, "'People's Instructions.'"

28. NA IRI RAN f. 2, r. I, op. 103, d.1, ll. 2ob.–3.

29. Irina Dunaevskaia, *Ot Leningrada do Kënigsberga: Dnevnik voennoi perevodchitsy (1942–1945)* (Moscow: ROSSPEN, 2010), 156; N. S. Frolov, ed., *Vse oni khoteli zhit': Frontovye pis'ma pogibshikh soldat, vospominaniia veteranov voiny* (Kazan: Tarikh, 2003), 34.

30. Charles David Shaw, "Making Ivan Uzbek: War, Friendship of the Peoples, and the Creation of Soviet Uzbekistan, 1941–1945" (PhD diss., University of California, Berkeley, 2015), iv, 13, 47–48, 198, 222.

31. RGASPI f. 88, op. 1, d. 964, l. 1ob.; Schechter, "'People's Instructions.'"

32. Tat'iana Repina (Atabek), *K biografii voennogo pokoleniia* (Moscow: Moskovskie uchebniki i kartolitografiia, 2004), 224. (Hereafter Atabek, her last name during the war.)

33. RGASPI f. 88, op. 1, d. 972, ll. 5–6. "Soveshchanie nachal'nikov otdelov agitatsii i propagandy politupravleniia voennykh okrugov," *Propagandist i agitator Krasnoi Armii*, no. 17 (1944): 36; Viktor Muradian, *Boevoe bratstvo* (Moscow: Voenizdat, 1978), 284–90.

34. Pomerants, *Zapiski gadkogo utënka*, 64.

35. Nison Shapiro, "Semero iz semnadtsati," *Neva*, no. 9 (2002): 223–25; Iurii Bondarev, *Tishina* (Moscow: Sovetskii pisatel', 1962), 70–73; NA IRI RAN f. 2, r. I, op. 27, l. 8.

36. NA IRI RAN f. 2, r. I, op. 57, d. 2 ll. 9–10; Baklanov, *Zhizn'*, 50.

37. Reese, *Why Stalin's Soldiers Fought,* 121–23; AP RF f. 2, op. 50, d. 266, ll. 27–34, in Kudriashov, *Voina,* 123; RGVA f. 4, op. 11, d. 73, ll. 148–53, in *Prikazy narodnogo komissara oborony SSSR 22 iiunia 1941 g.–1942 g.*, ed. A. I. Barsukov et al., vol. 13 (2–2) of *Russkii arkhiv: Velikaia Otechestvennaia* (Moscow: TERRA, 1997), 361–65.

38. Konstantin Simonov, *Dni i nochi: Povest'* (Moscow: Sovetskii pisatel', 1944), 77–78; *Zavetnoe slovo Fomy Smyslova russkogo byvalogo soldata* (Moscow: Voenizdat, 1943). See also "Prikaz komandira vypolnit' bystro i tochno," *Bloknot agitatora Krasnoi Armii*, no. 1 (1943): 20–22.

39. NA IRI RAN f. 2, r. I, op. 223, d. 10, ll. 1–1ob.

40. Vasilii Grossman, *Gody voiny* (Moscow: Pravda, 1989), 352.

41. See chapters 2 and 3 of Krylova, *Soviet Women in Combat*.

42. See, e.g., RGVA f. 4, op. 11, d. 70, ll. 251–52, in *Prikazy narodnogo komissara oborony SSSR 22 iiunia 1941 g.–1942 g.*, 213–14; RGVA f. 4, op. 11, d. 70, ll. 369–71, in *Prikazy narodnogo komissara oborony SSSR 22 iiunia 1941 g.–1942 g.*, 214–15; RGVA f. 4, op. 11, d. 74, ll. 7–8, in *Prikazy Narodnogo komissara oborony SSSR (1941–1945)*, 13–14.

43. NA IRI RAN f. 2, r. X, op. 7, d. 3, l. 7ob.; Iakupov, *Frontovye zarisovki*, 104; NA IRI RAN f. 2, r. I, op. 16, d. 4, l. 109. Gabriel Temkin, however, disagreed (*My Just War,* 202–3).

44. *Prikaz Narodnogo komissara oborony N. 336* (Moscow: Voenizdat, 1942), 60.

45. NA IRI RAN f. 2, r. X, op. 7, d. 13-b, ll. 97–98.

46. NA IRI RAN f. 2, r. I, op. 16, d. 4, l. 108ob.

47. RGVA f. 4, op. 12, d. 97, ll. 263–72, in *Prikazy narodnogo komissara oborony SSSR 1937–21 iiunia 1941 g.: Dokumenty i materialy*, ed. A. I. Barsukov et al., vol. 13 (2–1) of *Russkii arkhiv: Velikaia Otechestvennaia* (Moscow: TERRA, 1994), 260.

48. Mansur Abdulin, *160 stranits iz soldatskogo dnevnika* (Moscow: Molodaia gvardiia, 1985), 60; Valentina Chudakova, *Ratnoe schast'e* (Moscow: Voenizdat, 1980), 174–75; RGVA f. 4, op. 12, d. 106-a, l. 512, in *Prikazy narodnogo komissara oborony SSSR 22 iiunia 1941 g.–1942 g.*, 368.

49. RGVA f. 4, op. 12, d. 99, ll. 274–77, in *Prikazy narodnogo komissara oborony SSSR 22 iiunia 1941 g.–1942 g.*, 111–12.

50. RGVA f. 4, op. 12, d. 99, ll. 274–77, in *Prikazy narodnogo komissara oborony SSSR 22 iiunia 1941 g.–1942 g.*, 111–12.

51. TsAMO f. 32, op. 920265, d. 4, ll. 2–3, in *Glavnye politicheskie organy vooruzhennykh sil SSSR v Velikoi Otechestvennoi voine 1941–1945 gg.: Dokumenty i materialy*, ed. N. I. Borodin and N.V. Usenko, vol. 17–6 (1–2) of *Russkii arkhiv: Velikaia Otechestvennaia* (Moscow, TERRA, 1996), 78–79.

52. NA IRI RAN f. 2, r. X, op. 7, d. 13-b, ll. 97–98; Al'bert Baiburin, *Sovetskii pasport: istoriia-struktura-praktika* (St. Petersburg: Izdatel'stvo Evropeiskogo universiteta v Sankt-Peterburge, 2017).

53. Tsentral'nyi arkhiv Federal'noi sluzhby bezopasnosti Rossiiskoi Federatsii (TsA FSB) f. 14, op. 5, d. 23, ll. 386–91, in *"Ognennaia duga": Kurskaia bitva glazami Lubianki*, ed. A. T. Zhadobin, V. V. Markovchin, and B. S. Khristoforov (Moscow: Moskovskie uchebniki i Kartolitografiia, 2003), 25–28; A. Litvinov, "Vvod v boi novogo popolneniia," *Krasnaia zvezda*, April 20, 1943.

54. Pomerants, *Zapiski gadkogo utënka*, 64. Pomerants noted that education could have a tremendous impact on one's fate, particularly for illiterate and "non-Russian" soldiers.

55. V. P. Iampol'skii et al., *Krushenie "Blitskriga" 1 ianvaria–20 iiunia 1942 g.: Organy gosudarstvennoi bezopasnosti SSSR v Velikoi Otechestvennoi voine*, vol. 3, bk. 1 (Moscow: Rus', 2003), 517–18.

56. Oles' Honchar, *Chelovek i oruzhie: Proizvedeniia v trekh tomakh* (Moscow: Khudozhestvennaia literatura, 1990), 2:24–28.

57. Gonchar, *Chelovek i oruzhie*, 186–87; Merridale, *Ivan's War*, 56.

58. RGVA f. 4, op. 15, d. 27, ll. 105–13, in *Tyl Krasnoi Armii v Velikoi Otechestvennoi voine 1941–1945 gg.: Dokumenty i materialy*, ed. P. I. Veshchikov et al., vol. 25 (14) of *Russkii arkhiv: Velikia Otechestvennaia* (Moscow, TERRA, 1998), 43–44, 53; Aleksandr Lesin, *Byla voina: Kniga-dnevnik* (Simferopol: Tavriia, 1990), 43–44, 53.

59. Viktor Astaf'ev, *Prokliaty i ubity* (Moscow: TERRA, 1999), 19; Efraim Sevela, *Monia Tsatskes—znamenosets* (St. Petersburg: Kristall, 2000), 49.

60. Frolov, *Vse oni khoteli zhit'*, 26; Chekhovich, *Dnevnik ofitsera*, 28.

61. David Samoilov, *Podënnye zapisi* (Moscow: Vremia, 2002), 1:152–53.

62. Aleksandr Bek, *Volokolamskoe shosse* (Moscow: Pravda, 1988), 58.

63. V. P. Kiselëv, "Voina i zhizn' v predstavlenii 20-letnikh frontovikov (iz moego dnevnika)," in *Obshchestvo i vlast': Rossiiskaia provintsiia*, ed. A. A. Kulikov and A. N. Sakharov, vol. 3: *Iiun' 1941 g.–1953 g.* (Moscow: IRI RAN, 2005), 1008.

64. G. Gorov, "Agitatsiia sredi boitsov nerusskoi natsional'nosti," *Agitator i propagandist Krasnoi Armii*, no. 22 (1942): 26–27.

65. Tvardovskii, *Vasilii Terkin*, 25.

66. Lesin, *Byla voina*, 142.

67. Lesin, *Byla voina*, 92.

68. *Rukovodstvo dlia boitsa pekhoty* (Moscow: Voenizdat, 1940) (hereafter *RBP-40*), 89–92, 103–20.

69. Veniamin Tongur, *Frontovoi dnevnik (1941–1945)* (Simferopol: Sliperi Rok, 2006), 82–83.

70. Dunaevskaia, *Ot Leningrada do Kënigsberga*, 141.

71. RGASPI f. 17, op. 125, d. 85, ll. 66–67; NA IRI RAN f. 2, r. III, op. 5, d. 2-a, l. 51.

72. NA IRI RAN f. 2, r. X, op. 7 d. 13-b, l. 21.

73. Anatolii Genatulin, *Sto shagov na voine* (Moscow: RBP, 1995), 5.

74. Bek, *Volokolamskoe shosse*, 27.

75. *Ustav garnizonnoi sluzhby Krasnoi Armii* (Moscow: Voenizdat, 1942), 21.

76. RGASPI f. M-33, op. 1, d. 853, l. 179.

77. Anatolii Genatulin, "Strakh," *Sovetskii voin*, no. 4 (1990): 15.

78. Vasil' Bykov, *Dolgaia doroga domoi* (Moscow: AST-Kharvest, 2005), 56.

79. RGVA f. 4, op. 15, d. 31, ll. 57–65, in *Prikazy Narodnogo komissara oborony SSSR 1937–21 iiunia 1941 g.*, 180–83.

80. See, e.g., Dunaevskaia, *Ot Leningrada do Kënigsberga*, 73; and Bek, *Volokolamskoe shosse*, 19–27.

81. Vladimir Bushin, *Ia posetil sei mir: Iz dnevnikov frontovika* (Moscow: Algoritm, 2012), 47.

82. RGVA f. 4, op. 11, d. 62, l. 331, in *Prikazy Narodnogo komissara oborony SSSR 22 iiunia 1941 g.–1942 g.*, 84. This was later reexamined. See RGVA f. 4, op. 11, d.

66, ll. 149–52, in *Prikazy Narodnogo komissara oborony SSSR 22 iiunia 1941 g.–1942 g.*, 108–9.

83. Amir Weiner has argued that these orders are a sort of distillation of Stalinist thinking, particularly on the need for purity over numbers, while Amnon Sella claims that these orders were damaging to morale. Alexander Hill notes that Order 227 was one of the few times the Soviet regime was honest about how desperate the situation was. I argue, alongside Roger Reese, that these orders are a desperate response to the situation at the front and would add that there is a surprising degree of flexibility that these orders allow. See Reese, *Why Stalin's Soldiers Fought*, 63, 153, 160–75; Weiner, "Something to Die For," 107–8; Amnon Sella, *The Value of Life in Soviet Warfare* (New York: Routledge, 1992), 99–100; and Alexander Hill, *The Red Army and the Second World War* (New York: Cambridge University Press, 2016), 353–56.

84. RGVA f. 4, op. 12, d. 98, ll. 617–22, in *Prikazy Narodnogo komissara oborony SSSR 22 iiunia 1941 g.–1942 g.*, 58–60.

85. RGVA f. 4, op. 12, d. 105, ll. 122–28, in *Prikazy Narodnogo komissara oborony SSSR 22 iiunia 1941 g.–1942 g.*, 276–79.

86. "Programma politicheskoi podgotovki nachal'stvuiushchego sostava," *Propagandist Krasnoi Armii*, no. 20 (1941): 21; NA IRI RAN f. 2, r. I, op. 30, d. 1, ll. 235–36. See also Reese, *Why Stalin's Soldiers Fought*, 169.

87. See, e.g., Vera Inber, "Zhenshine," *Krasnoarmeets*, no. 15 (1942): 3.

88. M. Mironov, "O voinskom vospitanii ofitserskikh kadrov," *Agitator i propagandist Krasnoi Armii*, no. 7 (1944): 8.

89. Irakli Toidze, "Rodina mat' zovet!" (Moscow: Iskusstvo, 1941).

90. RGASPI f. 644, op. 1, d. 5, l. 176.

91. NA IRI RAN f. 2, r. I, op. 123, d. 13, l. 2ob; Larionov, *Frontovaia povsednevnost'*, 251–54.

92. Temkin, *My Just War*, 178; NA IRI RAN f. 2, r. X, op. 2, d. 3, l. 23; Reutov, *Gvardeets*, 84; TsAMO RF, f. 208, op. 2563, d. 48, ll. 64–75, in *Tyl Krasnoi Armii*, 239.

93. Vasilii Chekalov, *Voennyi dnevnik: 1941, 1942, 1943* (Moscow: Zdravyi smysl, 2004), 34–36.

94. Bek, *Volokolamskoe shosse*, 23, 26.

95. Tongur, *Frontovoi dnevnik*, 104–5.

96. NA IRI RAN f. 2, r. X, op. 2, d. 3, 11ob.; Belov, *Frontovoi dnevnik N. F. Belova. 1941–1944 gody*, entry from February 13, 1943, www.booksite.ru/fulltext/vol/ogd/atwo/19.htm#27.

97. Krivosheev, *Grif sekretnosti sniat*, 140.

98. Budnitskii, "Great Patriotic War and Soviet Society," 793–94.

99. Weiner, "Something to Die For," 108; *The Army Lawyer: A History of the Judge Advocate General's Corps, 1775–1975* (Washington, DC: The Corps, 1975), 192.

100. Iakov Aizenshtat, *Zapiski sekretaria voennogo tribunala* (London: Overseas Publications Interchange Limited, 1991), 6–7.

101. Reese, *Why Stalin's Soldiers Fought*, 168–71.

102. TsAMO RF f. 243, op. 2963, d. 26, ll. 85–86, in *Tyl Krasnoi Armii*, 580–81.

103. Krivosheev, *Grif sekretnosti sniat*, 146–47.

104. RGASPI f. 17, op. 125, d. 85, ll. 55–68; RGASPI f. 17, op. 125, d. 104, ll. 200–201.

105. NA IRI f. 2, r. I, op. 28, d. 30, l. 13.

106. Vasilii Glotov, *Vstrechi: Frontovoi dnevnik, ocherki* (Lviv: Kameniar, 1980), 47. For a discussion of desertion in the Red Army, see Mark Edele, *Stalin's Defectors: How Red Army Soldiers Became Hitler's Collaborators, 1941–1945* (New York: Oxford University Press, 2017).

107. Reutov, *Gvardeets*, 116. See Lesin, *Byla voina*, 86–87, for a similar account.

108. RGASPI f. 17, op. 125, d. 130, l. 33. He went on to claim that these soldiers were essentially worthless (l. 34).

109. Krivosheev, *Grif sekretnosti sniat*, 140.

110. NA IRI RAN f. 2, r. I, op. 16, d. 3b, ll. 242–43.

111. TsA FSB f. 14, on. S, d. 23, ll. 386–91, in *Kurskaia duga glazami Lubianki*, 25.

112. K. Kniazeva, "Vospitanie devushek-voennosluzhashchikh," *Agitator i propagandist Krasnoi Armii*, no. 15 (1944): 33.

113. Brandon Schechter, "'Girls' and 'Women': Love, Sex, Duty, and Sexual Harassment in the Ranks of the Red Army, 1941–1945," *Journal of Power Institutions in Post-Soviet Societies* 17 (2015), https://journals.openedition.org/pipss/4202.

114. RGASPI f. 88, op. 1, d. 948, ll. 12–14.

115. NA IRI RAN f. 2, r. X, op. 7, d. 2a, ll. 2–2ob.

116. A.A. Kovalevskii, "Nynche u nas peredyshka . . .," *Neva*, no. 5 (1995): 86–87.

117. On PPZh, see Grossman, *Gody voiny*, 242–43; Aizenshtat, *Zapiski sekretaria voennogo tribunala*, 116.

118. Lev Kopelev, *Khranit' vechno* (Moscow: TERRA, 2004), 1:90.

119. RGVA f. 4, op. 12, d. 110, ll. 279–80, in *Prikazy Narodnogo komissara oborony SSSR (1943–1945 gg.)*, 317–18.

120. Tikhonov, "O prostykh sovetskikh liudiakh," 1.

121. AP RF f. 3, op. 50, d. 270, ll. 146–56, in Kudriashov, *Voina*, 217: most divisions were reduced to 2,500 soldiers or less when brought to the rear for reformation in 1942. That is out of a projected strength of over 10,000 (RGASPI f. 644, op. 1, d. 48, l. 25).

122. IRI RAN f. 2, r. III, op. 5, d. 16, l. 7; Reutov, *Gvardeets*, 98; Belov, *Frontovoi dnevnik*, July 15, 1942; A. I. Erëmenko, *Dnevniki, zapiski, vospominaniia* (Moscow: ROSSPEN, 2013), 298.

123. I. Mints, "O traditsiiakh Krasnoi armii," *Agitator i propagandist Krasnoi Armii*, no. 13 (1944): 7. See also F. Gorokhov, "Komandir—dusha boevoi spaiki i sokhraneniia boevykh traditsii chasti," *Agitator i propagandist Krasnoi Armii*, no. 3–4 (1943): 33–39; N. Markovich, "Ofitserstvo—kostiak armii," *Agitator i propagandist Krasnoi Armii*, no. 9–10 (1943): 9–16; and K. Kulik, "Vospitanie na boevykh traditsiiakh," *Agitator i propagandist Krasnoi Armii*, no. 15 (1943): 14–22. The "kostiak" of one elite unit toward the war's end was 260 men left of the original 18,000 (NA IRI RAN f. 2, r. 1, op. 230, d. 6, l. 1ob.).

124. Kulik, "Vospitanie na boevykh traditsiiakh," 17.

125. Pomerants, *Zapiski gadkogo utënka*, 141; Petr Liubarov, *Inache ia ne mog: Voina* (Odessa: Maiak, 2004), 19.

126. N. N. Nikulin, *Vospominaniia o voine* (St, Petersburg: Izdatel'stvo Gosudarstvennogo Ermitazha, 2008), 39, 47–48, 99.

127. Lesin, *Byla voina*, 241.

128. Omer Bartov, *Hitler's Army: Soldiers, Nazis, and War in the Third Reich* (New York: Oxford University Press, 1992), 36.

129. Chudakova, *Ratnoe schast'e*, 9; see also A. B. Priadekhin, "Frontovye dorogi," in *Veteran,* ed. Ia. F. Potekhin, M.P. Streshinskii, and I. M. Frantishev (Leningrad: Lenizdat, 1977), 16–32. Even on the relatively static Leningrad Front Priadekhin was sent to a new unit every time he was wounded.

130. See, e.g., Dunaevskaia, *Ot Leningrada do Kënigsberga*; and RGVA f. 4, op. 126, d. 99, ll. 341–42, in *Prikazy narodnogo komissara oborony SSSR 22 iiunia 1941 g.–1942 g.*, 128–29.

131. Ivan Yakushin, *On the Roads of War*, trans. and ed. Bair Irincheev (Barnsley: Pen & Sword, 2005), 151; NA IRI RAN f. 2, r. I, op. 226, d. 2, l. 13.

132. Evgenii D. Moniushko, *From Leningrad to Hungary: Notes of a Red Army Soldier, 1941–1946*, trans. Oleg Sheremet, ed. David M. Glantz (New York: Frank Cass, 2005), 122–25.

133. RGVA f. 4, op. 12, d. 105, l. 677, in *Prikazy narodnogo komissara oborony SSSR 22 iiunia 1941 g.–1942 g.*, 310–11.

134. Dunaevskaia, *Ot Leningrada do Kënigsberga*, 307.

135. Litvinov, "Vvod v boi novogo popolneniia"; RGVA f. 4, op. 11, d. 62, ll. 310–11, in *Prikazy narodnogo komissara oborony SSSR 22 iiunia 1941 g.–1942 g.*, 70. Whether a unit would be disbanded depended on the presence or absence of "combat traditions,"—that is, combat record—even if the personnel were entirely new.

136. M. Garussiko and S. Gliazer, "Moi polk—moia sem'ia," *Bloknot agitatora Krasnoi Armii*, no. 10 (1943): 1, 5.

137. Boris Komskii, "Dnevnik 1943–1945 gg.," in *Arkhiv evreiskoi istorii*, vol. 6 (Moscow: ROSSPEN, 2011), 59.

138. See, e.g., Suris, *Frontovoi dnevnik,* 175, 357; and Gennadii Tokarev, *Vesti dnevnik na fronte zapreshchalos'* (Novosibirsk: Svin'in i synov'ia, 2005), 185.

139. NA IRI RAN f. 2, r. II, op. 28, d. 35, l. 8.

140. Mansur Abdulin, *Iz vospominanii soldata: Rasskazy* (St. Petersburg: Del'fa R.A., 1995), 44–46.

141. NA IRI RAN f. 2, r. III, op. 5, d. 116, l. 58.

142. Elaine Scarry, *The Body in Pain: The Making and Unmaking of the World* (New York: Oxford University Press, 1985), 21.

143. *Programmy uskorennoi boevoi podgotovki strelkovykh podrazdelenii* (Moscow: Voenizdat, 1941), 5; "Zakony sovetskoi gvardii," *Bloknot agitatora Krasnoi Armii*, no. 8 (1942): 21–22.

144. Sella, *Value of Life in Soviet Warfare*, 31. See also Scarry, *Body in Pain,* 76.

145. Nikulin, *Vospominaniia o voine*, 100.

146. NA IRI RAN f. 2, r. X, op. 2, d. 3, ll. 28ob.–29.

147. NA IRI RAN f. 2, r. I, op. 230, d. 1, l. 12.

148. Grigorii Baklanov, "Piad' zemli," in *Voennye povesti* (Moscow: Sovetskii pisatel', 1981), 341.

149. NA IRI RAN f. 2, r. I, op. 28, d. 27, l. 4ob.; Karel Berkhoff, *Motherland in Danger: Soviet Propaganda During World War II* (Cambridge, MA: Harvard University Press, 2012), 62–65.

150. Kiselev, "Voina i zhizn' v predstavlenii 20-letnikh frontovikov," 1020.

151. NA IRI RAN f. 2, r. II, op. 103, d. 6, l. 3.

152. Baklanov, *Zhizn'*, 64; Dunaevskaia, *Ot Leningrada do Kënigsberga*, 50, 168.

153. Grigorii Baklanov, "Naveki—deviatnadtsatiletnie," in *Voennye povesti*, 222; RGASPI f. M-33, op. 1, d. 853, l. 350.

154. Galuzevii derzhavnii arkhïv Sluzhba bezpeki Ukraïni (GDA SB Ukraïni) f. 9, spr. 216-sp, ark. 13–21 zv., quoted in V. Litvinenko and V. Ogorodnik, "Viddili viis'kovoï tsenzuri ta politichnogo kontroliu NKVD–NKGB SRSR u Chervonii armiï ta Viis'kovo-mors'komu floti (kin. 1930-kh—berezen' 1946 rr.)," *Z arkhiviv VChK–GPU–NKVD–KGB*, no. 1 (42) (2014): 220; GDA SB Ukraini f. 9, spr. 223-sp, ark. 38–38 zv., quoted in Litvinenko and Ogorodnik, "Viddili viis'kovoï tsenzuri," 274–75.

155. RGASPI f. M-33, op. 1, d. 1085, l. 96.

156. Loginov, *Eto bylo na fronte*, 10.

157. Belov, *Frontovoi dnevnik*, entry from August 8, 1943.

158. RGASPI f. M-33, op. 1, d. 853, l. 220.

159. *Boevoi ustav pekhoty Krasnoi Armii*, pt. 1 *(Boets, otdelenie, vzvod, rota)* (Moscow: Voenizdat, 1942) (hereafter *BUP-42*), 28. Bold print in original.

160. A. A. Vishnevskii, *Dnevnik khirurga: Velikaia Otechestvennaia Voina 1941–1945 gg.* (Moscow: Meditsina, 1970), 137.

161. A. I. Burnazian, *Bor'ba za zhizn' ranenykh i bol'nykh na Kalininskom-1-m Pribaltiiskom fronte 1941–1945* (Moscow: Meditsina, 1982), 125.

162. *Ukazaniia po meditsinskoi sortirovke i evakuatsii po naznacheniiu tak nazyvaemykh legko ranenykh v polevoi sanitarnoi sluzhbe* (Moscow: Medgiz, 1942), 2–4. For a harrowing account of what it was like to work in a head trauma ward, see Anatolii Genatulin, "Dve nedeli," in *Vot konchitsia voina* (Moscow: Pravda, 1988), 75–107.

163. Burnazian, *Bor'ba za zhizn'*, 39.

164. Komskii, "Dnevnik 1943–1945 gg.," 30.

165. Burnazian, *Bor'ba za zhizn'*, 116; NA IRI RAN f. 2, r. X, op. 7, d. 11, ll. 3ob.–4ob, Sella, *Value of Human Life in Soviet Warfare*, 26–27, 36, 38, 60; NA IRI RAN f. 2, r. I, op. 123, d. 13, l. 3.

166. TsAMO RF f. 449, op. 9957, d. 3, ll. 131–32, in *Tyl Krasnoi Armii*, 575–77.

167. IRI RAN f. 2, r. X, op. 2, d. 3, l. 7.

168. Sella, *Value of Human Life in Soviet Warfare*, 60.

169. NA IRI RAN f. 2, r. I, op. 123, d. 13, l. 2.

170. TsAMO RF f. 449, op. 9957, d. 4, l. 25, in *Tyl Krasnoi Armii*, 475–76.

171. Komskii, "Dnevnik 1943–1945 gg.," 31; TsAMO RF f. 208, op. 2563, d. 48, ll. 64–75, in *Tyl Krasnoi Armii*, 239.

172. RGASPI f. 17, op. 125, d. 130, l. 54.

173. TsAMO RF f. 67, op. 12020, d. 19, ll. 1–59, in *Tyl Krasnoi Armii*, 706; I. B. Rostotskii, *Boets v gospitale* (Moscow: Institut sanitarnogo prosveshcheniia, 1942), 40–41.

174. Gataulla Makhmutov, "Iz voennogo dnevnika 1941–1945 godov," *Gostinnyi dvor*, no. 16 (2005): 82; Nikulin, *Vospominaniia o voine*, 82; NA IRI RAN f. 2, r. III, op. 5, d. 4, l. 1.

175. Lesin, *Byla voina*, 222.

176. Bulat Okudzhava, *Bud' zdorov, shkoliar* (Frankfurt am Main: Posev-Verlag, 1964), 69–70.

177. TsAMO f. 2, op. 795437, d. 9, l. 274, in *Tyl Krasnoi Armii*, 233–34.

178. Rostotskii, *Boets v gospitale*, 39.
179. Samoilov, *Pamiatnye zapiski*, 204–5.

2. A Personal Banner

1. RGVA f. 4, op. 12, d. 98, ll. 617–22, in *Prikazy narodnogo komissara oborony SSSR 22 iiunia 1941 g.–1942 g.*, 58–60. Epigraph: Vasilii Subbotin, "Soldatskaia dusha," *Krasnaia zvezda*, February 19, 1943.
2. Iakupov, *Frontovye zarisovki*, 19–20, 26.
3. Dunaevskaia, *Ot Leningrada do Kënigsberga*, 170.
4. R. Moran, "O vruchenii pogonov," *Krasnaia zvezda*, January 28, 1943.
5. Subbotin, "Soldatskaia dusha."
6. The British, German, French, and US armies all had some device—such as distinctive patches, brass shoulder titles, collar tab, or cuff titles for individual military units—used as a point of pride and source of corporate identity.
7. Elizabeth Wilson, *Adorned in Dreams: Fashion and Modernity* (New Brunswick: Rutgers University Press, 2003), 11–12; Alison Lurie, *The Language of Clothes* (New York: Random House, 1981), 11–13.
8. Nathan Joseph, *Uniforms and Nonuniforms: Communication through Clothing* (New York: Greenwood, 1986), 66.
9. Jennifer Craik, *Uniforms Exposed: From Conformity to Transgression* (New York: Berg, 2005), 4; Joseph, *Uniforms and Nonuniforms*, 69.
10. *RBP-40*, 53–55; A. Krivitskii, "O voennom mundire i pogonakh," *Krasnaia zvezda*, January 17, 1943; Joseph, *Uniforms and Nonuniforms*, 3, 66–67.
11. In 1935 all ranks up to colonel returned, and in 1940 the rank of general was introduced to the Red Army. RGVA f. 4, op. 12-a, d. 82, ll. 562–628, in *Prikazy narodnogo komissara oborony SSSR 1937–21 iiunia 1941 g.*, 43; RGVA f. 4, op. 15, d. 31, l. 288, in *Prikazy narodnogo komissara oborony SSSR 1937–21 iiunia 1941 g.*, 133–34.
12. TsAMO RF f. 2, op. 920266, d. 1, ll. 463–96, in *Tyl Krasnoi Armii*, 105–8; TsAMO RF f. 2, op. 795437, d. 9, l. 254, 255, in *Tyl Krasnoi Armii*, 231–33; RGVA f. 4, op. 12, d. 108, l. 36a, in *Prikazy narodnogo komissara oborony SSSR 1943–1945 gg.*, 186–87.
13. TsAMO RF f. 2, op. 795437, d. 9, l. 158, in *Tyl Krasnoi Armii*, 216; RGVA f. 4, op. 11, d. 67, ll. 158–60, in *Prikazy narodnogo komissara oborony SSSR 22 iiunia 1941 g.–1942 g.*, 193–94.
14. TsAMO RF f. 4, op. 11, d. 70, l. 1, in *Tyl Krasnoi Armii*, 193–94.
15. TsAMO RF f. 2, op. 920266, d. 2, ll. 12–19, in *Tyl Krasnoi Armii*, 208–12; Abdulin, *160 stranits*, 60.
16. See, e.g., Sevela, *Monia Tsatskes*, 28; Baklanov, "Naveki–deviatnadtsatiletnie," 177; TsAMO RF f. 4, op. 11, d. 70, l. 1, in *Tyl Krasnoi Armii*, 193–94.
17. Lesin, *Byla voina*, 90–99.
18. RGVA f. 4, op. 12, d. 99, ll.128–43, 146–47, 151–52, in *Prikazy narodnogo komissara oborony SSSR 22 iiunia 1941 g.–1942 g.*, 97; RGVA f. 4, op. 12, d. 107, l. 677, in *Prikazy narodnogo komissara oborony SSSR 1943–1945 gg.*, 115.
19. Chekhovich, *Dnevnik ofitsera*, 43; Frolov, *Vse oni khoteli zhit'*, 69.
20. Chekhovich, *Dnevnik ofitsera*, 69; see also 21–22.
21. *RBP-40*, 75–76.

22. Ol'ga Gurova, *Sovetskoe nizhnee bel'ë: Mezhdu ideologiei i povsednevnost'iu* (Moscow: NLO, 2008), 38–64.

23. Grigorii Pernavskii, "Ekipazh mashiny boevoi," in *Ia dralsia na T-34*, ed. Artëm Drabkin (Moscow: Eksmo, 2005), 64.

24. TsAMO RF f. 239, d. 2204, op. 167, ll. 115–20, in *Tyl Krasnoi Armii*, 569; TsAMO RF f. 235, op. 2096, d. 103, ll. 84–86, in *Tyl Krasnoi Armii*, 600.

25. Genatulin, "Vot konchitsia voina," in *Vot konchitsia voina*, 210.

26. Lesin, *Byla voina*, 82.

27. Boris Antropov, "Daleko za rubezhom," *Gostinyi dvor* 16 (2005): 71–72.

28. V. P. Kiselev. "Voina i zhizn' v predstavlenii 20-letnikh frontovikov," 1008.

29. Suris, *Frontovoi dnevnik*, 77.

30. TsA FSB RF f. 14, op. 5, d. 96, ll. 83–91, in *Stalingradskaia epopeia: Materialy NKVD SSSR i voennoi tsenzury iz Tsentral'nogo arkhiva FSB RF*, ed. A.T. Zhadobin et al. (Moscow: Zvonnitsa MG, 2000), 141–47; RGVA f. 4, op. 11, d. 77, ll. 174–93, in *Prikazy narodnogo komissara oborony SSSR 1943–1945 gg.*, 257.

31. Dunaevskaia, *Ot Leningrada do Kënigsberga*, 76, 186–88.

32. Suris, *Frontovoi dnevnik*, 24.

33. Chekhovich, *Dnevnik ofitsera*, 46; Dunaevskaia, *Ot Leningrada do Kënigsberga*, 57.

34. Iudif' Vladimirovna Golubkova, interview by Artëm Drabkin, www.iremember.ru (March 14, 2007); Vera Ivanovna Malakhova, "Four Years a Frontline Physician," in *A Revolution of Their Own: Voices of Women in Soviet History*, ed. Barbara Alpern Engel and Anastasia Posadskaya-Vanderbeck (Boulder: Westview, 1998), 198.

35. Malakhova, "Four Years a Frontline Physician," 199.

36. NA IRI RAN f. 2, r. X, op. 7, d. 2a, ll. 6ob.–7; Dunaevskaia, *Ot Leningrada do Kënigsberga*, 98–99.

37. TsAMO RF f. 213, op. 2026, d. 1, l. 52, in *Tyl Krasnoi Armii*, 195; Samoilov, *Podënnye zapisi*, 1: 152–53.

38. TsAMO RF f. 32, op. 11289, d. 434, ll. 332–48, in *Preliudiia Kurskoi bitvy*, 369.

39. Abdulin, *160 stranits*, 73; RGASPI f. 88, op. 1, d. 958, l. 3.

40. Antropov, "Daleko za rubezhom," 72.

41. Irina Vladimirovna Iavorskaia, interview by Bair Irincheev, in *"A zori zdes' gromkie": Zhenskoe litso voiny*, ed. Artëm Drabkin and Bair Irincheev (Moscow: Eksmo, 2012), 102–3; Iuliia K. Zhukova, *Devushka so snaiperskoi vintovkoi: Vospominaniia vypusknitsy Tsentral'noi zhenskoi shkoly snaiperskoi podgotovki 1944–1945* (Moscow: Tsentrpoligraf, 2006), 77; Dunaevskaia, *Ot Leningrada do Kënigsberga*, 98–99.

42. Zhukova, *Devushka so snaiperskoi vintovkoi*, 73.

43. Baklanov, *Zhizn'*, 52; Genatulin, *Strakh*, 13; *Pamiatka krasnoarmeitsa o podgonke, noshenii i ukhode za obmundirovaniem i obuv'iu* (Moscow: Voenizdat, 1941), 26, 41.

44. Nikolai Litvin and Stuart Britton, *800 Days on the Eastern Front: A Russian Soldier Remembers World War II* (Lawrence: University Press of Kansas, 2007), 51; A. Kibovskii, A. Stepanov, and K. Tsyplenkov, *Uniforma rossiiskogo voennogo vozdushnogo flota*, 2 vols. (Moscow: Russkie vitiazi, 2007), 2, pt. 1:50.

45. Nicholas S. Timasheff, *The Great Retreat: The Growth and Decline of Communism in Russia* (New York: E. P. Dutton, 1946), 404–8; N. A. Antipenko, *Na glavnom napravlenii* (Moscow: Nauka, 1967), 303.

46. Slëzkin, *Do voiny i na voine*, 309.
47. Kibovskii et al., *Uniforma rossiiskogo voennogo vozdushnogo flota*, 50.
48. Yakushin, *On the Roads of War*, 133–34.
49. Suris, *Frontovoi dnevnik*, 32; Zhukova, *Devushka so snaiperskoi vintovkoi*, 170.
50. Belov, *Frontovoi dnevnik*, January 4, 1943.
51. TsAMO RF f. 208, op. 14139, d. 1, ll. 188–90, in *Tyl Krasnoi Armii*, 166–68; Antipenko, *Na glavnom napravlenii*, 302.
52. Chekhovich, *Dnevnik ofitsera*, 65, 68.
53. TsAMO RF f. 32, op. 11289, d. 434, ll. 332–48, in *Preliudiia Kurskoi bitvy*, 369.
54. Zhukova, *Devushka co snaiperskoi vintovkoi*, 74–75.
55. However, as Nikolai Inozemtsev wrote in his diary, having ragged boots that spoke to one coming from the front could earn favors among civilians (*Frontovoi dnevnik*, 92).
56. Rakhimzhan Koshkarbaev, *Shturm: Den' 1410* (Alma-Ata: Zhalyn, 1983), 128.
57. Loginov, *Eto bylo na fronte*, 93; Litvin, *800 Days on the Eastern Front*, 51.
58. Tvardovskii, *Vasilii Tërkin*, 12.
59. *Pamiatka krasnoarmeitsa o podgonke*, 16, 38–39; *RBP-40*, 76.
60. Kibovskii et al., *Uniforma rossiiskogo voennogo vozdushnogo flota*, 150.
61. TsAMO RF f. 208, op. 14703 c, d. 2, l. 339–43, in *Tyl Krasnoi Armii*, 136–37.
62. TsAMO RF f. 2, op. 920266, d. 2, ll. 12–19, in *Tyl Krasnoi Armii*, 208–12; Abdulin, *160 stranits*, 60.
63. RGVA f. 4, op. 14, d. 2737, ll. 58–70, in *"Zimniaia voina": Rabota nad oshibkami aprel'–mai 1940 g.: Materialy Komissii Glavnogo voennogo soveta Krasnoi Armii po obobshcheniiu opyta finskoi kampanii*, ed. N. S. Tarkhova et al. (St. Petersburg: Letnii sad, 2004), 119; RGVA f. 4, op. 12, d. 97, l. 384, in *Tyl Krasnoi Armii*, 103.
64. "'Eto to, chto nabolelo, chto prositsia naruzhu': Pis'ma I. V. Stalinu. 1941–1942 gg.," *Istoricheskii arkhiv*, no. 2 (2005): 28.
65. In her diary Dunaevskaia referred to the uniform as the "new-old model" (*Ot Leningrada do Kënigsberga*, 170).
66. NA IRI RAN f. 2, r. III, op. 5, d. 2a, l. 16.
67. Gosudarstvennyi arkhiv Rossiiskoi Federatsii (GARF) f. R-5446, op. 44a, d. 9427, l. 41.
68. Kibovskii et al., *Uniforma rossiiskogo voennogo vozdushnogo flota*, 132, 218.
69. David Ortenberg, *Stalin, Shcherbakov, Mekhlis i drugie* (Moscow: Kodeks, 1995), 96–97.
70. Moran, "O vruchenii pogonov."
71. B. I. Kolonitskii, *Pogony: Bor'ba za vlast' v 1917 godu* (St. Petersburg: Ostrov, 2001), 79–82.
72. NA IRI RAN f. 2, r. III, op. 5, d. 2a, ll. 7ob.–8.
73. A. Krivitskii, "O voennom mundire i pogonakh," *Krasnaia zvezda*, January 7, 1943.
74. On the failure of nationalizing projects before the Bolsheviks, see Brandenberger, *National Bolshevism*, 10, 226; and Sanborn, *Drafting the Russian Nation*, 203–6. On prewar propaganda promoting Soviet patriotism, see Brandenberger, *Propaganda State in Crisis*, 98–119.
75. RGVA f. 4, op. 12, d. 106-a, ll. 566–68, in *Prikazy narodnogo komissara oborony SSSR 22 iiunia 1941 g.–1942 g.*, 383–84.

76. RGVA f. 4, op. 12-a, d. 82, ll. 562–628, in *Prikazy narodnogo komissara oborony SSSR 1937–21 iiunia 1941 g.*, 43.

77. RVGA f. 4, op. 12, d. 90, l. 5, in *Prikazy narodnogo komissara oborony SSSR 22 iiunia 1941 g.–1942 g.*, 64; RVGA f. 4, op. 12, d. 106-a, ll. 242–43, in *Prikazy narodnogo komissara oborony SSSR 22 iiunia 1941 g.–1942 g.*, 360–61.

78. Dunn, *Hitler's Nemesis*, xviii–xix; General-maior V. N. Nesmelov, "Krasnaia armiia stala kadrovoi armiei," *Agitator i propagandist Krasnoi Armii*, no. 7 (1943): 1–2.

79. IRI RAN f. 2, r. 1, op. 124, d. 1, l. 1.

80. TsAMO RF f. 2, op. 729266, d. 6, l. 72, in *Tyl Krasnoi Armii*, 314.

81. Moran, "O vruchenii pogonov"; Lesin, *Byla voina*, 130.

82. Brychev, "Novye znaki razlichiia," 29.

83. "Perekhod na novye znaki razlichiia—pogony," *Krasnaia zvezda*, January 31, 1943; "Voennosluzhashchie vne stroia," *Krasnaia zvezda*, July 18, 1943.

84. See, e.g., Lesin, *Byla voina*, 246–47; and Temkin, *My Just War*, 129–30.

85. Dunaevskaia, *Ot Leningrada do Kënigsberga*, 243.

86. General-leitenant A. Ignat'ev, "Ofitsery," *Krasnoarmeets*, no. 17–18 (1943): 5.

87. TsAMO RF f. 32, op. 795436, d. 12, l. 19, in *Glavnye politicheskie organy vooruzhennykh sil SSSR*, 271–72.

88. Suris, *Frontovoi dnevnik*, 74.

89. N. V. Pupyshev, *V pamiati i v serdtse* (Moscow: Voenizdat, 1986), 110; Bulatov, *Budni frontovykh let*, 45.

90. Frolov, *Vse oni khoteli zhit'*, 59.

91. Slëzkin, *Do voiny i na voine*, 456–57.

92. NA IRI RAN f. 2, r. III, op. 5, d. 4, l. 30ob.

93. TsA FSB RF f. 14, op. 4, d. 913, ll. 151–53, in *Stalingradskaia epopeia*, 388–89.

94. Samoilov, *Podënnye zapisi*, 1: 178.

95. TsA FSB RF f. 14, op. 4, d. 913, ll. 151–53, in *Stalingradskaia epopeia*, 390–91. See also Merridale, *Ivan's War*, 164.

96. Dunaevskaia, *Ot Leningrada do Kënigsberga*, 237; Slutskii, *O drugikh i o sebe*, 120–23.

97. Suris, *Frontovoi dnevnik*, 137–38.

98. Litvin, *800 Days on the Eastern Front*, 29. See also Lesin, *Byla voina*, 185; Genatulin, "Vot konchitsia voina," 250.

99. RGASPI f. 17, op. 125, d. 242, l. 101.

100. "Gimn SSSR" and "The Internationale."

101. Slutskii, *O drugikh i o sebe*, 132.

102. Berkhoff, *Motherland in Danger*, 60–61. Jochen Hellbeck has dubbed this "the hero strategy" (*Stalingrad: The City that Defeated the Third Reich* [New York: Public Affairs, 2015], 35, 47–50). David Brandenberger, among others, has tracked the obsession with heroes in 1930s popular culture (*Propaganda State in Crisis*, 67–97). Samuel Clark has argued that medals serve as a coordinating device that communicate expectations about exemplary behavior and that new regimes often create new awards in order to foster new hierarchies (*Distributing Status: The Evolution of State Honours in Western Europe* [Montreal: McGill-Queens' University Press, 2016], 90, 189, 192, 273).

103. "Nagrada v boiu," *Krasnaia zvezda*, March 6, 1943. See also M. Rubinshtein, "O voinskom vospitanii," *Krasnaia zvezda*, October 18, 1942.

104. RGASPI f. 17, op. 125, d. 78, l. 122, contains a bibliography, "The System of Military Decorations in Russia," submitted to the Propaganda Section of the Party on September 24, 1942; N. Markovin, "Boevye nagrady v staroi russkoi armii," *Agitator i propagandist Krasnoi Armii*, no. 7 (1943): 42–48.

105. Joseph, *Uniforms and Nonuniforms*, 97–98.

106. "Orden i medal'—slava sovetskogo voina," *Krasnaia zvezda*, June 20, 1943.

107. RGVA f. 4, op. 12, d. 108, l. 141, in *Prikazy narodnogo komissara oborony SSSR 1943–1945 gg.*, 148–49; Oleg Smyslov, *Istoriia sovetskikh nagrad: Vo slavu otechestva* (Moscow: Veche, 2007), 214.

108. For example, Suris received his Stalingrad medal almost a year after it was instituted (*Frontovoi dnevnik*, 163).

109. RGASPI f. 17, op. 125, d. 78, l. 57.

110. RGASPI f. 17, op. 125, d. 78, l. 123.

111. RGVA f. 4, op. 11, d. 71, ll. 386–89, in *Prikazy narodnogo komissara oborony SSSR 22 iiunia 1941 g.–1942 g.*, 269–70; "Nagrada v boiu," *Krasnaia zvezda*, March 6, 1943; *Ukaz ot 10 noiabria 1942 goda "O predostavlenii prava nagrazhdeniia ordenami i medaliami SSSR i nagrudnymi znakami komanduiushchim frontami, flotami, armiiami i flotiliiami, komandiram korpusov, divizii, brigad, polkov."*

112. Smyslov, *Istoriia sovetskikh nagrad*, 204–5.

113. Alexandra Orme, *Comes the Comrade* (New York: William Morrow, 1950), 255. One author has estimated that in 1941, 1 percent of all Red Army personnel had been awarded a medal, in 1942, 7 percent, and in 1945, 86 percent (Smyslov, *Istoriia sovetskikh nagrad*, 133–34, 207). In 1947 the state was paying benefits to roughly 5.6 million decorated veterans (Edele, *Soviet Veterans*, 192). Edele's figures would lead to about one in six soldiers having a decoration. The differences in numbers here can probably be ascribed to the fact that many soldiers awarded medals were killed in action and that only those in the active army who survived their initial period at the front received medals.

114. "Polozhenie o nagrudnom znake 'otlichnyi povar,'" *Krasnaia zvezda*, July 9, 1943.

115. NA IRI RAN f. 2, r. X, op. 7, d. 13-b, l. 130.

116. RGVA f. 4, op. 12, d. 99, ll. 110–12, in *Prikazy narodnogo komissara oborony SSSR 22 iiunia 1941 g.–1942 g.*, 85–86; RGVA f. 4, op. 12, d. 104, ll. 389–91, in *Prikazy narodnogo komissara oborony SSSR 22 iiunia 1941 g.–1942 g.*, 243–45; Temkin, *My Just War*, 138–39; AP RF f. 3, op. 50, d. 311, ll. 13–30, in Kudriashov, *Voina*, 156–57.

117. Frolov, *Vse oni khoteli zhit'*, 64, 66; Tsentral'nyi gosudarstvennyi arkhiv istoriko-politicheskoi dokumentatsii Respubliki Tatarstan (TsGA IPD RT) f. 319, op. 1, d. 20, l. 25.

118. Aleksandr Kuznetsov, *Entsiklopediia russkikh nagrad* (Moscow: Golos-Press, 2001), 370–73, 287; RGVA f. 4, op. 12, d. 111, ll. 365–68, in *Prikazy narodnogo komissara oborony SSSR 1943–1945 gg.*, 371–72.

119. Lidzhi Indzhiev, *Frontovoi dnevnik* (Elista: Kalmytskoe knizhnoe izdatel'stvo, 2002), 72. The author was a Kalmyk who celebrated Victory Day in an internment camp, because his entire people had been deported. He waited forty years to get his Defense of the Caucasus campaign medal.

120. "Vvedenie otlichitel'nykh znakov dlia voennosluzhashchikh, ranenykh na frontakh Otechestvennoi voiny," *Krasnaia zvezda*, July 16, 1942; TsAMO RF f. 32, op.

920265, d. 5, l. 539, in *Glavnye politicheskie organy vooruzhennykh sil SSSR*, 151–52; L. Vol'fovskii, "Pochemu do sikh por net otlichitel'nykh znakov dlia ranenykh?" *Krasnaia zvezda*, August 2, 1942; Subbotin, "Soldatskaia dusha."

121. Clark, *Distributing Status*, 265–66. This is a variation of what Clark has called "status inflation"—when a regime begins awarding more decorations, their value often decreases.

122. Vladimir Gel'fand, ed. and intro. Oleg Budnitskii, *Dnevnik 1941–1946* (Moscow, ROSSPEN, 2015), 317.

123. Abdulin, *160 stranits*, 77.

124. Lesin, *Byla voina*, 336.

125. Sonke Neitzel and Harald Welzer, *Soldiers: German POWs on Fighting, Killing, and Dying* (New York: Vintage, 2013), 40–43.

126. RGASPI f. 17, op. 125, d. 78, ll. 122–25.

127. "Pravila nosheniia ordenov, medalei, ordenskikh lent i znakov otlichiia," *Krasnaia zvezda*, June 20, 1943.

128. G. A. Kolesnikov and A. M. Rozhkov, *Ordena i medali SSSR* (Moscow: Voenizdat, 1983), 95.

129. "Statut Ordena Otechestvennoi voiny," *Pravda*, May 21, 1942.

130. RGVA f. 4, op. 12, d. 107, l. 590, in *Prikazy narodnogo komissara oborony SSSR 1943–1945 gg.*, 100–101.

131. Kolesnikov and Rozhkov, *Ordena i medali SSSR*, 68–69.

132. Brandon M. Schechter, "Embodied Violence: A Red Army Soldier's Journey as Told by Objects," in *Objects of War: The Material Culture of Conflict and Displacement*, ed. Leora Auslander and Tara Zahra (Ithaca: Cornell University Press, 2018), 152–54.

133. V. F. Loboda and I. P. Kargal'tsev, *Pravila nosheniia ordenov, medalei SSSR, nagrudnykh znakov i ordenskikh lent* (Moscow: Voenizdat, 1948), 26.

134. A. Krivitskii, "O naimenovanii voinskikh chastei i soedinenii," *Krasnaia zvezda*, February 7, 1943.

135. Kolesnikov and Rozhkov, *Ordena i medali SSSR*, 28–29.

136. Kolesnikov and Rozhkov, *Ordena i medali SSSR*, 31.

137. RGVA f. 4, op. 12, d. 105, ll. 689–96, in *Prikazy narodnogo komissara oborony SSSR 22 iiunia 1941 g.–1942 g.*, 315.

138. Abdulin, *160 stranits*, 77.

139. TsAMO RF f. 243, op. 2963, d. 98, l. 21, in *Tyl Krasnoi Armii*, 645–46; "Sobirat' pamiatniki i relikvii Otechestvennoi voini," *Krasnaia zvezda*, April 7, 1943.

140. *Ustav garnizonnoi sluzhby*, 86–87.

141. Kotkin, *Magnetic Mountain*, 198–238; *Obshchestvo i vlast': Rossiiskaia provintsiia*, vol. 3: *Iiun' 1941 g.–1953g* (Moscow: IRI RAN, 2005), 977–78.

142. Baklanov, "Naveki–deviatnadtsatiletnie," 263.

143. GARF f. R-7523, op. 13. d. 66, ll. 1–2, 38–39.

144. GARF f. R-7523, op. 13, d. 66, ll. 66–69.

145. AP RF f. 3, op. 50, d. 464, ll. 131–33, in Kudriashov, *Voina*, 158.

146. Frolov, *Vse oni khoteli zhit'*, 82.

147. Tvardovskii, *Vasilii Tërkin*, 45–48.

148. Lesin, *Byla voina*, 217.

149. Dunaevskaia, *Ot Leningrada do Kënigsberga*, 257; G. V. Slavgorodskii, *Frontovoi dnevnik 1941–1945* (Moscow: Politicheskaia entsiklopediia, 2017), 120.

150. RGVA f. 4, op. 11, d. 78, ll. 53–54, in *Prikazy narodnogo komissara oborony SSSR 1943–1945 gg.*, 299–300.

151. Aizenshtat, *Zapiski sekrataria voennogo tribunala*, 120. See also Malakhova, "Four Years a Frontline Physician," 215; and Kopelev, *Khranit' vechno*, 1:90.

152. Dunaevskaia, *Ot Leningrada do Kënigsberga*, 257–58; NA IRI RAN f. 2, r. I, op. 30, d. 23, l. 2ob.

153. Suris, *Frontovoi dnevnik*, 175, 357.

154. For a full description of Damcheev's medals and accomplishments, see Schechter, "Embodied Violence."

155. RGASPI f. 17, op. 125, d. 130, ll. 55–56.

156. RGASPI f. 88, op. 1, d. 972, l. 3.

157. Prezidium Verkhovnogo Soveta SSSR, *Ukaz ot 16 dekabria 1947 goda: O vnesenii izmenenii v zakonodatel'stvo SSSR v sviazi s izdaniem ukaza Prezidiuma Verkhovnogo soveta SSSR ot 10 sentiabria 1947 goda "O l'gotakh i preimushchestvakh, predostavliaemykh nagrazhdënnym ordenami i medaliami SSSR"*; Edele, *Soviet Veterans*, 192–93.

158. Pomerants, *Zapiski gadkogo utënka*, 130.

159. Once veterans became central to the regime's legitimacy, the issuance of medals on anniversaries became standard.

160. Aleksandr Valer'evich Pecheikin, "Iz istorii voennogo obmundirovaniia i snariazheniia," http://history.milportal.ru/2011/06/voennaya-shinel/.

161. Lesin, *Byla voina*, 307.

162. Samoilov, *Pamiatnye zapiski*, 193.

163. TsAMO RF f. 208, op. 14703 c, d. 2, ll. 339–43, in *Tyl Krasnoi Armii*, 136.

164. Genatulin, "Ataka," in *Vot konchitsia voina*, 27–30; Inozemtsev, *Frontovoi dnevnik*, 155.

165. *Instruktsiia po ukladke, prigonke, sborke i nadevaniiu pokhodnogo snariazheniia boitsa pekhoty Krasnoi Armii* (Moscow: Voenizdat, 1941), 27–32; Pecheikin, "Iz istorii voennogo obmundirovaniia i snariazheniia."

166. Bek, *Volokolamskoe shosse*, 62–65.

167. Geroi Sovetskogo Soiuza V. Galakhov, "Tridtsatoe noiabria," in *Boi v Finliandii: Vospominaniia uchastnikov* (Moscow: Voenizdat, 1941), 1:36.

168. Lesin, *Byla voina*, 65.

169. Anatolii Genatulin, "Bessonnaia pamiat'," *Znamia*, no. 5 (2005), http://magazines.russ.ru/znamia/2010/5/ge2.html.

170. Vladimir Karpov, *Russia at War, 1941–1945* (London: Vendome, 1987), 201.

171. Lesin, *Byla voina*, 178.

172. Bulatov, *Budni frontovykh let*, 219.

173. See, e.g., Kharis Iakupov, *Khater. Pamiat'. Memory: Al'bom* (Kazan: Tatknigizdat, 2002), 62, 66, 70; Frolov, *Vse oni khoteli zhit'*, 56; and Suris, *Frontovoi dnevnik*, 96–97.

174. *Pamiatka krasnoarmeitsa o podgonke*, 35.

175. Joanne Finkelstein, *The Fashioned Self* (Philadelphia: Temple University Press, 1991), 107–29.

176. *Ustav garnizonnoi sluzhby*, 126.

177. *RBP-40*, 48, 71–72, 98.

178. Slëzkin, *Do voiny i na voine*, 326; Dunaevskaia, *Ot Leningrada do Kënigsberga*, 307.

179. Lesin, *Byla voina*, 66.

180. T. K. Strizhenova, *Iz istorii sovetskogo kostiuma* (Moscow: Sovetskii khudozhnik, 1972), 24–25. There is widely circulated apocryphal story that these hats were designed during the Great War for the victory parade in Berlin.

181. Mints, "O traditsiiakh Krasnoi Armii," 15.

182. Chekhovich, *Dnevnik ofitsera*, 68.

183. Lesin, *Byla voina*, 277–79.

184. Baklanov, "Mertvye sramu ne imut," in *Voennye povesti*, 139; *Pamiatka krasnoarmeitsa o podgonke*, 26.

185. Tvardovskii, *Vasilii Tërkin*, 126.

186. Aleksandr Tvardovskii, *"Ia v svoiu khodil ataku . . ." Dnevniki. Pis'ma 1941–1945* (Moscow: Vagrius, 2005), 90.

187. RGVA f. 4, op. 12, d. 107, l. 209, in *Prikazy narodnogo komissara oborony SSSR 22 iiunia 1941 g.–1942 g.*, 40.

188. Yakushin, *On the Roads of War*, 139.

189. *RBP-40*, 48; *Pamiatka krasnoarmeitsa o podgonke*, 31–34.

190. Orme, *Comes the Comrade*, 19–20.

191. Loginov, *Eto bylo na fronte*, 24; Samoilov, *Pamiatnye zapiski*, 206.

192. Samoilov, *Pamiatnye zapiski*, 280.

3. The State's Pot and the Soldier's Spoon

1. Lesin, *Byla voina*, 146. Epigraph: Lesin, *Byla voina*, 76.

2. TsAMO RF f. 2, op. 795437, d. 11, ll. 546–49, in *Tyl Krasnoi Armii*, 401–5. This order is also mentioned in memoirs by provisioning officers and histories of the war. See F. S. Saushin, *Khleb i sol'* (Iaroslavl: Verkhne-Volzhskoe knizhnoe izdatel'stvo, 1983), 56; Anastas Mikoian, *Tak bylo: Razmyshleniia o minuvshem* (Moscow: Vagrius, 1999), 431; S. K. Kurkotkin, *Tyl sovetskikh vooruzhennykh sil v Velikoi Otechestvennoi voine 1941–1945 gg.* (Moscow: Voenizdat, 1977), 202–3.

3. TsAMO RF f. 2, op. 795437, d. 9, l. 696, in *Tyl Krasnoi Armii*, 291–92; TsAMO RF f. 2, op. 795437, d. 11, ll. 66–68, in *Tyl Krasnoi Armii*, 306–8; TsAMO RF f. 47, op. 1029, d. 83, ll. 53–55, in *Tyl Krasnoi Armii*, 321–25; TsAMO RF f. 2, op. 795437, d. 11, ll. 293–95; TsAMO RF f. 47, op. 1029, d. 84, ll. 23–24, in *Tyl Krasnoi Armii*, 380–82.

4. *1936 Constitution of the USSR*, http://www.departments.bucknell.edu/russian/const/36cons01.html.

5. A note on historiography: in English-language historiography, provisioning often receives treatment in larger works on the Red Army more generally, particularly in the works of military historians interested in combat effectiveness and military science. Colonel David Glantz provides a brief but very good soldiers' eye view of provisioning (*Colossus Reborn*, 555–60). William Moskoff's *The Bread of Affliction: The Food Supply in the USSR during World War II* is a pioneering overview but was written before many of the relevant primary sources became available (New York: Cambridge University Press, 1990). Nicholas Ganson provides a concise overview of the provisioning situation in the country and makes many astute observations, particularly on ways in which the state often waited until situations became disastrous before getting involved in provisioning ("Food Supply, Rationing, and Living Standards," 69–92). Lizzie Collingham's thorough treatment of food during the war provides an excellent

global context and a fair overview of the situation in the Soviet Union. However, given its immense scope, this work inevitably suffers from a limited source base in its treatment of the Soviet Union. That being said, she provides a good summary of the role of Lend-Lease food aid in provisioning the Soviet army, something that I barely touch on. See Lizzie Collingham, "Fighting on Empty," in *The Taste of War: World War Two and the Battle for Food* (New York: Penguin, 2012), 317–46. Soviet historiography on the subject tended to be written by participants with an eye for improving provisioning in the future, often taking the form of memoirs. One of the few Soviet authors interested in the cultural dimensions of rations was the veteran and food historian Vil'iam Pokhlëbkin, whose work has inspired my own attempt to write an ethnography of Red Army rations. For issues of provisioning in the Soviet Union as a whole, see Wendy Goldman and Donald Filtzer, eds., *Hunger and War: Food Provisioning in the Soviet Union during World War II* (Bloomington: Indiana University Press, 2015).

6. V. P. Zotov, *Pishchevaia promyshlennost'* (Moscow: Pishchevaia promyshlennost', 1967), 24–25.

7. RGVA f. 4, op. 11, d. 72, l. 270, in *Prikazy narodnogo komissara SSSR 22 iiunia 1941 g.–1942 g.*, 277.

8. See, e.g., Joseph Stalin, *The Great Patriotic War of the Soviet Union* (New York: International Publishers, 1945), 15.

9. Zotov, *Pishchevaia promyshlennost'*, 136–38.

10. For further discussion of the semantics of rationing, see my chapter in *Hunger and War*.

11. P. Ia. Chernykh, *Istoriko-etimologicheskii slovar' sovremennogo russkogo iazyka* (Moscow: Russkii iazyk, 1993), 1:615.

12. Aleksandr Grigor'evich Preobrazhenskii, *Etimologicheskii slovar' russkogo iazyka*, vol. 2: *P–S* (Moscow: Gosudarstvennoe izdatel'stvo inostrannykh i natsional'nykh slovarei, 1959), 725.

13. This was in line with earlier provisioning policies. See, for example, E. A. Osokina, *Za fasadom "Stalinskogo izobiliia": Raspredelenie i rynok v snabzhenii naseleniia v gody industrializatsii, 1927–1941* (Moscow: ROSSPEN, 1998), 99.

14. Sanborn, *Drafting the Russian Nation*, 107–10.

15. S. Gurov, *Boets i otdelenie na pokhode* (Moscow: Voenizdat, 1941), 11.

16. F. G. Krotkov, ed., *Gigiena: Opyt sovetskoi meditsiny v Velikoi Otechestvennoi voine 1941–1945 gg.* (Moscow: MEDGIZ, 1955), 33:3, 134; I. Ia. Moreinis, *Uchebnik pishchevoi gigieny dlia sanitarno-fel'dsherskikh shkol* (Moscow: MEDGIZ, 1940), 126.

17. TsAMO RF f. 2, op. 920266, d. 1, ll. 718–929, in *Tyl Krasnoi Armii*, 147–48.

18. Moreinis, *Uchebnik pishchevoi gigieny*, 28, 203.

19. Kurkotkin, *Tyl vooruzhennykh sil*, 191; Krotkov, *Gigiena*, 139.

20. See, e.g., TsAMO RF f. 47, op. 1029, d. 84, ll. 23–24, in *Tyl Krasnoi Armii*, 381.

21. S. Gurov, *Pokhod i otdykh pekhoty* (Moscow: Voenizdat, 1940), 72–73.

22. Matthew Payne, "The Forge of the Kazakh Proletariat?" in *A State of Nations*, ed. Ronald Grigor Suny and Terry Martin (New York: Oxford University Press, 2001), 234–35; Richard S. Fogarty, *Race and War in France: Colonial Subjects in the French Army, 1914–1918* (Baltimore: John Hopkins University Press, 2008), 169, 173, 183–89; Tarak Barkawi, "Peoples, Homelands, and Wars? Ethnicity, the Military, and Battle among British Imperial Forces in the War against Japan," *Comparative Studies in Society and History* 46 (2004): 134–63.

23. TsAMO RF f. 2, op. 920266, d. 1, ll. 718–929, in *Tyl Krasnoi Armii*, 148; Krotkov, *Gigiena*, 140. According to Boris Slutskii, soldiers resented the *doppaëk* (*O drugikh i o sebe*, 28).

24. Bulatov, *Budni frontovykh let*, 251.

25. TsAMO RF, f. 2, op. 920266, d. 1, ll. 718–929, in *Tyl Krasnoi Armii*, 148; Krotkov, *Gigiena*, 140.

26. See, e.g., NA IRI RAN f. 2, r. I, op. 123, d. 13, l. 2ob.

27. TsAMO RF f. 2, op. 920266, d. 1, ll. 718–929, in *Tyl Krasnoi Armii*, 149–55.

28. See Goldman, "Not by Bread Alone: Food, Workers, and the State," in *Hunger and War*, 56–60.

29. TsAMO RF f. 2, op. 920266, d. 1, ll. 718–929, in *Tyl Krasnoi Armii*, 148; TsAMO RF f. 2, op. 795437, d. 10, l. 276, in *Tyl Krasnoi Armii*, 414; Loginov, *Eto bylo na fronte*, 10.

30. TsAMO RF f. 2, op. 920266, d. 1, ll. 718–929, in *Tyl Krasnoi Armii*, 149–56.

31. See, e.g., Slutskii, *O drugikh i o sebe*, 29–31.

32. Lesin, *Byla voina*, 53.

33. Baklanov, *Zhizn'*, 47–48.

34. Lesin, *Byla voina*, 64–65, 80.

35. L. P. Grachev, *Doroga ot Volkhova* (Leningrad: Lenizdat, 1983), 213.

36. See, e.g., Saushin, *Khleb i sol'*, 41; and Antipenko, *Na glavnom napravlenii*, 60.

37. Antipenko, *Na glavnom napravlenii*, 7–8, 117; TsAMO RF f. 2, op. 795437, d. 5, ll. 545–47, in *Tyl Krasnoi Armii*, 90–100; TsAMO RF f. 208, op. 224922-c, d. 1, ll. 139–40, in *Tyl Krasnoi Armii*, 123–24; TsAMO RF f. 2, op. 795437, d. 4, l. 378, in *Tyl Krasnoi Armii*, 173–74; TsAMO RF f. 244, op. 3017, d. 2, ll. 20–23, in *Tyl Krasnoi Armii*, 345–47; TsAMO RF f. 2, op. 795437, d. 11, l. 370, in *Tyl Krasnoi Armii*, 384. See also TsAMO RF f. 2, op. 920266, d. 6, l. 47, in *Tyl Krasnoi Armii*, 313–14; TsAMO RF f. 2, op. 920266, d. 6, l. 405, 406, in *Tyl Krasnoi Armii*, 388–89; TsAMO RF f. 236, op. 2719, d. 76, ll. 14–15, in *Tyl Krasnoi Armii*, 659–60; Lesin, *Byla voina*, 319; Mikoian, *Tak bylo*, 469–70; and Grachev, *Doroga ot Volkhova*, 77, 208–13. For an English-language treatment of subsidiary agriculture, see Moskoff, *Bread of Affliction*, 94–113; Goldman, "Not by Bread Alone," 46, 53–55, 74–78, 92–93.

38. John Samuels, ed., *Ration Development* (Washington, DC: Quartermaster Food & Container Institute for the Armed Forces, 1947), 24–48.

39. The Quartermaster School for the Quartermaster General, "Rations Conference Notes," January 1949, US Army Quartermaster Foundation, Fort Lee, VA, http://www.qmfound.com/history_of_rations.htm.

40. Collingham, *Taste of War*, 37–39, 184–87.

41. See Adam Tooze, *The Wages of Destruction: The Making and Breaking of the Nazi Economy* (New York: Penguin, 2006), 467, 469, 476–80, 538–49.

42. TsAMO RF f. 2, op. 920266, d. 1, l. 480, in *Tyl Krasnoi Armii*, 91–92; TsAMO RF f. 208, op. 3031, d. 2, ll. 501, 502, in *Tyl Krasnoi Armii*, 172–73.

43. Temkin, *My Just War*, 115; Genatulin, "Ataka," in *Vot konchitsia voina*, 36.

44. RGVA f. 4, op. 11, d. 76, ll. 70–75, in *Prikazy narodnogo komissara oborony SSSR 1943–1945 gg.*, 168; Loginov, *Eto bylo na fronte*, 9–10; TsAMO RF f. 47, op. 1029, d. 83, ll. 53–55, in *Tyl Krasnoi Armii*, 324; Krotkov, *Gigiena*, 145–50.

45. D. P. Vorontsov, *Prodovol'stvennoe snabzhenie strelkovogo batal'ona i polka v deistvuiushchei armii* (Moscow: Voennaia Akademiia tyla i snabzheniia Krasnoi armii imeni Molotova V. M., 1943), 12.

46. Kurkotkin, *Tyl sovetskikh vooruzhennykh sil*, 190; see also TsAMO RF f. 208, op. 14703c, d. 2, ll. 339–43, in *Tyl Krasnoi Armii*, 137–38; TsAMO RF f. 47, op. 1029, d. 83, ll. 53–55, in *Tyl Krasnoi Armii*, 324.

47. See, e.g., TsAMO RF f. 235, op. 2096, d. 104, l. 9, in *Tyl Krasnoi armii*, 479–80; and Khisam Kamalov, *U kazhdogo zhizn' odna*, trans. from Tatar by V. Mal'tsev (Kazan': Tatknigizdat, 1983), 203–5.

48. Vorontsov, *Prodovol'stvennoe snabzhenie*, 12.

49. RGASPI f. 88, op. 1, d. 958, l. 7.

50. Vil'iam Pokhlëbkin, *Kukhnia veka* (Moscow: Polifakt, 2000), 209.

51. See *Pamiatka voiskovomu povaru* (Moscow: Voenizdat, 1943), 14–15.

52. Pokhlëbkin, *Kukhnia veka*, 210; *Vospominaniia frontovikov: Sbornik No. 1* (Moscow: Voennaia akademiia tyla i snabzheniia Krasnoi Armii imeni Molotova V. M., 1943), 16.

53. Lesin, *Byla voina*, 88; Loginov, *Eto bylo na fronte*, 9–10. See also TsAMO RF f. 208, op. 14703c, d. 2, ll. 339–43, in *Tyl Krasnoi Armii*, 138; and NA IRI RAN f. 2, r. I, op. 223, d. 9, l. 1ob.

54. *RBP-40*, 45; Vorontsov, *Prodovol'stvennoe snabzhenie*, 16. The increased use of dry rations was a lesson the Red Army took away from the Winter War (RGASPI f. 74, op. 2, d. 121, ll. 11, 32).

55. Antipenko, *Na glavnom napravlenii*, 92.

56. Vorontsov, *Prodovol'stvennoe snabzhenie*, 49; RGASPI f. 74, op. 2, d. 121, l. 11.

57. TsAMO RF f. 208, op. 14703c, d. 2, ll. 339–43, in *Tyl Krasnoi Armii*, 138; GARF f. R5446, op. 43a, d. 8627, l. 7.

58. Antipenko, *Na glavnom napravlenii*, 92; Kurkotkin, *Tyl sovetskikh vooruzhennykh sil*, 206; Krotkov, *Gigiena*, 150–53.

59. Slëzkin, *Do voiny i na voine*, 278–79, 404; Antipenko, *Na glavnom napravlenii*, 92; Collingham, *Taste of War*, 339.

60. Dunaevskaia, *Ot Leningrada do Kënigsberga*, 94; *Instruktsiia po ukladke pokhodnykh kukhon' i nalivnogo kipatil'nika* (Moscow: Voenizdat, 1942), 18–19.

61. Vil'iam Pokhlëbkin, *Moia kukhnia i moe meniu* (Moscow: Tsentrpoligraf, 1999), 278–79.

62. TsGA IPD RT f. 8288, op. 1, d. 14, l. 22, in D. I. Ibragimov et al., *Pis'ma s fronta 1941–1945 gg.: Sbornik dokumentov* (Kazan': Gasyr, 2010), 81–82.

63. Slutskii, *O drugikh i o sebe*, 29.

64. See, e.g., Nikita Lomagin, *Neizvestnaia blokada: Dokumenty, prilozheniia* (St. Petersburg: Neva, 2004), 2:185.

65. Stalin, *Great Patriotic War of the Soviet Union*, 22–23.

66. TsAMO RF f. 2, op. 795437, d. 11, ll. 546–49, in *Tyl Krasnoi Armii*, 403.

67. Quoted in Antipenko, *Na glavnom napravlenii*, 299.

68. Lesin, *Byla voina*, 82; TsA FSB RF f. 14, op. 4, d. 943, l. 327, in *Stalingradskaia epopeia*, 379; Chekhovich, *Dnevnik ofitsera*, 73; NA IRI RAN f. 2, r. I, op. 28, d. 30, l. 10.

69. TsA FSB RF f. 14, op. 4, d. 777, ll. 40–44, in *Stalingradskaia epopeia*, 259–60; RGASPI f. 84, op. 1, d. 84, l. 2.

70. RGASPI f. 88, op. 1, d. 958, l. 3. See also Lesin, *Byla voina*, 146; and TsAMO RF f. 2, op. 795437, d. 9, ll. 394–95, in *Tyl Krasnoi Armii*, 237–41.

71. TsAMO RF f. 2, op. 795437, d. 11, ll. 546–49 in *Tyl Krasnoi Armii*, 403.

72. RGVA f. 4, op. 12, d. 98, ll. 210–14, in *Prikazy narodnogo komissara oborony SSSR 22 iiunia 1941 g.–1942 g.*, 11–13.

73. RGASPI f. 88, op. 1, d. 958, l. 10. This problem continued throughout the war. See RGASPI f. 84, op. 1, d. 86, ll. 223–30.

74. See GARF f. R-5446, op. 46a, d. 7395, ll. 27–28; RGASPI f. 88, op. 1, d. 958, l. 13.

75. RGASPI f. 88, op. 1, d. 958, ll. 11–13; see also Larionov, *Frontovaia povsednevsnost'*, 122–62.

76. TsAMO RF f. 241, op. 2618, d. 12, ll. 131–33, in *Tyl Krasnoi Armii*, 591–93.

77. Antipenko, *Na glavnom napravlenii*, 125.

78. TsAMO RF f. 208, op. 2563, d. 47, ll. 212–14, in *Tyl Krasnoi Armii*, 258–61; TsAMO RF f. 2, op. 795437, d. 9, l. 527, in *Tyl Krasnoi Armii*, 261–62; TsAMO RF, f. 2, op. 795437, d. 9. l. 696, in *Tyl Krasnoi Armii*, 291–92; and RGASPI f. 88, op. 1, d. 958, l. 15.

79. TsAMO RF f. 208, op. 2563, d. 47, ll. 212–14, in *Tyl Krasnoi Armii*, 258–61; TsAMO RF f. 2, op. 795437, d. 9, l. 696, in *Tyl Krasnoi Armii*, 291–92, TsAMO RF f. 47, op. 1029, d. 83, ll. 53–55, in *Tyl Krasnoi Armii*, 321–25; RGVA f. 4, op. 11, d. 73, ll. 299–301, in *Prikazy narodnogo komissara oborony SSSR 22 iiunia 1941 g.–1942 g.*, 372–74; RGVA f. 4, op. 11, d. 75, ll. 38–40, in *Prikazy narodnogo komissara oborony SSSR 1943–1945 gg.*, 24–26; RGVA f. 4, op. 11, d. 75, ll. 41–46, in *Prikazy narodnogo komissara oborony SSSR 1943–1945 gg.*, 26–28; RGVA, f. 4, op. 11, d. 75, ll. 52–54, in *Prikazy narodnogo komissara oborony SSSR 1943–1945 gg.*, 29–30; RGVA, f. 4, op. 12, d. 107, ll. 307, in *Prikazy narodnogo komissara oborony SSSR 1943–1945 gg.*, 70–71.

80. RGVA f. 4, op. 11, d. 75, ll. 38–40, in *Prikazy narodnogo komissara oborony SSSR 1943–1945 gg.*, 24–25.

81. See, e.g., RGVA f. 4, op. 11, d. 71, ll. 472–75, in *Prikazy narodnogo komissara oborony SSSR 22 iiunia 1941 g.–1942 g.*, 273–75.

82. TsAMO RF f. 2, op. 795437, d. 11, ll. 546–49, in *Tyl Krasnoi Armii*, 404.

83. Pokhlëbkin, *Kukhnia veka*, 209. See also "Pshënnye dni," *Krasnaia zvezda*, June 8, 1943.

84. TsAMO RF f. 2, op. 795437, d. 11, ll. 546–49, in *Tyl Krasnoi Armii*, 402.

85. TsAMO RF f. 2, op. 795437, d. 11, ll. 546–49, in *Tyl Krasnoi Armii*, 404; RGVA f. 4, op. 11, d. 75, ll. 94–96, in *Prikazy narodnogo komissara oborony SSSR 1943–1945 gg.*, 38. See also "Krasnoarmeiskaia kukhnia," *Krasnaia zvezda*, April 11, 1943; and RGVA f. 4, op. 11, d. 71, ll. 472–75, in *Prikazy narodnogo komissara oborony SSSR 22 iiunia 1941 g.–1942 g.*, 274.

86. Kurkotkin, *Tyl vooruzhennykh sil*, 203.

87. Pokhlëbkin, *Kukhnia veka*, 212, 227, 230–31; Saushin (*Khleb i sol'*, 59) agreed.

88. *Novye vidy produktov, postupaiushchikh na dovol'stvie Krasnoi Armii* (Moscow: Voenizdat, 1944).

89. *Pamiatka voinskovomu povaru*, 4; Antipenko, *Na glavnom napravlenii*, 131; TsAMO RF f. 217, op. 1250, d. 183, l. 188, in *Tyl Krasnoi Armii*, 417; Saushin, *Khleb i sol'*, 44, 59–61.

90. "Soveshchanie nachal'nikov otdelov agitatsii i propagandy Politupravlenii frontov i okrugov," *Agitator i propagandist Krasnoi Armii*, no. 5–6 (1943): 22.

91. Benjamin Zajicek, "Scientific Psychiatry in Stalin's Soviet Union: The Politics of Modern Medicine and the Struggle To Define 'Pavlovian' Psychiatry, 1939–1953" (PhD diss., University of Chicago, 2009), 153–54.

92. Grossman, *Gody voiny*, 362.

93. RGASPI f. 88, op. 1, d. 958, ll. 1–17.

94. NA IRI RAN f. 2, r. I, op. 223, d. 9, ll. 1–1ob.

95. Loginov, *Eto bylo na fronte*, 33–34.

96. Loginov, *Eto bylo na fronte*, 9–10.

97. TsAMO RF f. 67, op. 12001, d. 5, ll. 202–17, in *Tyl Krasnoi Armii*, 36; Suris, *Frontovoi dnevnik*, 65.

98. Tvardovskii, *Vasilii Tërkin*, 124.

99. Abdulin, *160 stranits*, 40.

100. Loginov, *Eto bylo na fronte*, 10.

101. On the meat crisis, see Zotov, *Pishchevaia promyshlennost'*, 128; Saushin, *Khleb i sol'*, 115–16; and Pokhlëbkin, *Kukhnia veka*, 209.

102. Lesin, *Byla voina*, 85, 89, 99, 102, 149–50; RGASPI f. 88, op. 1, d. 958, l. 2; Slutskii, *O drugikh i o sebe*, 29.

103. Dunaevskaia, *Ot Leningrada do Kënigsberga*, 158, 296.

104. Malakhova, "Four Years a Frontline Physician," 209–11; Iakupov, *Frontovye zarisovki*, 30. Conversely, Abdulin (*160 stranits*, 22) gave his bread ration to horses to keep them alive.

105. TsAMO RF f. 2, op. 795437, d. 9, ll. 394–95, in *Tyl Krasnoi Armii*, 237.

106. See TsA FSB RF f. 14, op. 4, d. 418, ll. 19–20, in *Stalingradskaia epopeia*, 246–48.

107. "Dolg voennykh khoziaistvennikov," *Krasnaia zvezda*, July 4, 1943. The *starshina* was a figure often derided and assumed to be corrupt. See, for example, Astaf'ev, *Prokliaty i ubity*, 109.

108. Loginov, *Eto bylo na fronte*, 9–10. This method goes back to at least the eighteenth century (Ilya Berkovich, *Motivation in War: The Experience of Common Soldiers in Old-Regime Europe* [New York: Cambridge University Press, 2017], 216).

109. Antipenko, *Na glavnom napravlenii*, 148–49.

110. Grigorii Baklanov, "Naveki deviatnadtsatiletnie," 280.

111. Abdulin, *160 stranits*, 134.

112. As late as August 1944, General Andrei Khrulëv, in charge of the rear area services for the entire Red Army, complained that there was a deficit of 2.7 million mess tins at the front (GARF f. R-5446, op. 46a, d. 7161, l. 2).

113. This model was deemed more useful (TsAMO RF f. 208, op. 14703 s, d. 2, ll. 339–43, in *Tyl Krasnoi Armii*, 137).

114. Slëzkin, *Do voiny i na voine*, 347.

115. Golubkova interview; and Malakhova, "Four Years a Frontline Physician," 199, 204–5.

116. Galimzhan Valiev, *Soldat khatlar* (Yar Chally, 2000), 57.

117. Tokarev, *Vesti dnevnik na fronte zapreshchalos'*, 137.

118. Loginov, *Eto bylo na fronte*, 9.

119. Temkin, *My Just War*, 104; Loginov, *Eto bylo na fronte*, 24.

120. "Mobilizatsionnoe predpisanie (oborot)," Soldat.ru, http://www.soldat.ru/doc/original/original.html?img=mobpredpis&id=2.

121. Kats interview.

122. *Vospominaniia frontovikov*, 7.

123. RGASPI f. 84, op. 1, d. 83, l. 172.

124. Malakhova, "Four Years a Frontline Physician," 201.

125. Slëzkin, *Do voiny i na voine*, 328.

126. Aleksandr Ustinov, "'Zavtra uedem v armiiu' (Iz frontovogo dnevnika fotozhurnalista)," *Rodina*, no. 6 (2011): 23.

127. Slutskii, *O drugikh i o sebe*, 29.

128. Koshkarbaev, *Shturm*, 109.

129. A. Lukovnikov, *Druz'ia-odnopolchane: Rasskazy o pesniakh, rozhdennykh voinoi, melodii i teksty* (Moscow: Muzyka, 1985), 32–33.

130. Saushin, *Khleb i sol'*, 87–93. See also RGVA f. 4, op. 12, d. 98, ll. 507–8, in *Prikazy narodnogo komissara oborony SSSR 22 iiunia 1941 g.–1942 g.*, 48; and RGVA f. 4, op. 11, d. 75, ll. 16–17, in *Prikazy narodnogo komissara oborony SSSR 1943–1945 gg.*, 18–19. However, it appears that appealing to the center for items in serious deficit was a standard part of how provisioning worked. A front seemed more or less invisible to the center while its provisioning was in order and became visible in moments of crisis.

131. Zotov, *Pishchevaia promyshlennost'*, 483.

132. Richard Klein, *Cigarettes Are Sublime* (Durham: Duke University Press, 1993), especially chapter 5, "The Soldier's Friend."

133. Moreinis, *Uchebnik pishchevoi gigieny*, 146; S. Gurov, *Boets i otdelenie na pokhode* (Moscow: Voenizdat, 1941), 22; Krotkov, *Gigiena*, 39.

134. Sidney W. Mintz, *Sweetness and Power: The Place of Sugar in Modern History* (New York: Penguin, 1986), 108–9, 110, 114, 122.

135. RGVA f. 4, op. 14, d. 2737, ll. 58–70, in *Zimniaia voina*, 118; Antipenko, *Na glavnom napravlenii*, 149; RGVA f. 4, op. 11, d. 65, ll. 413–14, in *Prikazy narodnogo komissara oborony SSSR 22 iiunia 1941 g.–1942 g.*, 73; RGVA f. 4, op. 11, d. 70, ll. 548–49, in *Prikazy narodnogo komissara oborony SSSR 22 iiunia 1941 g.–1942 g.*, 228–29; RGVA f. 4, op. 11, d. 71, ll. 191–92, in *Prikazy narodnogo komissara oborony SSSR 22 iiunia 1941 g.–1942 g.*, 252–53; RGVA f. 4, op. 11, d. 73, ll. 154–55, in *Prikazy narodnogo komissara oborony SSSR 22 iiunia 1941 g.–1942 g.*, 365–66; RGVA f. 4, op. 11, d. 75, l. 51, in *Prikazy narodnogo komissara oborony SSSR 1943–1945 gg.*, 28; RGVA, f. 4, op. 11, d. 75, l. 649, in *Prikazy narodnogo komissara oborony SSSR 1943–1945 gg.*, 145.

136. Nina Ivanovna Kunitsina, interview by Artëm Drabkin, *Ia pomniu*, http://iremember.ru/letno-tekh-sostav/kunitsina-nina-ivanovna.html; Klavdiia Andreevna Deriabina (Ryzhkova), interview by Artëm Drabkin, *Ia pomniu*, http://iremember.ru/letchiki-bombardirov/deryabina-rizhkova-klavdiya-andreevna-letchitsa-po-2.html.

137. Koshkarbaev, *Shturm*, 109.

138. NART f. R-3610, op. 1, d. 327, l. 40.

139. Abdulin, *160 stranits*, 105.

140. Slutsky, *O drugikh i o sebe*, 30; Temkin, *My Just War*, 197; Nikulin, *Vospominaniia o voine*, 144, 169, 187, 199.

141. Suris, *Frontovoi dnevnik*, 204–5, 234, 236; Inozemtsev, *Frontovoi dnevnik*, 199, 208–9, 226.

142. Vil'iam Pokhlëbkin, *Istoriia vazhneishikh pishchevykh produktov* (Moscow: Tsentrpoligraf, 2001), 272.

143. TsAMO RF f. 208, op. 14703c, d. 2, ll. 339–43, in *Tyl Krasnoi Armii*, 137; Krotkov, *Gigiena*, 92.

144. RGASPI f. 84, op. 1, d. 83, l. 173; GARF f. R5446, op. 44a, d. 9410, l. 13. There were constant problems with realizing these orders. See, e.g., GARF f. R-5446, op. 44a, d. 9410, l. 28.

145. See Krotkov, *Gigiena*, 49–50, 110–11; Gurov, *Boets and otdelenie na pokhode*, 23.

146. *Nastavlenie po polevomu vodosnabzheniiu voisk* (Moscow: Voenizdat, 1941), 3, 6, 71.

147. Krotkov, *Gigiena*, 44. See also *Nastavlenie po inzhenernomu delu dlia pekhoty (INZh-43)* (Moscow: Voenizdat, 1943), 7, 227–31.

148. Koshkarbaev, *Shturm*, 158.

149. Gurov, *Boets i otdelenie na pokhode*, 23.

150. Slutskii, *O drugikh i o sebe*, 29.

151. TsAMO RF f. 217, op. 1305, d. 17, ll. 37, 38, in *Tyl Krasnoi Armii*, 197.

152. F. G. Krotkov, "Problemy pitaniia voisk v gody Velikoi Otechestvennoi voiny," *Voprosy pitaniia*, no. 3 (1975): 6; Vorontsov, *Prodovol'stvennoe snabzhenie*, 24–25.

153. Saushin, *Khleb i sol'*, 53.

154. Saushin, *Khleb i sol'*, 52.

155. TsAMO RF f. 217, op. 1305, d. 17, ll. 37, 38, in *Tyl Krasnoi Armii*, 197.

156. Temkin, *My Just War*, 115.

157. Temkin, *My Just War*, 104.

158. Slutskii, *O drugikh i o sebe*, 28.

159. Koshkarbaev, *Shturm*, 81.

160. Abram Efimovich Shoikhet, interview by Grigorii Koifman, *Ia pomniu*, http://iremember.ru/pulemetchiki/shoykhet-abram-efimovich.html; Loginov, *Eto bylo na fronte*, 9–10; Frolov, *Vse oni khoteli zhit'*, 61–62.

161. Kamalov, *U kazhdogo zhizn'—odna*, 203–4.

162. Pëtr Fëdorovich Bazhenov, interview by I. Trifonov, *Ia pomniu*, http://iremember.ru/pekhotintsi/bazhenov-petr-fedorovich.html; Aizenshtat, *Zapiski sekretaria voennogo tribunala*, 116.

163. Pokhlëbkin, *Kukhnia veka*, 227–28; Collingham, *Taste of War*, 70.

164. Stefanovskii, *Poslednie pis'ma s fronta, 1941*, 1:31–32.

165. Inozemtsev, *Frontovoi dnevnik*, 107–8, 195. Akhmetov was a Crimean Tatar who was not deported but served through the end of the war.

166. D. D. Petrov, ed., *Frontovaia pechat' o voinakh iz Iakutii* (Iakutsk: Iakutskoe knizhnoe izdatel'stvo, 1982), 40–41.

167. RGASPI f. 17, op. 125, d. 85, l. 60.

168. RGASPI f. 88, op. 1, d. 958, l. 7.

169. Adamskii interview.

170. Mikhail Fëdorovich Borisov, Hero of the Soviet Union, interview by Artëm Drabkin, *Ia pomniu*, http://iremember.ru/artilleristi/borisov-mikhail-fedorovich-geroy-sovetskogo-soiuza-artillerist.html#comment-963.

171. Astaf'ev, *Prokliaty i ubity*, 82, described how Kazakhs in his unit slowly came to eat pork, initially vomiting it back up.

172. Genatulin, *Strakh*, 11–12.

173. Frolov, *Vse oni khoteli zhit'*, 38–39, 43, 63.

174. Slutskii, *O drugikh i o sebe*, 30. This is not to say that there were no problems with food after 1943, as a variety of sources attest. See, e.g., TsAMO RF f. 240,

op. 2824, d. 123, ll. 62–65, in *Tyl Krasnoi Armii*, 471; and NA IRI RAN f. 2, r. I, op. 28, d. 33, l. 12.

175. NA IRI RAN f. 2, r. I, op. 30, d. 23, l. 4.

176. Grossman, *Gody voiny*, 444.

177. Lesin, *Byla voina*, 287.

178. Slutskii, *O drugikh i o sebe*, 30.

179. Graft, particularly around major holidays, remained a problem late in the war, however; see, e.g., GARF f. R-5446, op. 46a, d. 7395, ll. 20–21, 26–28.

180. N. A. Antipenko, *Front i tyl* (Moscow: Znanie, 1977), 59.

181. TsAMO RF f. 236, op. 2719, d. 76, ll. 62–65, in *Tyl Krasnoi Armii*, 674–75.

182. Slutskii, *O drugikh i o sebe*, 30.

4. Cities of Earth, Cities of Rubble

1. Valentina Chudakova, *Chizhik—ptichka s kharakterom* (Leningrad: Lenizdat, 1980), 114. David Samoilov made a similar comparison (*Pamiatnye zapiski*, 204). Epigraph: Oles' Gonchar, *Shchodenniki, 1943–1967* (Kiev: Veselka, 2002), 1:52 (diary entry, July 13, 1944).

2. S. Gliazer, "Lopata—vernyi drug soldata," *Bloknot agitatora Krasnoi Armii*, no. 14 (1943): 15–16.

3. P. Mozgovoi, "Lopata—podruga i zashchitnitsa boitsa," *Bloknot agitatora Krasnoi Armii*, no. 4 (1942): 18–20.

4. Stephen Kotkin, "The Search for the Socialist City," *Russian History* 23 (1996): 231–61; Kotkin, *Magnetic Mountain*, 144, 160, 165–75, 180–83, 191, 356, 360; Milyausha Zakirova, "The City as a Genuine Place: The Paradoxes of Soviet Urbanization. The Search for the Genuine Soviet City," in *The City in Russian Culture*, ed. Pavel Lyssakov and Stephen M. Norris (New York: Routledge, 2018), 141–65.

5. This was nothing new, dating back at least to World War I. See, e.g., Ross J. Wilson, *Landscapes of the Western Front: Materiality during the Great War* (New York: Routledge, 2012), 136.

6. Lesin, *Byla voina*, 224; Chekalov, *Voennyi dnevnik*, 35.

7. Baklanov, "Piad' zemli," 326.

8. Atabek, *K biografii voennogo pokoleniia*, 252.

9. Nikulin, *Vospominaniia o voine*, 49.

10. Astaf'ev, *Prokliaty i ubity*, 311.

11. Kovalevskii, "Nynche u nas peredyshka," 71 (entry from February 29, 1944).

12. This figure is cited on monuments scattered around the center of St. Petersburg, where the damage done by the Germans to landmarks such as St. Isaac's Cathedral and the Anichkov Bridge has been left unrepaired.

13. TsAMO f. 132a, op. 2642, d. 13, ll. 131–34, in *Russkii arkhiv: Velikaia Otechestvennaia. Stavka VGK. Dokumenty i materialy, 1942 god*, vol. 16 (5–2), ed. A. M. Sokolov, Iu. N. Semin, et al. (Moscow: Terra, 1996), 430. See also Atabek, *K biografii voennogo pokoleniia*, 117.

14. NA IRI RAN f. 2, r. III, op. 5, d. 4, l. 16.

15. Chekalov, *Voennyi dnevnik*, 124.

16. Kiselëv, "Voina i zhizn' v predstavlenii 20-letnikh frontovikov," 1012.

17. N. Rubinshtein, "Zametki ob agitatsii," *Propagandist Krasnoi Armii*, no. 14 (1942): 47.

18. V. Olizerenko, "Mitingi mesti," *Agitator i Propagandist Krasnoi Armii*, no. 9–10 (1943): 29. As Karel Berkhoff has pointed out, the solid basis in reality for the claims of Nazi atrocities was increasingly clear by the middle of the war (*Motherland in Danger,* 133). Agitators were encouraged to spread information about German crimes in newly liberated towns. See G. Rumiantsev, "Zametki frontovogo agitatora," *Agitator i Propagandist Krasnoi Armii*, no. 15 (1943): 32–34.

19. RGASPI f. 17, op. 125, d. 171, l. 37; Gel'fand, *Dnevnik 1941–1945*, 154.

20. See, e.g., NA IRI RAN f. 2, r. I, op. 28, d. 33, ll. 6–6ob.; RGASPI f. 17, op. 125, d. 171, ll. 32–43.

21. See, e.g., Ia. Gertsovich, "Vospitanie nenavisti," *Agitator i Propagandist Krasnoi Armii*, no. 19–20 (1942): 33–35; NA IRI RAN f. 2, r. I, op. 28, d. 33, ll. 6–6ob.

22. *Pis'mo tatarskomu narodu ot tatar-boitsov 1-go Ukrainskogo fronta* (Kazan: Tatgosizdat, 1944), 4.

23. RGASPI f. 17, op. 125, d. 171, l. 40.

24. RGASPI f. 17, op. 125, d. 171, ll. 32–43.

25. RGASPI f. 17, op. 125, d. 171, l. 39. The overlap between the personal and public apparent here is discussed in detail by Lisa Kirschenbaum ("'Our City, Our Hearths, Our Families': Local Loyalties and Private Life in Soviet World War II Propaganda," *Slavic Review* 59 [2000]: 825–47.)

26. Komskii, "Dnevnik 1943–1945 gg.," 33.

27. RGASPI f. 17, op. 125, d. 171, ll. 40–43.

28. Tvardovskii, *Vasilii Tërkin*, 58–64, 146, 158–65, 174–80; Viktor Kurochkin, "Na voine kak na voine," in *Povesti i rasskazy* (Leningrad: Khudozhestvennaia literatura, 1978), 41–47; Temkin, *My Just War*, 100–102; Atabek, *K biografii voennogo pokoleniia*, 166; Samoilov, *Podënnye zapisi*, 1: 188, 201–2; Gel'fand, *Dnevnik 1941–1945*, 131, 133, 155, 272; Budnitskii, "Great Patriotic War and Soviet Society."

29. S. Gurov, *Boets i otdelenie na pokhode* (Moscow: Voenizdat, 1941), 9, 53.

30. Grossman. *Gody voiny*, 284.

31. *Pamiatka boitsam-razvedchikam v osnovnykh vidakh boia* (Moscow: Voenizdat, 1943), 23.

32. Iakupov, *Frontovye zarisovki*, 23

33. Nikulin, *Vospominaniia o voine*, 37.

34. Evgenii Petrov, *Frontovoi dnevnik* (Moscow: Sovetskii pisatel', 1942), 25.

35. *Boets pomni*, listovka from NA IRI RAN f. 2, r. XIII Z7, op. 4, d. 71; *Bloknot agitatora Krasnoi Armii*, no. 15 (1943), 20; another, less glamorous version stated, "Wherever a nanny goat has passed, a Russian soldier will also pass."

36. *Nastavlenie po inzhenernomu delu dlia pekhoty* (hereafter *Inzh. P-43*) (Moscow: Voenizdat, 1943), 189–90.

37. *Inzh. P-43*, 154–68.

38. Chekalov, *Voennyi dnevnik*, 299.

39. NA IRI RAN f. 2, r. X, op. 7, d. 13-b, l. 95.

40. *Boevoi ustav pekhoty* (hereafter *BUP-42*) (Moscow: Voenizdat, 1942), 19.

41. *Pamiatka komandiru strelkovogo otdeleniia* (Moscow: Voenizdat 1943), 20.

42. *BUP-42*, 14–15, 226–30.

43. Petrov, *Frontovoi dnevnik*, 29–30.

44. Viktor Nekrasov, *V okopakh Stalingrada: Povest', rasskazy* (Moscow: Khudozhestvennaia literatura, 1990), 58–59.

45. The Russian term for this was *ognevaia otchetnaia kartochka*, often shortened to either *ognevaia kartochka* or *otchetnaia kartochka*.

46. *Pamiatka boitsu pekhoty v oborone* (Moscow: Voenizdat, 1943), 4.

47. *BUP-42*, 14–15, 226–30.

48. NA IRI RAN f. 2, r. III, op. 5, d. 14, l. 52.

49. Suris, *Frontovoi dnevnik*, 254.

50. Gliazer, "Lopata–vernyi drug soldata," 14–17; *Instruktsiia po ukladke*, 39.

51. *Inzh. P-43*, 8.

52. RGASPI f. 84, op. 1, d. 91, l. 256.

53. *Instruktsiia po ukladke*, 17.

54. Mozgovoi, "Lopata—podruga i zashchitnitsa boitsa," 18–20.

55. *Inzh. P-43*, 33.

56. *Inzh. P-43*, 32–36.

57. *BUP-42, 235–57*, *Inzh. P-43*, 8–9, 33–39.

58. Daniel Giblin, "Digging for Victory: Mobilization of Civilian Labor for the Battle of Kursk, 1943" (PhD diss., University of North Carolina at Chapel Hill, 2016).

59. *Inzh. P-43*, 10.

60. *BUP-42*, 252.

61. *Inzh. P-43*, 40.

62. *Inzh. P-43*, 8, 37–47.

63. Abdulin, *Iz vospominanii soldata*, 14.

64. V. Kaplin, "V transhee," *Gvardiia*, June 15, 1944.

65. A. Shnyrkevich, "Oborona v gorakh Severnogo Kavkaza," *Krasnaia zvezda*, September 3, 1942; V. Shterenberg, "Okapyvanie v stepnykh raionakh," *Krasnaia zvezda*, September 24, 1942.

66. Dunaevskaia, *Ot Leningrada do Kënigsberga*, 296; Samoilov, *Pamiatnye zapiski*, 204; Lesin, *Byla voina*, 97–98.

67. *Inzh. P-43*, 102–4.

68. *BUP-42*, 252–57.

69. *Inzh. P-43*, 27, 75–80.

70. *Inzh. P-43*, 132–54.

71. *BUP-42*, 235–57; D. Ushakov, *Voenno-inzhenernoe delo* (Moscow: Voenizdat, 1939), 26.

72. Gliazer, "Lopata—vernyi drug soldata," 15–16.

73. Mozgovoi, "Lopata—podruga i zashchitnitsa boitsa," 18.

74. *Inzh. P-43*, 12–15.

75. Grossman, *Gody voiny*, 408; *BUP-42*, 49.

76. Indzhiev, *Frontovoi dnevnik*, 20.

77. Pomerants, *Zapiski gadkogo utënka*, 74.

78. TsAMO RF f. 243, op. 2963, d. 26, ll. 85–86, in *Tyl Krasnoi Armii*, 580–81.

79. V. Pichuzhin, "Kaska da lopata—druz'ia soldata," *Gvardiia*, September 28, 1944.

80. TsAMO RF, f. 2, op. 795437, d. 9, l. 158, in *Tyl Krasnoi Armii*, 216. In 1940 a new helmet was designed with a three-pad liner.

81. Genatulin, "Dve nedeli," 76; TsAMO RF f. 208, op. 14703s, d. 2, ll. 339–43, in *Tyl Krasnoi Armii*, 137. A report from 1941 pointed out issues of visibility with the paint used on helmets: "The steel helmet's current paint is highly reflective and can be pierced by bullets. It is necessary to change the paint."

82. *Frontovoi tovarishch* (s.l.: Voenizdat, 1942), 77–78; RGASPI f. 644, op. 1, d. 28, l. 56.

83. Dotsenko "Kaska spasla mne zhizn'," *Gvardiia*, June 17, 1944.

84. Nikulin, *Vospominaniia o voine*, 79.

85. M. I. Kalinin, "Slovo agitatora na fronte," *Bloknot agitatora Krasnoi Armii*, no. 13 (1943): 12.

86. AP RF f. 3, op. 50, d. 226, ll. 184–88, in Kudriashov, *Voina*, 150.

87. NA IRI RAN f. 2, r. III, op. 1, d. 7, l. 93.

88. NA IRI RAN f. 2, r. III, op. 5, d. 8, l. 52.

89. TsA FSB Rossii, f. 41, on. 51, d. 11, ll. 286–87, in *"Ognennaia duga,"* 90–91.

90. V. I. Ermolenko, *Voennyi dnevnik starshego serzhanta* (Belgorod: Otchii krai, 2000), 96.

91. A. Simukov, "Kak Kuz'ma s zemlei-matushkoi dogovarivalsia," *Krasnoarmeets*, no. 1 (1943): 14.

92. Atabek, *K biografii voennogo pokoleniia*, 102.

93. Nikulin, *Vospominaniia o voine*, 81.

94. *Inzh. P-43*, 15.

95. *Inzh. P-43*, 16.

96. *Inzh. P-43*, 18.

97. *Inzh. P-43*, 68; Inozemtsev, *Frontovoi dnevnik*, 40; Sero Khanzadian, *Povesti i rasskazy* (Moscow: Izvestiia, 1986), 155–56.

98. *Inzh. P-43*, 21–23.

99. *BUP-42*, 235–57; *Posobie komandiru strelkovogo otdeleniia*, 80–96.

100. Grossman, *Gody voiny*, 356.

101. K. Zlygostev, "Umelo maskirovat' ognevye pozitsii," *Gvardiia*, December 20, 1943.

102. S. Gurov, *Pokhod i otdykh pekhoty* (Moscow: Voenizdat, 1940), 53.

103. *Frontovoi tovarishch*, 70.

104. Ia. Malakhov, "Voiskovoe nabliudenie," *Agitator i propagandist Krasnoi Armii*, no. 15 (1943): 42–45.

105. AP RF f. 3, op. 58, d. 451, ll. 111–16, in Kudriashov, *Voina*, 446–47.

106. Gliazer, "Lopata–vernyi drug soldata," 15–16.

107. Chudakova, *Chizhik—ptichka s kharakterom*, 525–26; *Frontovoi tovarishch*, 75–77; Grossman, *Gody voiny*, 349.

108. Loginov, *Eto bylo na fronte*, 8.

109. Abdulin, *160 stranits*, 33–34.

110. *Inzh. P-43*, 98–101.

111. Nikulin, *Vospominaniia o voine*, 151.

112. NA IRI RAN f. 2, r. I, op. 30, d. 13, l. 4ob.

113. RGASPI f. M-33, op. 1, d. 853, l. 129.

114. Loginov, *Eto bylo na fronte*, 8.

115. Tokarev, *Vesti dnevnik na fronte zapreshchalos'*, 91–92.

116. Chekhovich, *Dnevnik ofitsera*, 38.

117. Grossman, *Gody voiny*, 339; Abdulin, *160 stranits*, 45–46; Samoilov, *Pamiatnye zapiski*, 204.

118. NA IRI RAN f. 2, r. X, op. 7, d. 7-b, l. 3.

119. RGASPI f. 88, op. 1, d. 958, l. 5.

120. RGASPI f. 88, op. 1, d. 958, l. 4.

121. "Dnevnik Lutovinina Semëna Mikhailovicha," in *S veroi v pobedu: Pis'ma, dnevniki, vospominaniia frontovikov*, ed. N. A. Cherepanov (Tiumen': Vektor Buk, 2002), 131–32; RGASPI f. 88, op. 1, d. 958, ll. 2–5.

122. Abdulin, *160 stranits*, 33–34; Chudakova, *Chizhik—ptichka c kharakterom*, 273–76; Dunaevskaia, *Ot Leningrada do Kënigsberga*, 115.

123. Grossman, *Gody voiny*, 349.

124. Vishnevskii, *Dnevnik khirurga*, 229–30.

125. "Dnevnik Lutovinina Semëna Mikhailovicha," 131–32.

126. Grossman, *Gody voiny*, 349.

127. Tokarev, *Vesti dnevnik na fronte zapreshchalos'*, 140.

128. *Instruktsiia po rabote pokhodnoi banno-dezinfekstionno-prachechnoi ustanovki (BDPU) Volkhovskogo fronta* (Moscow: Medgiz, 1943); "Polevye bani-prachechnye," *Krasnaia zvevda*, July 1, 1942.

129. Dunaevskaia, *Ot Leningrada do Kënigsberga*, 187.

130. *Nastavlenie po inzhenernomu delu dlia pekhoty* (Moscow: Voenizdat, 1939), 94–96.

131. Suris, *Frontovoi dnevnik*, 34.

132. Dunaevskaia, *Ot Leningrada do Kënigsberga*, 178.

133. Malakhova, "Four Years a Frontline Physician," 200; Dunaevskaia, *Ot Leningrada do Kënigsberga*, 333.

134. Tokarev, *Vesti dnevnik na fronte zapreshchalos'*, 40; Lesin, *Byla voina*, 91.

135. Boris Slutskii, *"Ia istoriiu izlagaiu. . ."* (Moscow: Pravda, 1990), 70.

136. For a rare look at the work of SMERSh informants, see TsA FSB Rossii f. 14, on. 5, d. 13, ia. 291, 299–309, and TsA FSB Rossii f. 41, on. 102, d. 268, il. 159–161, in *"Ognennaia duga,"* 126–31.

137. NA IRI RAN f. 2, r. III, op. 5, d. 16, ll. 73–74.

138. Amanzholov, *Iz opyta politiko-vospitatel'noi raboty*; A. Shipov., "Biblioteka agitatora tov. Muzafarova," *Bloknot agitatora Krasnoi Armii*, no. 25 (1943): 21–24; Dunaevskaia, *Ot Leningrada do Kënigsberga*, 93.

139. NA IRI RAN f. 2, r. III, op. 5, d. 2-a, l. 58.

140. Slëzkin, *Do voiny i na voine*, 420–21.

141. Samoilov, *Pamiatnye zapiski*, 206–10.

142. NA IRI RAN f. 2, r. X, op. 7, d. 13-b, ll. 86–87, 104; Dunaevskaia, *Ot Leningrada do Kënigsberga*, 71, 137.

143. Samoilov, *Podennye zapisi*, 1:177.

144. "Kak deistvovat' v boiu," *Bloknot agitatora Krasnoi Armii*, no. 14 (1942): 8–10; Dunaevskaia, *Ot Leningrada do Kënigsberga*, 112; Kopelev, *Khranit' vechno*, 113–15.

145. Lukovnikov, *Druz'ia-odnopolchane*, 54–56.

146. NA IRI RAN f. 2, r. X, op. 7, d. 13-b, ll. 86–87.

147. Suris, *Frontovoi dnevnik*, 67.

148. Slutskii, *Ia istoriiu izlagaiu*, 83, see also 63.

149. Kopelev, *Khranit' vechno*, 1:14.

150. Kiselёv, "Voina i zhizn' v predstavlenii 20-letnikh frontovikov," 1008.

151. Dunaevskaia, *Ot Leningrada do Kёnigsberga*, 94–95.

152. Dunaevskaia, *Ot Leningrada do Kёnigsberga*, 112.

153. NA IRI RAN f. 2, r. III, op. 1, d. 7, l. 20.

154. Komskii, "Dnevnik 1943–1945 gg.," 60; Atabek, *K biografii voennogo pokoleniia*, 85, 90, 97, 98, 102, 107, 134–35.

155. Atabek, *K biografii voennogo pokoleniia*, 237.

156. See, e.g., Kunitsina interview.

157. NA IRI RAN f. 2, r. X, op. 7, d. 3, l. 7ob; Dunaevskaia, *Ot Leningrada do Kёnigsberga*, 55.

158. Kovalevskii, "Nynche u nas peredyshka," 104–5.

159. See, e.g., NA IRI RAN f. 2, r. X, op. 7, d. 8, ll. 5, 9, 12ob.–13.

160. NA IRI RAN f. 2, r. X, op. 7, d. 4, l. 7ob.

161. K. Kniazeva, "Vospitanie devushek-voennosluzhashchikh," *Agitator i propagandist Krasnoi Armii*, no. 15–16 (1944): 23.

162. Iakupov, *Frontovye zarisovki*, 104.

163. Nikulin, *Vospominaniia o voine*, 79.

164. TsAMO RF f. 208, op. 2563, d. 48, ll. 64–75, in *Tyl Krasnoi Armii*, 239.

165. Nikulin, *Vospominaniia o voine*, 56.

166. Grossman, *Gody voiny*, 334.

167. TsAMO f. 32, op. 920265, d. 4, ll. 275–76, in *Glavnye politicheskie organy vooruzhennykh sil SSSR v Velikoi Otechestvennoi voine 1941–1945 gg.*, 97–98.

168. Kalinin, "Slovo agitatora na fronte," 13.

169. L. Zheleznov, "Mogila voina," *Krasnaia zvezda*, October 21, 1943.

170. Nikulin, *Vospominaniia o voine*, 76.

171. TsAMO RF f. 214, op. 1470, d. 308, l. 340, in *Tyl Krasnoi Armii*, 616–17.

172. Zheleznov, "Mogila voina."

173. TsAMO RF f. 243, op 2963, d. 98, ll. 21–21ob., in *Tyl Krasnoi Armii*, 645–46.

174. TsA FSB Rossii f. 40, op. 16, d. 3, ll. 175–77, in *"Ognennaia duga,"* 102; RGASPI f. 644, op. 1, d. 26, ll. 16–17.

175. Reutov, *Gvardeets*, 87.

176. Kiselёv, "Voina i zhizn' v predstavlenii 20-letnikh frontovikov," 1023. *Natsmen*—national minorities—was a generally derogatory term for "non-Russians."

177. Genatulin, "Dve nedeli," 92.

178. Chekalov, *Voennyi dnevnik*, 233.

179. Chekhovich, *Dnevnik ofitsera*, 81.

180. TsGA IPD RT f. 4034, op. 25, d. 226, ll. 57, 58, 65, in Ibragimov et al., *Pis'ma s fronta*, 68–69.

181. Chekalov, *Voennyi dnevnik*, 349–50.

182. Atabek, *K biografii voennogo pokoleniia*, 113.

183. Loginov, *Eto bylo na fronte*, 25–26.

184. Inozemtsev, *Frontovoi dnevnik*, 218.

185. Loginov, *Eto bylo na fronte*, 22.

186. See, e.g., Nina Tumarkin, *The Living and the Dead: The Rise and Fall of the Cult of World War II in Russia* (New York: Basic Books, 1994).

187. Gliazer, "Lopata—vernyi drug soldata," 15–16.

5. "A Weapon Is Your Honor and Conscience"

1. Abdulin, *160 stranits*, 4, 11–13. Chapter title from *Frontovoi tovarishch*, 37–38. Epigraph: NA IRI RAN f. 2, r. I, op. 16, d. 1, l. 21ob.

2. Grossman, *Gody voiny*, 339.

3. Peter Fritzsche and Jochen Hellbeck, "The New Man in Stalinist Russia and Nazi Germany," in *Beyond Totalitarianism: Stalinism and Nazism Compared*, ed. Michael Geyer and Sheila Fizpatrick (New York: Cambridge University Press, 2008), 321.

4. Hellbeck, *Stalingrad*, 20, 33.

5. Baurdzhan Momysh-uly, *Psikhologiia voiny: Kniga-Khronika* (Alma-Ata: Kazakhstan, 1990), 42.

6. S. L. A. Marshall, *Men against Fire: The Problem of Battle Command* (Norman: University of Oklahoma Press, 2000), 54, 71, 78; Dave Grossman, *On Killing: The Psychological Cost of Learning to Kill in War and Society*, rev. ed. (New York: Back Bay Books, 2009), 18–36; Joanna Bourke, *An Intimate History of Killing: Face to Face Killing in Twentieth Century Warfare* (New York: Basic Books, 1999), 64.

7. Krylova, *Soviet Women in Combat*, 30, 175–83.

8. *Programmy uskorennoi boevoi podgotovki strelkovykh podrazdelenii* (Moscow: Voenizdat, 1941), 5, pt. 1:7; 5, pt. 2:5.

9. See, e.g., David L. Hoffmann, *Peasant Metropolis: Social Identities in Moscow, 1929–1941* (Ithaca: Cornell University Press, 1994), 73–82.

10. RGASPI f. 644, op. 1, d. 24, ll. 160–61.

11. AP RF f. 3, op. 50, d. 263, ll. 145–53, in Kudriashov, *Voina*, 60.

12. Reutov, *Gvardeets*, 68.

13. AP RF f. 3, op. 50, d. 266, ll. 82–99, in Kudriashov, *Voina*, 142; AP RF f. 3, op. 50, d. 266, ll. 184–88, in Kudriashov, *Voina*, 150–53. This continued: RGVA f. 4, op, 11, d. 72, ll. 482–85, in *Prikazy narodnogo komissara oborony SSSR 22 iiunia 1941 g.–1942 g.*, 318–20.

14. See Ronald M. Smelser and Edward J. Davies, *The Myth of the Eastern Front: The Nazi-Soviet War in American Popular Culture* (New York: Cambridge University Press, 2008); Roger Reese, *Why Stalin's Soldiers Fought*, 153–55.

15. NA IRI RAN f. 2, r. III, op. 5, d. 11, l. 16.

16. Chekalov, *Voennyi dnevnik*, 187–88.

17. AP RF f. 3, op. 50, d. 264, ll. 144–47, in Kudriashov, *Voina*, 76.

18. AP RF f. 3, op. 50, d. 266, ll. 82–89, in Kudriashov, *Voina*, 142.

19. Krylova, *Soviet Women in Combat*, 176–79.

20. Daniil Granin, Lev Slëzkin, and Boris Komskii all shared this fate.

21. NA IRI RAN f. 2, r. 3, op. 14, d. 2b, l. 11.

22. Chekalov, *Voennyi dnevnik*, 84; Kiselëv, "Voina i zhizn′ v predstavlenii 20-letnikh frontovikov," 1017; Erëmenko, *Dnevniki*, 176.

23. Baklanov, "Mertvye sramu ne imut," 134.

24. Grossman, *On Killing*, 146; Bourke, *Intimate History of Killing*, 75.

25. Hannah Arendt, *On Violence* (New York: Harcourt, 1969), 56; Reese, *Why Stalin's Soldiers Fought*, 151–75.

26. Bourke, *Intimate History of Killing*, 75; Grossman, *On Killing*, 153; Marshall, *Men against Fire*, 57. Notice that Grossman's interviewee at the beginning of this chapter said that only crew-served weapons could be relied on during a fight.

27. RGVA f. 4, op. 12, d. 106, ll. 8–16, in *Prikazy narodnogo komissara oborony SSSR 22 iiunia 1941 g.–1942 g.*, 324–25; G. Morozov, "Deistvennost' ognia v boiu i mery dlia umen'sheniia poter'," *Agitator i propagandist Krasnoi Armii*, no. 12 (1943): 44.

28. Abdulin, *160 stranits*, 5.

29. Pavel Nilin, "Sovest'," *Krasnoarmeets*, no. 10 (1942): 8–10.

30. Konstantin Simonov, "Ubei ego," *Krasnaia zvezda*, July 18, 1942.

31. Elena Kononenko, "Tin-Tinych," *Krasnoarmeets*, no. 1 (1943): 6–8.

32. Weiner, "Something to Die For," 114–15.

33. G. Lomidze, "Druzhba, skreplennaia krov'iu," *Agitator i propagandist Krasnoi Armii*, no. 18 (1944): 44; Amanzholov, *Opyt politiko-vospitatel'noi raboty v deistvuiushchei armii*, 12, 21.

34. Amanzholov, *Opyt politiko-vospitatel'noi raboty v deistvuiushchei armii*, 21–22.

35. *Frontovoi tovarishch*, 89–91; *Metkuiu puliu v serdtse vraga* (Unknown frontline publisher, 1942), 34.

36. RGVA f. 4, op. 11, d. 71, ll. 320–22, in *Prikazy narodnogo komissara oborony SSSR 22 iiunia 1941 g.–1942 g.*, 264–65.

37. *Frontovoi tovarishch*, 89–91. Dramatic overcounting of casualties inflicted on the enemy was also part of this culture, as multiple units could claim having downed the same plane or destroyed the same tank. See, e.g., Nikulin, *Vospominaniia o voine*, 129.

38. This is a universal trope in total war (Bourke, *Intimate History of Killing*, 20–24).

39. John Keegan, *The Face of Battle* (New York: Penguin, 1978), 320–22; Scarry, *Body in Pain*, 65–68.

40. "'Na voine rozhdaetsia samaia sil'naia liubov'. . .': Perepiska V. P. Stroevoi i B. Momysh-uly 1944–1965 gody," *Kinovedcheskie zapiski*, no. 72 (2005): 56.

41. Momysh-uly, *Psikhologiia voiny*, 48.

42. Lukovnikov, *Druz'ia-odnopolchane*, 12–14.

43. NA IRI RAN f. 2, r. III, op. 1, d. 7, l. 41; Pomerants, *Zapiski gadkogo utënka*, 136; Tokarev, *Vesti dnevnik na fronte zapreshchalos'*, 169–70.

44. RGASPI f. M-33, op. 1, d. 853, l. 337.

45. Slëzkin, *Do voiny i na voine*, 468.

46. NA IRI RAN f. 2, r. X, op. 7, d. 11, l. 5ob.

47. AP RF f. 3, op. 50, d. 263, ll. 145–53, in Kudriashov, *Voina*, 60; N. D'iakonov, "Zametki o roli ofitsera kak vospitatelia," *Agitator i propagandist Krasnoi Armii*, no. 15–16 (1944): 26.

48. NA IRI RAN f. 2, r. I, op. 71, d. 2, l. 1ob.

49. Momysh-uly, *Psikhologiia voiny*, 40.

50. NA IRI RAN f. 2, r. I, op. 201, d. 1, ll. 3–3ob.

51. On war as posturing, see Momysh-uly, *Psikhologiia voiny*, 40; and Grossman, *On Killing*, 5–17.

52. N. D'iakonov, "Zametki o roli ofitsera kak vospitatelia," 28–29; "Kak deistvovat' v boiu (beseda komvzvoda A. Panova s mladshimi boitsami)," *Bloknot agitatora Krasnoi Armii*, no. 14 (1942): 8–9.

53. Morozov, "Deistvennost' ognia v boiu i mery dlia umen'sheniia poter'," 44; D'iakonov, "Zametki o roli ofitsera kak vospitatelia," 28–29; N. Rubenshtein, "Zametki ob agitatsii," *Agitator i propagandist Krasnoi Armii*, no. 14 (1942): 47–48.

54. NA IRI RAN f. 2, r. II, op. 16, d. 4, l. 31; Amanzholov, *Opyt politiko-vospitatel'noi raboty v deistvuiushchei armii*, 43.

55. Stalin, *Great Patriotic War of the Soviet Union*, 55.

56. Dunn, *Hitler's Nemesis*, xviii–xix.

57. Krylova, *Soviet Women in Combat*, 30–31, 173–203; M. Protsenko, "Osobennosti boia v naselënnom punkte," *Propagandist Krasnoi Armii*, no. 17 (1941): 51.

58. Loginov, *Eto bylo na fronte*, 9; Scarry, *Body in Pain*, 67.

59. Nikolai Bogdanov, *Beregi oruzhie, kak zenitsu oka* (Moscow: Voenizdat, 1942), 19–20.

60. *Boevoi ustav bronetankovykh i mekhanizirovannykh voisk Krasnoi Armii*, pt. 1: *Tank, tankovyi vzvod, tankovaia rota* (hereafter *BUBiMV-44*) (Moscow: Voenizdat, 1944), 164–66.

61. "Moia vintovka," *Bloknot agitatora Krasnoi Armii*, no. 19 (1943): 24; Bogdanov, *Beregi oruzhie, kak zenitsu oka*, 6.

62. P. Burlaka, "Kak zenitsu oka, berech' boevuiu tekhniku i vooruzhenie," *Propagandist Krasnoi Armii*, no. 19 (1941): 6–7.

63. *BUP-42*, 28.

64. I. Stepanenko, "Utrennie osmotry i vechernie proverki," *Krasnaia zvezda*, July 14, 1943.

65. NA IRI RAN f. 2, r. I, op. 16, d. 1, l. 21ob.; Krylova, *Soviet Women in Combat*, 251–52.

66. NA IRI RAN f. 2, r. III, op. 14, d. 2b, l. 31.

67. Krylova, *Soviet Women in Combat*, 250–51; Khazratkul Faiziev, *Ognennye vërsty: Iz frontovoi tetradi*, trans. from Tajik by V. S. Zhukovskii (Dushanbe: Irfon, 1980), 158–69.

68. Grossman, *Gody voiny*, 422.

69. Indzhiev, *Frontovoi dnevnik*, 26.

70. RGASPI f. M-7, op. 1, d. 6387, l. 118.

71. Samoilov, *Podennye zapisi*, 1: 180; see also Indzhiev, *Frontovoi dnevnik*, 22.

72. RGASPI f. M-7, op. 2, d. 650, ll. 87, 97, 139. This is the famed partisan Zoia Kosmodem'ianskaia's younger brother. He was killed in March 1945 in Germany.

73. NA IRI RAN f. 2, r. I, op. 30, d. 32, ll. 1–1ob.

74. *BUP-42*, 19, 39–48.

75. RGVA f. 4, op. 14, d. 2736, ll. 152–72, in *"Zimniaia voina,"* 183–84.

76. D. N. Bolotin, *Sovetskoe strelkovoe oruzhie* (Moscow: Voenizdat, 1983), 68.

77. John Barber and Mark Harrison, *The Soviet Home Front 1941–1945: A Social and Economic History of the USSR in World War II* (London: Longman, 1991), 180.

78. *Nastavlenie po strelkovomu delu (NSD-38)* (Alma-Ata, 1941), 120–21; V. Glazatov, *Vintovka i eë primenenie* (Moscow: Voenizdat, 1941), 4.

79. Glazatov, *Vintovka i eë primenenie*, 5; *Instruktsiia po ukladke, prigonke, sborke i nadevaniiu pokhodnogo snariazheniia boitsa Krasnoi Armii* (Moscow: Voenizdat, 1941), 13–14.

80. "Moia vintovka," 23.

81. *O komsomol'skoi rabote na fronte* (Moscow: Voenizdat, 1942), 8.

82. Stalin, *Great Patriotic War of the Soviet Union*, 55.

83. See, e.g., NA IRI RAN f. 2, r. III, op. 5, d. 26; NA IRI RAN f. 2, r. I, op. 103, d. 6, l. 8ob. Oleg Budnitskii has pointed out that this movement, like the Stakhanovite movement before it, was instituted from above and for many participation was mandatory ("Harvard Project in Reverse," 190).

84. RGASPI f. 644, op. 1, d. 34, l. 203; *Sbornik zakonov, postanovlenii pravitel'stva i prikazov NKO po voprosam prokhozhdeniia sluzhby i forme odezhdy lichnogo sostava Krasnoi armii* (Moscow: NKO, 1946), 52. A sniper held the rank of *efreitor* (corporal) and could rise to the rank of sergeant and earn up to 200 rubles a month in pay (regular soldiers received 11 rubles, 20 kopecks).

85. *Metkuiu puliu v serdtse vraga*, 3–5. Tallies were often exaggerated (Budnitskii, "Harvard Project in Reverse," 188–91).

86. Dem'ian Bednyi, "Semën Nomokonov," *Krasnoarmeets*, no. 21 (1942): 16.

87. NA IRI RAN f. 2, r. I, op. 103, d. 6, ll. 1–2, 8ob.

88. Elena Kononenko, "Devushka v shineli," *Krasnoarmeets*, no. 12 (1943): 20–22; Vera Inber, "Zhenshchine," *Krasnoarmeets*, no. 15 (1942): 3; *Nakaz naroda* (Moscow: Voenizdat, 1943), 26, 82.

89. Kononenko, "Devushka v shineli"; GARF f. P-7523, op. 13, d. 66, ll. 66–69; *V boiakh za Rodinu* (s.l.: Izdanie gazety *Ataka*, 1942), 48.

90. G. Morozov, "Snaipery v boiu," *Agitator i propagandist Krasnoi Armii*, no. 14 (1942), 26–29.

91. Bourke, *Intimate History Of Killing*, 42, 79.

92. "Pogovorki russkikh soldat," *Bloknot agitatora Krasnoi Armii*, no. 15 (1943): 20; Morozov, "Deistvennost' ognia v boiu i mery dlia umen'sheniia poter'," 42.

93. *Frontovoi tovarishch*, 68–69; Pomerants, *Zapiski gadkogo utënka*, 142.

94. *BUP-42*, 10.

95. NA IRI RAN f. 2, r. III, op. 5, d. 4, l. 11ob.

96. Pomerants, *Zapiski gadkogo utënka*, 116.

97. Temkin, *My Just War*, 112; Koshkarbaev, *Shturm*, 123.

98. Bolotin, *Sovetskoe strelkovoe oruzhie*, 117, 125.

99. RGASPI f. 74, op. 2, d. 121, l. 5; Bolotin, *Sovetskoe strelkovoe oruzhie*, 114; Erëmenko, *Dnevniki*, 144.

100. Bolotin, *Sovetskoe strelkovoe oruzhie*, 118–20, 126–28, 134; Glantz, *Colossus Reborn*, 192. Their full names were Pistolet-pulemët Degtiarëva obraztsa 1940 goda, Pistolet-pulemët Shpagina obraztsa 1941 goda and Pistolet-pulemët Sudaeva obraztsa 1943 goda, respectively. Each was named after its inventor and year of adoption.

101. RGVA f. 4, op. 11, d. 66, ll. 104–5, in *Prikazy narodnogo komissara oborony SSSR 22 iiunia 1941 g.–1942 g.*, 107; RGVA f. 4, op. 11, d. 66, ll. 182–83, in *Prikazy narodnogo komissara oborony SSSR 22 iiunia 1941 g.–1942 g.*, 117.

102. *Deistviia roty avtomatchikov v boiu* (Moscow: Voenizdat, 1942).

103. P. Lymarev, "Bei vraga granatoi," *Bloknot agitatora Krasnoi Armii*, no. 13 (1943): 19.

104. *Nastavlenie po strelkovomu delu: Ruchnye, oskolochnye i protivotankovye granaty i zazhigatel'nye butylki* (Moscow, Voenizdat, 1946), 6.

105. NA IRI RAN f. 2, r. III, op. 5, d. 2a, l. 5ob.

106. Molotov cocktails were actually invented by the Finns during the 1939–1940 Winter War. To fight Red Army tanks, the Finns, who were in dire straits, filled bottles with fuel and dubbed them "Molotov cocktails" in reference to Soviet Foreign Minister Viacheslav Molotov.

107. *Nastavlenie po strelkovomu delu: Ruchnye*, 87–94.

108. Dem'ian Bednyi, *Nesokrushimaia uverennost'* (Moscow, Goslitizdat, 1943), 24.

109. Inozemtsev, *Frontovoi dnevnik*, 37.

110. Baklanov, "Naveki-deviatnadtsatiletnie," 196.

111. Samoilov, *Pamiatnye zapiski*, 205–8, 213.

112. NA IRI RAN f. 2, r. III, op. 5, d. 36, l. 12.

113. Bolotin, *Sovetskoe strelkovoe oruzhie*, 161–62.

114. *V pomoshch' agitatoru* (Dushanbe: Tadzhikgosizdat, 1942), 101; RGASPI f. 644, op. 1, d. 34, ll. 104–5.

115. *Pamiatka komandiru pulemëtnogo otdeleniia* (Moscow: Voenizdat, 1943), 15, 21, 28.

116. *BUP-42*, 76.

117. NA IRI RAN f. 2, r. I, op. 71, d. 7, l. 22.

118. David M. Glantz, *Companion to Collossus Reborn: Key Documents and Statistics* (Lawrence: University Press of Kansas, 2005), 141–45.

119. *Boevoi ustav artillerii RKKA*, pt. 1, bk. 1: *Voiskovaia artilleriia* (Moscow: Voenizdat, 1938). *Navodchik*, which refers to the soldier who aimed the piece, was a sought-after position because it was easy to gain praise serving as one (NA IRI RAN f. 2, r. I, op. 30, d. 1, l. 120).

120. RGVA f. 4, op. 11, d. 71, ll. 320–22, in *Prikazy narodnogo komissara oborony SSSR 22 iiunia 1941 g.–1942 g.*, 264–65. These privileges included double pay for commanders, a rank one above the norm for crew members, and a guarantee to be returned to their unit after being wounded.

121. NA IRI RAN f. 2, r. I, op. 30, d. 23, ll. 1–1ob.

122. Dunn, *Hitler's Nemesis*, 150, 196–203.

123. *Posobie dlia boitsa-tankista*, 1; RGVA f. 4, op. 12, d. 106, ll. 112–22, in *Prikazy narodnogo komissara oborony SSSR 22 iiunia 1941 g.–1942 g.*, 334–37.

124. Dunn, *Hitler's Nemesis*, 119, 121.

125. *Tank T-34 v boiu* (Mosow: Voenizdat, 1942), 1–6.

126. *BUBiMV-44*, 14–25.

127. Dunn, *Hitler's Nemesis*, 150.

128. *BUBiMV-44*, 68; *Posobie dlia boitsa-tankista*, 72–83.

129. *Posobie dlia boitsa-tankista*, 118.

130. *Posobie dlia boitsa-tankista*, 7, 127; Grossman, *Gody voiny*, 315–16.

131. *BUBiMV-44*, 63; *Posobie dlia boitsa-tankista*, 198–201.

132. *Posobie dlia boitsa-tankista*, 145–75.

133. *BUBiMV-44*, 185–89.

134. *Tank v boiu* (Moscow: Voenizdat, 1946), 42–47; *Tank T-34 v boiu*, 31–32; *BUBiMV-44*, 70–72.

135. *BUBiMV-44*, 34–38, 59–61; *Posobie dlia boitsa-tankista*, 18.

136. *Boevye priëmy tankistov* (Moscow: Voenizdat, 1942), 36–41.

137. *BUBiMV-44*, 95.

138. *Posobie dlia boitsa-tankista*, 46; RGVA f. 4, op. 12, d. 106, ll. 112–22 in *Prikazy narodnogo komissara oborony SSSR 22 iiunia 1941 g.–1942 g.*, 334–38.

139. *BUBiMV-44*, 97–98, 108–13.

140. *BUBiMV-44*, 94–95.

141. RGVA f. 4, op. 12, d. 106, ll. 112–22, in *Prikazy narodnogo komissara oborony SSSR 22 iiunia 1941 g.–1942 g.*, 334–36.

142. RGVA f. 4, op. 11, d. 73, ll. 341–42, in *Prikazy narodnogo komissara oborony SSSR 22 iiunia 1941 g.–1942 g.*, 379–80.

143. RGVA f. 4, op. 11, d. 73, ll. 53–54, in *Prikazy narodnogo komissara oborony SSSR 22 iiunia 1941 g.–1942 g.*, 338; RGVA f. 4, op. 11, d. 73, l. 340, in *Prikazy narodnogo komissara oborony SSSR 22 iiunia 1941 g.–1942 g.*, 378.

144. RGVA f. 4, op. 11, d. 78, ll. 376–77, in *Prikazy narodnogo komissara oborony SSSR 1943–1945 gg.*, 328–29.

145. *Tank v boiu*, 26; *Tank T-34 v boiu*, 37.

146. *BUBiMV-44*, 92.

147. Stalin, *Great Patriotic War of the Soviet Union*, 32.

148. M. Vistin, "Uchit'sia voevat' tak, kak etogo trebuet delo pobedy," *Bloknot agitatora Krasnoi Armii*, no. 15 (1943): 12–13.

149. NA IRI RAN f. 2, r. I, op. 120, d. 3, l. 2ob.

150. NA IRI RAN f. 2, r. X, op. 7, d. 13-b, l. 141.

151. A. Strelkov, "Sviazisty v boiu," *Bloknot agitatora Krasnoi Armii*, no. 22 (1943): 6.

152. Iakushin, *On the Roads of War*, 139.

153. Momysh-uly, *Psikhologiia voiny*, 11–12.

154. Kopelev, *Khranit' vechno*, 1:114.

155. Kopelev, *Khranit' vechno*, 1:113.

156. NA IRI RAN f. 2 r. 1, op. 74, d. 4, ll. 2–3.

157. See Kamalov, *U kazhdogo zhizn' odna*, 7–8. This novel opens with an artillery unit complaining to infantry about their slovenly trenches.

158. In connection with this event, an entire issue of *Bloknot agitatora Krasnoi Armii* (no. 28 [1944]) was dedicated to the artillery. See Serhy Yekelchyk, *Stalin's Citizens: Everyday Politics in the Wake of Total War* (New York: Oxford University Press, 2014), 41–42.

159. E. Burdzhalov, "Zabota tovarishcha Stalina ob ukreplenii Krasnoi Armii," *Bloknot agitatora Krasnoi Armii*, no. 35 (1944): 28–29.

160. NA IRI RAN f. 2, r. I, op. 230, d. 6, l. 1ob. They also make up a major portion of the diarists, memoirists, interviewees, and novelists whose works are available to researchers (e.g., Inozemtsev, Kiselëv, Baklanov, Nikulin, Abdulin, and Komskii).

161. RGVA f. 4, op. 11, d. 66, ll. 232–34, in *Prikazy narodnogo komissara oborony SSSR 22 iiunia 1941 g.–1942 g.*, 122–23.

162. RGASPI f. M-7, op. 2, d. 650, ll. 87; NA IRI RAN f. 2, r. I, op. 120, d. 3, l. 4.

163. AP RF f. 3, op. 50, d. 264, l. 166, in Kudriashov, *Voina*, 84.

164. Dunn, *Hitler's Nemesis*, 150.

165. Ivan Pyr'ev, dir., *Traktoristy* (Mosfilm, 1939).

166. Kurochkin, "Na voine kak na voine," 19.

167. Kurochkin, "Na voine kak na voine," 71–72.

168. "Voennaia istoriia zhenshchiny i eë tanka 'Boevaia podruga,'" *RIA Novosti*, April 14, 2010. http://ria.ru/ocherki/20100414/222121628.html.

169. *BUBiMV-44*, 202. These letters were written in white during the summer and red during the winter.

170. Grossman, *Gody voiny*, 423.

171. F. Gorokhov, "Komandir—dusha boevoi spaiki i sokhraneniia boevykh traditsii chasti," *Agitator i propagandist Krasnoi Armii*, no. 3–4 (1943): 34–36.

172. Grossman, *Gody voiny*, 354.

173. A. Burdzhalov, "Rost Krasnoi Armii v khode Otechestvennoi voiny," *Agitator i propagandist Krasnoi Armii*, no. 2 (1944): 16-28, here 22. For a more detailed and

skeptical discussion of the Red Army's abilities from 1943 on, see Hill, *Red Army and the Second World War.*

174. Dunn, *Hitler's Nemesis*, xxi, 37–40, 68; Inozemtsev, *Frontovoi dnevnik*, 159; Atabek, *K biografii voennogo pokoleniia*, 98.

175. Krylova, *Soviet Women in Combat*, 211–13; Grossman, *Gody voiny*, 423. Michael Geyer and Mark Edele point to continued heavy casualties but note that these losses were "recoverable"—sick and wounded, rather than killed, captured or invalided out of the army ("States of Exception," 388).

176. Chudakova, *Chizhik—ptichka s kharakterom*, 110–13; Nikulin, *Vospominaniia o voine*, 132–33.

177. See, e.g., Dunn, *Hitler's Nemesis*, 12–13; TsAMO f. 202, op. 5, d. 1329, ll. 27–28, in *Preliudiia Kurskoi bitvy*, 140–42.

178. A typical storm group included "two rifle squads; 1–2 heavy machine guns; 1 antitank rifle squad; 1 platoon of 50mm mortars; 1–2 cannons; a squad of sappers with plastic explosives, mine detectors, fuel containers, and the means to cut wire; 2 flamethrowers; 2–3 medium or 1–2 heavy tanks" (M. M. Kozhevnikov, *Deistviia tankov v sostave shturmovoi gruppy pri atake DOT i DZOT* [Moscow: Voenizdat, 1944], 12).

179. NA IRI RAN f. 2, r. I, op. 226, l. 25ob.

180. Antipenko, *Front i tyl*, 4.

181. AP RF f. 3, op. 50, d. 311, ll. 13–30, in Kudriashov, *Voina*, 156–57.

182. Grossman, *Gody voiny*, 387.

183. NA IRI RAN f. 2, r. III, op. 1, d. 7, l. 62.

184. NA IRI RAN f. 2, r. III, op. 5, d. 2a, l. 50.

185. Slutskii, *O drugikh i o sebe*, 19–20.

186. Loginov, *Eto bylo na fronte*, 8.

187. NA IRI RAN f. 2, r. I, op. 28, d. 35, l. 8; Anatolii Genatulin, "Tri protsenta," in *Nas ostaetsia malo: Rasskazy i povest'* (Moscow: Sovremennik, 1988), 10–11.

188. Baklanov, *Zhizn'*, 77.

6. The Thing-Bag

1. S. G. Gurov, *Inzhenernoe obespechenie boevykh deistvii pekhoty: Vzvod. Rota. Batal'on* (Moscow: Voenizdat, 1939), 10; Abdulin, *Iz vospominanii soldata*, 11; Genatulin, "Dve nedeli," 75–76. Epigraph: General Grigorii Maliukov to the Mints Commission, May 1944 (NA IRI RAN f. 2, r. I, op. 104, d. 4, ll. 3ob.–4).

2. N. D'iakonov, "Zametki o roli ofitsera kak vospitatelia," *Agitator i propagandist Krasnoi Armii*, no. 15 (1944): 25–29.

3. See, e.g., V. G. Perov's paintings *Starik-Strannik* (ca. 1869) and *Strannik* (1870); and I. I. Shishkin's sketch *Krest'ianka s kotomkoi za spinoi* (1852–1855).

4. TsAMO f. 208, op. 14703 s, d. 2, ll. 339–43, in *Tyl Krasnoi Armii*, 137.

5. RGVA f. 4, op. 14, d. 2737, ll. 58–70, in *"Zimniaia voina,"* 120.

6. *Instruktsiia po ukladke*, 7, 25–26; Gel'fand, *Dnevnik*, 73.

7. N. Losikov, "Ob ispol'zovanii politprosvetimushchestva na fronte," *Agitator i propagandist Krasnoi Armii*, no. 7–8 (1944): 36.

8. Inozemtsev, *Frontovoi dnevnik*, 164; Suris, *Frontovoi dnevnik*, 177; Gel'fand, *Dnevnik*, 73.

9. See, e.g., TsGA IPD RT f. 15, op. 5, d. 559, ll. 33–33ob. in *Pis'ma s fronta*, 173.

10. TsAMO f. 32, op. 920265, d. 3, l. 325, in *Glavnye politicheskie organy vooruzhennykh sil SSSR v Velikoi Otechestvennoi voine*, 74.

11. Grossman, *Gody voiny*, 362.

12. RGASPI f. M-33, op. 1, d. 853, ll. 192–93.

13. "'Na voine rozhdaetsia samaia sil'naia liubov','" 59.

14. M. Kalinin, "Slovo agitatora na fronte," *Bloknot agitatora Krasnoi Armii*, no. 13 (1943): 9–10; Tokarev, *Vesti dnevnik na fronte zapreshchalos'*, 140.

15. Liubarov, *Inache ia ne mog*, 130.

16. Abdulin, *160 stranits*, 60.

17. Suris, *Frontovoi dnevnik*, 37. See also *Vospominaniia frontovikov*, 8. Here supply officers discuss tying strikers and flint to soldiers' rain capes.

18. Glotov, *Vstrechi*, 22.

19. Boris Likhachev, "Mundshtuchok," *Krasnoarmeets*, no. 20 (1943): 5.

20. Tvardovskii, *Ia v svoiu khodil ataku*, 123; Tvardovskii, *Vasilii Tërkin*, 65–70.

21. Baklanov, "Naveki—deviatnadtsatiletnie," 269.

22. Dunaevskaia, *Ot Leningrada do Kënigsberga*, 139.

23. NART f. 4821, op. 1, d. 2, l. 304.

24. NA IRI RAN f. 2, r. I, op. 28, d. 3, l. 6.

25. TsAMO RF f. 208, op. 14703 c, d. 2, ll. 339–43, in *Tyl Krasnoi Armii*, 136–37; Nikulin, *Vospominaniia o voine*, 37.

26. RGASPI f. M-33, op. 1, d. 853, ll. 192–93.

27. "'Eto to, chto nabolelo, chto prositsia naruzhu,'" 35.

28. RGASPI f. 88, op. 1, d. 973, l. 8.

29. N. P. Popov and N. A. Gorokhov, *Sovetskaia voennaia pechat' v gody Velikoi Otechestvennoi voiny 1941–1945* (Moscow: Voenizdat, 1981), 261, 5.

30. R. P. Ovsepian, *Istoriia noveishei otechestvennoi zhurnalistiki (fevral' 1917–nachalo 90-kh godov)* (Moscow: Izdatel'stvo MGU, 1996), 92; Matthew Lenoe, *Closer to the Masses: Stalinist Culture, Social Revolution, and Soviet Newspapers* (Cambridge, MA: Harvard University Press, 2004), 67.

31. Popov and Gorokhov, *Sovetskaia voennaia pechat'*, 54–57.

32. Atabek, *K biografii voennogo pokoleniia*, 294–96.

33. Popov and Gorokhov, *Sovetskaia voennaia pechat'*, 50.

34. V. Panov, "O zadachakh frontovoi pechati," *Propagandist Krasnoi Armii*, no. 18 (1941): 15.

35. I. Shikin, "Krasnoarmeiskaia gazeta—vazhneishii tsentr partiino-politicheskoi raboty," *Agitator i propagandist Krasnoi Armii*, no. 16 (1943): 6.

36. Shikin, "Krasnoarmeiskaia gazeta," 6.

37. See Schechter, "People's Instructions"; TsAMO f. 32, op. 920265, d. 8, ll. 42–43, in *Glavnye politicheskie organy vooruzhennykh sil SSSR v Velikoi Otechestvennoi voine*, 266–67; Popov and Gorokhov, *Sovetskaia voennaia pechat'*, 261. For a treatment of Tatar-language newspapers, see K. Ainutdinov, *Letopis' podviga (Istoricheskii ocherk po stranitsam tatar. frontovykh gazet)* (Kazan: Tatknigizdat, 1984).

38. See, e.g., TsGA IPD RT f. 15, op. 5, d. 553, l. 21; and RGASPI f. 17. op. 125, d. 85, ll. 69–76.

39. See, e.g., NA IRI RAN f. 2, r. I, op. 28, d. 33, l. 5ob.

40. Lenoe, *Closer to the Masses*, 20.

41. "Soveshchanie redaktorov frontovykh i armeiskikh gazet," *Agitator i propagandist Krasnoi Armii*, no. 16 (1943): 24.

42. TsAMO f. 32, op. 920265, d. 5, t. 2, ll. 534–35, in *Glavnye politicheskie organy vooruzhennykh sil SSSR v Velikoi Otechestvennoi voine*, 151; "Soveshchanie redaktorov frontovykh i armeiskikh gazet," 34–35; A. Shipov, "Patrioticheskie pis'ma," *Agitator i propagandist Krasnoi Armii*, no. 13–14 (1943): 40.

43. See, e.g., the regularly appearing column "Kavalery Ordena slavy" in the divisional newspaper *Gvardiia*.

44. See, e.g., Ia. Miletskii, "Pshënnye dni," *Krasnaia zvezda*, June 8, 1943; T. Mikhailov, "Komsorg bezdeistvuet," *Gvardiia*, March 2, 1944; and I. Perebeinos, "Trus pogibaet pervym!," *Gvardiia*, March 2, 1944.

45. N. Shafranovich, "Rabota s gazetoi," *Propagandist Krasnoi Armii*, no. 3 (1942): 24.

46. N. Rubinshtein, "Zametki ob agitatsii," *Agitator i propagandist Krasnoi Armii*, no. 14 (1942): 43.

47. Anderson, *Imagined Communities*, 30–36. This often took the form of published letters that cast the Soviet Union as a family. See Kirschenbaum, "'Our City, Our Hearths, Our Families,'" 828, 831, 838, 844.

48. V. Kuznetsov, "Pis'ma voenkorov v krasnoarmeiskoi gazete," *Agitator i propagandist Krasnoi Armii*, no. 7–8 (1945): 36–42.

49. RGVA f. 4, op. 11, d. 76, ll. 354–58, in *Prikazy narodnogo komissara oborony SSSR 1943–1945 gg.*, 237.

50. N. Gagulin, "Okopnaia gazeta," *Gvardiia*, November 27, 1944; G. Gorov, "Agitatsiia sredi boitsov nerusskoi natsional'nosti," *Agitator i propagandist Krasnoi Armii*, no. 22 (1942): 27; Dunaevskaia, *Ot Leningrada do Kënigsberga*, 116, 179.

51. NA IRI RAN f. 2, r. I, op. 195, d. 14, l. 31.

52. S. Blokhin, "Agitatsiia v nashei divizii," *Bloknot agitatora Krasnoi Armii*, no. 32 (1944): 26–29.

53. *Zakony Sovetskoi Gvardii: Na krasnoarmeiskom mitinge v N-skoi Gvardeiskoi Divizii* (Moscow: Voenizdat, 1942), 19–20.

54. Brandenberger, *National Bolshevism*, 144–57, 160–80.

55. Stezhenskii, *Soldatskii dnevnik*, 101, 106, 157, 158, 185.

56. See Katerina Clark, *The Soviet Novel: History as Ritual*, 3rd rev. ed. (Bloomington: University of Indiana Press, 2000).

57. See, e.g., L. Timofeev, "Patrioticheskaia sila russkoi literatury," *Agitator i propagandist Krasnoi Armii*, no. 12 (1943): 13–20; P. Rod'kin, "Iz praktiki moei vospitatel'noi raboty," *Agitator i propagandist Krasnoi Armii*, no. 18 (1944): 29–31.

58. Weiner, *Making Sense of War*, 43.

59. P. Prokhorov, "Ob ideino-politicheskom vospitanii kommunistov-frontovikov," *Agitator i propagandist Krasnoi Armii*, no. 15 (1943): 29–32.

60. I. Khodiakov, "Rotnaia bibliotechka," *Gvardiia*, January 24, 1944; "Knizhnaia polka," *Gvardiia*, January 24, 1944; A. Shipov, "Biblioteka agitatora tovarishcha Muzafarova," *Bloknot agitatora Krasnoi Armii*, no. 25 (1943): 22–24; Shikin, "Krasnoarmeiskaia gazeta," 9.

61. TsAMO f. 32, op. 920265, d. 5, t. 2, l. 461, in *Glavnye politicheskie organy vooruzhennykh sil SSSR v Velikoi Otechestvennoi voine*, 140; NA IRI RAN f. 2, r. I, op. 30, d. 23, l. 4.

62. Dunaevskaia, *Ot Leningrada do Kënigsberga*, 86; N. Bas'ko, "Kak ia rabotaiu (zametki chtetsa)," *Gvardiia*, June 12, 1944; Shipov, "Patrioticheskie pis'ma," 39.

63. On the paper shortage under Stalin in general, see Lenoe, *Closer to the Masses*, 179.

64. RGASPI f. M-33, op. 1, d. 1085, l. 14; Gel'fand, *Dnevnik*, 73; Baklanov, "Naveki-deviatnadtsatiletnie," 269.

65. John Bushnell, "Peasants in Uniform: The Tsarist Army as a Peasant Society," *Journal of Social History* 13 (1980): 565–576, here 566.

66. Tvardovskii, *Ia v svoiu khodil ataku*, 247.

67. Momysh-uly, *Psikhologiia voiny*, 24–30.

68. Shikin, "Krasnoarmeiskaia gazeta," 6–10.

69. Nikulin, *Vospominaniia o voine*, 152; Suris, *Frontovoi dnevnik*, 176. Dismissal of the official press and its celebration of heroes was also common before the war (Brandenberger, *Propaganda State in Crisis*, 235–39).

70. Stezhenskii, *Soldatskii dnevnik*, 218.

71. Nikulin, *Vospominaniia o voine*, 152.

72. Tvardovskii, *Ia v svoiu skhodil ataku*, 48.

73. Liubarov, *Inache ia ne mog*, 179; NA IRI RAN f. 2, r. III, op. 5, d. 8, l. 92.

74. William G. Rosenberg, "Reading Soldiers' Moods: Russian Military Censorship and the Configuration of Feeling in World War I," *American Historical Review* 119 (2014): 714–740, here 716. While the majority of soldiers may have been literate during World War I, their families would have been largely illiterate.

75. For a much more detailed discussion and theorization of wartime letters as a genre, see S. Ushakin and A. Golubev, eds., *XX vek: Pis'ma* (Moscow: Novoe literaturnoe obozrenie, 2016). This volume goes into detail about a variety of issues—including death, love, material conditions, and war as work—that I only touch on. See also the work of Irina Tadzhinova, who has systematically analyzed the personal lives of soldiers via letters (cited below), as well as Jochen Hellbeck (cited below) and Lisa Kirschenbaum (cited earlier). Steven Jug applies a systematic chronology to the content of soldiers' letters, tracking shifts in their themes as the war progressed ("All Stalin's Men? Soldierly Masculinities in the Soviet War Effort" [PhD diss., University of Illinois at Urbana-Champaign, 2013]). For an early and fascinating attempt to interpret letters to and from Red Army soldiers, see V. Zenzinov, *Vstrecha s Rossiei: Kak i chem zhivut v Sovetskom Soiuze. Pis'ma v Krasnuiu Armiiu 1939–1940* (New York: L. Rausen, 1944).

76. Lesin, *Byla voina*, 123.

77. TsGA IPD RT f. 8250, op. 2, d. 6, l. 16. Grammatical error in the original. It is worth noting that these are two ethnically "non-Russian" soldiers corresponding in Russian, and that the letter shows a mix of profanity and Stalinist stock phrases.

78. Tvardovskii, *Vasilii Tërkin*, 116–21.

79. Indzhiev, *Frontovoi dnevnik*, 105.

80. I. Shikin, "Za dal'neishii pod"ëm politicheskoi agitatsii i propagandy," *Agitator i propagandist Krasnoi Armii*, no. 1 (1943): 2–5.

81. NA IRI RAN f. 2, r. I, op. 30, d. 1, l. 235.

82. A. Shipov, "Patrioticheskie pis'ma," 38–41; N. D'iakonov. "Zametki o roli ofitsera kak vospitatelia."

83. Ushakin and Golubev, *XX vek*, 11.

84. TsA FSB Rossii f. 40, on. 28, d. 29, ll. 70–72, in "*Ognennaia duga*," 32.

85. David M. Glantz and Jonathan M. House, *The Battle of Kursk* (Lawrence: University of Kansas Press, 1999), 343.

86. *Pamiatka krasnoarmeitsu i krasnoflottsu ob adresovanii pochtovoi korrespondentsii* (Moscow: Voenizdat, 1941).

87. RGVA f. 4, op. 11, d. 73, ll. 329–32, in *Prikazy narodnogo komissara oborony SSSR 22 iiunia 1941 g.–1942 g.*, 375–77; TsAMO f. 32, op. 920265, d. 5, t. 2, l. 700, in *Glavnye politicheskie organy vooruzhennykh sil SSSR v Velikoi Otechestvennoi voine*, 175.

88. E. Nikiforova, *Rozhdënnaia voinoi* (Moscow: Molodaia gvardiia, 1985), 80; Frolov, *Vse oni khoteli zhit'*, 61; RGASPI f. M-33, op. 1, d. 853, l. 434.

89. A. Gupalo, "Rabota agitatora," *Agitator i propagandist Krasnoi Armii*, no. 24 (1942): 39; "Individual'nyi podkhod," *Bloknot agitatora Krasnoi Armii*, no. 22 (1943): 33.

90. "Pishite pravil'no adres," *Gvardiia*, May 30, 1944; *RBP-40*, 48–49.

91. Jochen Hellbeck, "'The Diaries of Fritzes and the Letters of Gretchens': Personal Writings from the German–Soviet War and Their Readers," *Kritika: Explorations in Russian and Eurasian History* 10 (2009): 571–606, here 598.

92. Iampol'skii et al., *Nachalo 22 iiunia–31 avgusta 1941 goda: Organy gosudarstvennoi bezopasnosti SSSR v Velikoi Otechestvennoi voine: Sbornik dokumentov* (Moscow: Rus', 2000), 2, bk. 1:308–9.

93. *RBP-40*, 48–49; GDA SB Ukraïni f. 9, spr. 220-sp, ark. 3–6, in "Viddili viis'kovoi tsenzuri," 251.

94. GDA SB Ukraïni f. 9, spr. 213-sp, ark. 18–26, in "Viddili viis'kovoi tsenzuri," 203.

95. GDA SB Ukraïni f. 9, spr. 15-sp, ark. 55–62, in "Viddili viis'kovoï tsenzuri," 157–64; GDA SB Ukraïni f. 16, op. 1, spr. 508, ark. 209–11, in "Viddili viis'kovoï tsenzuri," 175–76; GDA SB Ukraïni f. 16, op. 1, spr. 508, ark. 185–88, in "Viddili viis'kovoï tsenzuri," 172–74.

96. Hellbeck, "Diaries of Fritzes and the Letters of Gretchens," 598.

97. Lesin, *Byla voina*, 49. See also TsAMO f. 32, op. 920625, d. 5, t. 1, ll. 69–70, in *Glavnye politicheskie organy vooruzhennykh sil SSSR*, 111–12; and TsAMO f. 32, op. 920265, d. 5, t. 1, l. 353, in *Glavnye politicheskie organy vooruzhennykh sil SSSR*, 131.

98. Loginov, *Eto bylo na fronte*, 62.

99. Iampol'skii et al., *Nachalo 22 iiunia–31 avgusta 1941 goda*, 309.

100. Il'ia Pystin, "Dnevnik I. Pystina (1941–1943 gg.)," *ART*, no. 1 (2005): 52.

101. M. T. Sergeev, ed., *Pis'ma s voiny 1941–1945* (Ioshkar-Ola: Mariiskii poligrafichesko-izdatel'skii kombinat, 1995), 99.

102. TsA FSB f. 40, op. 28, d. 29, ll. 70–72, in "*Ognennaia duga*," 32.

103. Loginov, *Eto bylo na fronte*, 62.

104. GDA SB Ukraïni f. 9, spr. 213-sp, ark. 18–26, in "Viddili viis'kovoi tsenzuri," 203; GDA SB Ukraïni f. 9, spr. 220-sp, ark. 76–76 zv., in "Viddili viis'kovoi tsenzuri," 272–74. Unsurprisingly, a parallel process took place in published letters (Kirschenbaum, "'Our City, Our Hearths, Our Families,'" 830.

105. GDA SB Ukraïni f. 9, spr. 216-sp, ark. 71–74 zv., in "Viddili viis'kovoï tsenzuri," 242–47; GDA SB Ukraïni f. 9, spr. 18-sp, ark. 41–41 zv., in "Viddili viis'kovoi tsenzuri," 231–32.

106. GDA SB Ukraïni f. 9, spr. 213-sp, ark. 18–26, in "Viddili viis'kovoi tsenzuri," 201.

107. Sergeev, *Pis'ma s voiny 1941–1945*, 102–3; *Pis'ma s Velikoi Otechestvennoi: Sbornik frontovykh pisem, 1941–1945* (Ivanovo: Ivanovskii gosudarstvennyi universitet, 2005), 32.

108. Gel'fand, letter from November 7, 1944, at "Vladimir Gel'fand, Pis'ma 1941–1946: Skaner," http://www.gelfand.de/a/3/a.html.

109. See, e.g., Akhtiamov's letters in Frolov, *Vse oni khoteli zhit'*, 23–68.

110. Iakupov, *Frontovye zarisovki*, 8; RGASPI f. M-7, op. 2, d. 650, ll. 87, 97, 139.

111. GDA SB Ukraïni f. 9, spr. 220-sp, ark. 3–6, in Viddili viis'kovoï tsenzuri," 248–52.

112. GDA SB Ukraïni f. 9, spr. 208-sp, ark. 78–81, in "Viddili viis'kovoï tsenzuri," 166; GDA SB Ukraïni f. 9, spr. 216-sp, ark. 11–12 zv., in "Viddili viis'kovoi tsenzuri," 206–8.

113. GDA SB Ukraïni f. 9, spr. 216-sp, ark. 11–12 zv., in "Viddili viis'kovoi tsenzuri," 206–8; GDA SB Ukraïni f. 9, spr. 220-sp, ark. 21–22 zv., in "Viddili viis'kovoi tsenzuri," 255–58.

114. GDA SB Ukraïni f. 9, spr. 18-sp, ark. 39–39 zv., in "Viddili viis'kovoi tsenzuri," 229–30.

115. TsGA IPD RT f. 15, op. 5, d. 562, l. 115.

116. GDA SB Ukraïni f. 9, spr. 216-sp, ark. 11–12 zv., in "Viddili viis'kovoï tsenzuri," 206–8; GDA SB Ukraïni f. 9, spr. 220-sp, ark. 21–22 zv., in "Viddili viis'kovoi tsenzuri," 255–58; GDA SB Ukraïni f. 9, spr. 220-sp, ark. 7–8. in "Viddili viis'kovoi tsenzuri," 252–54.

117. "Na voine rozhdaetsia samaia sil'naia liubov'," 61; A. S. Smykalin, *Perliustratsiia korrespondentsii i pochtovaia voennaia tsenzura v Rossii i SSSR* (St. Petersburg: Iuridicheskii tsentr, 2008), 179.

118. Charles Shaw, "Soldiers' Letters to Inobatxon and O'g'ulxon: Gender and Nationality in the Birth of a Soviet Romantic Culture," *Kritika: Explorations in Russian and Eurasian History* 17 (2016): 517–52.

119. Frolov, *Vse oni khoteli zhit'*, 24–25, 42, 59, 63, 64–65.

120. GDA SB Ukraïni f. 9, spr. 216-sp, ark. 13–21 zv., in "Viddili viis'kovoï tsenzuri," 208–22.

121. See, e.g., TsAMO f. 32, op. 920625, d. 2, l. 381 in Borodin et al., *Glavnye politicheskie organy vooruzhennykh sil SSSR*, 35–36; TsAMO f. 32, op. 920625, d. 3, ll. 373–75, in *Glavnye politicheskie organy vooruzhennykh Sil SSSR*, 76–77.

122. See, e.g., TsAMO f. 32, op. 920265, d. 5, t. 1, ll. 380–81, in *Glavnye politicheskie organy vooruzhennykh sil SSSR*, 134–35.

123. Lesin, *Byla voina*, 122–23.

124. *Instruktsiia voiskovomu pochtal'onu* (Moscow: Voenizdat, 1944), 3.

125. Suris, *Frontovoi dnevnik*, 67.

126. Inozemtsev, *Frontovoi dnevnik*, 90.

127. Zenzinov, *Vstrecha s Rossiei*, 73–74.

128. TsAMO f. 32, op. 920265, d. 5, t. 2, ll. 574–75, in *Glavnye politicheskie organy vooruzhennykh sil SSSR*, 158.

129. TsAMO f. 32, op. 795436, d. 10, l. 44, in *Glavnye politicheskie organy vooruzhennykh sil SSSR*, 258. For a full collection of one soldier's letters, including the variety of paper he used, see "Vladimir Gel'fand, Pis'ma 1941–1946: Skaner," http://www.gelfand.de/a/3/a.html.

130. V. Kuznetsov, "Pis'ma voenkorov v krasnoarmeiskoi gazete," *Agitator i propagandist Krasnoi Armii*, no. 7–8 (1945): 36.

131. GDA SB Ukraïni f. 9, spr. 18-sp, ark. 32–32 zv., in "Viddili viis'kovoï tsenzuri," 227–28.

132. *Pis'ma s Velikoi Otechestvennoi*, 20.

133. Atabek, *K biografii voennogo pokoleniia*, 71.

134. Viktor Nekrasov, *I zhiv ostalsia* (Moscow: Kniga, 1991), 436.

135. Iakupov, *Frontovye zarisovki*, 32.

136. RGASPI f. M-33, op. 1, d. 360, ll. 8–9.

137. Shakirov, "Individual'nyi podkhod," *Bloknot agitatora Krasnoi Armii*, no. 22 (1943): 32.

138. Gorov, "Agitatsiia sredi boitsov nerusskoi natsional'nosti," 26–27.

139. Shipov, "Patrioticheskie pis'ma," 38–41; see also Shikin, "Za dal'neishii pod"ëm politicheskoi agitatsii i propagandy," 2–5.

140. RGASPI f. M-33, op. 1, d. 853, ll. 191–92; RGASPI f. M-33, op. 1, d. 1386, l. 48; NA IRI RAN f. 2, r. II, op. 28, d. 26, ll. 32–32ob. For an analysis of this phenomenon, see I. G. Tadzhinova, "Perepiska sovetskikh frontovikov s zaochno znakomymi zhenshchinami kak iavlenie voennogo vremeni (1941–1945 gg.)," *Golos minuvshego*, no. 3–4 (2012): 66–84.

141. Shaw, "Soldiers' Letters to Inobatxon and O'g'ulxon," 545.

142. See Schechter, "People's Instructions," on the genre of the *nakaz*.

143. "Gvardeitsy proslavili vashu pushku," *Gvardiia*, February 1, 1945. See also "Gromite vraga do kontsa, nashi osvoboditeli," and "Vash nakaz budet vypolnen! (Otvet gvardeitsev-artilleristov)," both in *Gvardiia*, April 7, 1945.

144. Ushakin and Golubev, *XX vek*, 416–17.

145. Cited in Hellbeck, *Stalingrad*, 47. See also NA IRI RAN f. 2, r. III, op. 5, d. 2-a, l. 49; V. Kalinin, "Perepiska s sem'iami geroev," *Gvardiia*, February 10, 1944.

146. Tsentral'nyi arkhiv obshchestvennykh dvizhenii Moskvy (TsAODM) f. 3, op. 52, d. 124, ll. 50–62, in *Aleksandr Shcherbakov: Stranitsy biografii*, ed. A. N. Ponomarev (Moscow: Glavarkhiv Moskvy, 2004), 319.

147. NA IRI RAN f. 2, r. I, op. 30, d. 1, l. 235.

148. Kurochkin, "Na voine kak na voine," 32. This theme is also common in the articles about soldiers' correspondences from *Agitator i propagandist Krasnoi Armii* cited elsewhere in this chapter.

149. Shipov, "Patrioticheskie pis'ma"; Shikin, "Za dal'neishii pod"ëm politicheskoi agitatsii i propagandy"; Gupalo, "Rabota agitatora"; Rubinshtein, "Zametki ob agitatsii."

150. Nekrasov, *I zhiv ostalsia*, 439.

151. Sergeev, *Pis'ma s voiny 1941–1945*, 99.

152. Samoilov, *Podënnye zapisi*, 1: 196

153. Inozemtsev, *Frontovoi dnevnik*, 292–93.

154. Baklanov, *Zhizn'*, 67.

155. Zenzinov noted that the contents of soldiers' letters in 1939–1940 were, at least at first glance, of little interest (*Vstrecha s Rossiei*, 102).

156. See Ushakin and Golubev, *XX vek*, 635–75; and Zenzinov, *Vstrecha s Rossiei*, for examples of letters to the front.

157. Slëzkin, *Do voiny i na voine*, 410.

158. Akhtiamov's letters, for example, repeat information frequently.

159. Martyn Lyons, *The Writing Culture of Ordinary People in Europe, 1860–1920* (Cambridge: Cambridge University Press, 2013), 86.

160. RGASPI f. M-33, op. 1, d. 316.

161. Sergeev, *Pis'ma s voiny 1941–1945*, 100. Catherine Merridale posits that some soldiers hid behind censorship as a way to avoid disturbing topics in their letters (*Ivan's War*, 233).

162. RGASPI f. M-33, op. 1, d. 853, ll. 177–78.

163. RGASPI f. M-33, op. 1, d. 853, l. 370.

164. This is a recurring theme in Ushakin and Golubev's *XX vek*; see 18 for first reference. Irina Tadzhinova has also written on this subject ("Problema ottsovstva frontovikov Velikoi Otechestvennoi voiny 1941–1945 gg. [Issledovanie privatnoi kul'tury sovetskogo cheloveka v voennye gody po epistoliarnym istochnikam]," *Vestnik arkhivistka*, no. 4 (2012): 71–83). This situation was by no means exclusive to the Red Army. As David Henkin has shown, men separated from their families during the California Gold Rush and US Civil War relied on letters to maintain connections with friends and authority in their family (*The Postal Age: The Emergence of Modern Communications in Nineteenth-Century America* [Chicago: University of Chicago Press, 2006], chap. 5). Martyn Lyons has noted the same phenomenon among French and Italian soldiers in World War I (*Writing Culture of Ordinary People in Europe*, 74, 85, 127, 151).

165. RGASPI f. M-33, op. 1, d. 853, l. 281.

166. RGASPI f. M-33, op. 1, d. 853, l. 298.

167. Frolov, *Vse oni khoteli zhit'*, 24, 37, 39.

168. Nikiforova, *Rozhdënnaia voinoi*, 83–84.

169. Igor' Sinani, *V poiskakh Mira: Povestvovanie po sledam pisem. Vospominaniia. Etiudy* (Moscow: ROSSPEN, 2005), 150; Loginov, *Eto bylo na fronte*, 120; Ushakin and Golubev, *XX vek*, 702–13.

170. Frolov, *Vse oni khoteli zhit'*, 24, 25, 36, 39. He went back and forth on this, giving them permission to sell it, possibly out of fear that their prolonged silence indicated economic hardship (32–33), but eventually telling them to keep it.

171. Frolov, *Vse oni khoteli zhit'*, 29.

172. Frolov, *Vse oni khoteli zhit'*, 24.

173. "Na voine rozhdaetsia samaia sil'naia liubov'," 59–60.

174. RGASPI f. M-33, op. 1, d. 853, l. 281.

175. Pystin, "Dnevnik I. Pystina," 62, 66, 75–76.

176. Genatulin, *Strakh*, 46; Shaw, "Soldiers' Letters to Inobatxon and O'g'ulxon," 536, 543.

177. Loginov, *Eto bylo na fronte*, 120.

178. Inozemtsev, *Frontovoi dnevnik*, 138.

179. Chekhovich, *Dnevnik ofitsera*, 96.

180. Nikiforova, *Rozhdënnaia voinoi*, 86.

181. Elena Andreeva, "Ia gorzhus' svoei docher'iu," *Gvardiia*, January 26, 1944; V. Shabanov, "Perepiska s sem'iami geroev," *Gvardiia*, February 10, 1944; "Voin! Otomsti za nashu Taniu!" *Gvardeets*, January 15, 1945; Shipov, "Patrioticheskie pis'ma," 40–41. For fascinating examples of running correspondences between fallen soldiers' families and their comrades, see Ushakin and Golubev, *XX vek*, 792–803.

182. Nikiforova, *Rozhdënnaia voinoi*, 86; Liubarov, *Inache ia ne mog*, 70–72, 120–23.

183. Lyons, *Writing Culture of Ordinary People in Europe*, 71, 84–86.

184. Lyons, *Writing Culture of Ordinary People in Europe*, 81, 138.

185. Lyons, *Writing Culture of Ordinary People in Europe*, 73, 74, 88, 127, 250.

186. Lyons, *Writing Culture of Ordinary People in Europe*, 89, 250; Elena Kozhina, *Through the Burning Steppe: A Wartime Memoir* (New York: Riverhead, 2000), 85–86.

187. RGASPI f. M-33, op. 1, d. 316; Atabek, *K biografii voennogo pokoleniia*, 203.

188. Hellbeck, "Galaxy of Black Stars," 620–21; Lyons, *Writing Culture of Ordinary People in Europe*, 157.

189. Oleg Budnitskii, "Jews at War: Diaries from the Front," trans. Dariia Kabanova, in *Soviet Jews in World War II: Fighting, Witnessing, Remembering*, ed. Harriet Murav and Gennady Estraikh (Boston: Academic Studies Press, 2014), 60–62.

190. Atabek, *K biografii voennogo pokoleniia*, 20–22.

191. Gel'fand, *Dnevnik*, 115–19.

192. Lesin, *Byla voina*, 367–68.

193. S. K. Bernev and S. V. Chernov, eds., *Blokadnye dnevniki i dokumenty* (St. Petersburg: Evropeiskii dom, 2007), 268–70, 309; Pystin, "Dnevniki I. Pystina," 53.

7. Trophies of War

1. NA IRI RAN f. 2, r. X, op. 6, d. 1, ll. 3–4. Epigraph: Suris, *Frontovoi dnevnik*, 227–28.

2. A note on historiography. Previous works on the Red Army in Europe have interrogated the end of the war, discussing trophies as simply part of the chaos of 1945. Most authors (e.g., Norman Naimark) have been interested in political or institutional histories, while those who have been interested in cultural history have yet to concentrate on the objects themselves (e.g., Oleg Budnitskii, Catherine Merridale). Filip Slaveski has provided a detailed assessment of the chaos caused by overlapping competencies during the initial period of occupation. See Budnitskii, "Intelligentsia Meets the Enemy"; Merridale, *Ivan's War*; Norman Naimark, *The Russians in Germany: A History of the Soviet Zone of Occupation, 1945–1949* (Cambridge, MA: Harvard University Press, 1995); and Filip Slaveski, *The Soviet Occupation of Germany*. My contribution is to provide a thick description and interpretation of how Red Army soldiers perceived and interacted with trophies. Closest to my approach is that of Natalie Moine, who also looks closely at objects, but whose interest in destruction and trophies is broader than mine, focusing on much more on civilians and the government. See Natalie Moine, "'Fascists Have Destroyed the Fruit of My Honest Work': The Great Patriotic War, International Law, and the Property of Soviet Citizens," *Jahrbücher für Geschichte Osteuropas* 61 (2013): 172–95; and Natalie Moine, "Of Loss and Loot: Stalin-Era Culture, Foreign Aid, and Trophy Goods in the Soviet Union during the 1940s," *Annales* 68 (2013): 221–58.

3. Hellbeck, "Diaries of Fritzes and the Letters of Gretchens," 585–98.

4. Michael David-Fox, *Showcasing the Great Experiment: Cultural Diplomacy and Western Visitors to the Soviet Union, 1921–1941* (New York: Oxford University Press, 2012), 1, 8, 18, 23, 286.

5. On meshchanstvo, see Svetlana Boym, *Common Places: Mythologies of Everyday Life in Russia* (Cambridge, MA: Harvard University Press, 1994), especially

"Introduction: Theoretical Common Places" and chap. 1 "Mythologies of Everyday Life"; Vera Dunham, *In Stalin's Time*; Timo Vihavainen, *The Inner Adversary: The Struggle against Philistinism as the Moral Mission of the Russian Intelligentsia* (Washington, DC: New Academic Publishing, 2005). Due to its specificity, I will use the Russian term throughout this chapter rather than translate it.

6. Volkov, "Concept of *Kul'turnost'*," 212–14, 216–21, 226.

7. This same tension was present when the Soviet Union annexed the Baltics and parts of Eastern Poland in 1939–1940. See Jan Gross, *Revolution from Abroad: The Soviet Conquest of Poland's Western Ukraine and Western Belorussia* (Princeton: Princeton University Press, 1988), 28, 46–47; and Alfreds Bernzish, *I Saw Vishinsky Bolshevize Latvia* (Washington, DC: Latvian Legation, 1948), 13–14.

8. Viola, *Peasant Rebels under Stalin,* 28; Lynne Viola, *Stalinist Perpetrators on Trial: Scenes from the Great Terror in Soviet Ukraine* (New York: Oxford University Press, 2017), 91–92, 97, 122, 129, 134; on 1918, see the forthcoming monograph by Anne O'Donnell, "Taking Stock: Power, Possession, and the Value of Things in Revolutionary Russia" (book manuscript).

9. P. Charles Hachten, "Property Relations and the Economic Organization of Soviet Russia, 1941–1948" (PhD diss., University of Chicago, 2005), 307.

10. Wolfgang Schivelbush, *The Culture of Defeat: On National Trauma, Mourning and Recovery*, trans. Jefferson Chase (New York: Metropolitan Books, 2003), 6.

11. D. N. Ushakov, *Tolkovyi slovar' russkogo iazyka* (Moscow: Gosudarstvennoe izdatel'stvo inostrannykh i natsional'nykh slovarei, 1940), 809. According to Maks Fasmer, this word entered the Russian language under Peter the Great (*Etimologicheskii slovar' russkogo iazyka* [Moscow: Progress, 1987], 4:107).

12. Erëmenko, *Dnevniki*, 217.

13. TsAMO RF f. 2, op. 920266, d. 1, ll. 483–86, in *Tyl Krasnoi Armii*, 94.

14. Kurkotkin, *Tyl sovetskikh vooruzhennykh sil*, 373–84.

15. TsAMO RF f. 219, op. 692, d. 1, ll. 352–54, in *Tyl Krasnoi Armii*, 160–62.

16. TsAMO RF f. 2, op. 795437, d. 9, ll. 229, 230, in *Tyl Krasnoi Armii*, 226–28.

17. TsAMO RF f. 2, op. 795137, d. 11, ll. 37, 38, in *Tyl Krasnoi Armii*, 300–301.

18. TsAMO RF f. 2, op. 795437, d. 11, l. 380, in *Tyl Krasnoi Armii*, 384.

19. TsAMO RF f. 236, op. 2719, d. 22, l. 18, in *Tyl Krasnoi Armii*, 507. Erëmenko, *Dnevniki*, 280.

20. TsAMO RF f. 2, op. 795437, d. 12, ll. 208–11, in *Tyl Krasnoi Armii*, 585–89.

21. RGASPI f. M-33, op. 1, d. 853, l. 358.

22. TsAMO RF f. 244, op. 3017, d. 45, ll. 4, 5, in *Tyl Krasnoi Armii*, 447, 462–63; TsAMO RF f. 233, op. 2342, d. 10, ll. 55–59, in *Tyl Krasnoi Armii*, 565.

23. TsAMO RF f. 233, op. 2342, d. 10, ll. 47, 48, in *Tyl Krasnoi Armii*, 521–22; RGVA f. 4, op. 11, d. 77, ll. 482–84, in *Prikazy narodnogo komissara oborony SSSR 1943–1945 gg.*, 290–92.

24. TsAMO RF f. 2, op. 795437, d. 12, l. 132, in *Tyl Krasnoi Armii*, 532–33.

25. TsA FSB Rossii f. 3, on. 10, d. 118, l. 62, in *"Ognennaia duga,"* 150–51.

26. Grigorii Pomerants, "Stanovlenie lichnosti skvoz' terror i voinu," *Vestnik Evropy*, no. 28–29 (2010), http://magazines.russ.ru/vestnik/2010/28/po30-pr.html.

27. Schechter, "'People's Instructions,'" 111.

28. Aizenshtat, *Zapiski sekretaria voennogo tribunala*, 21–22.

29. Vishnevskii, *Dnevnik khirurga*, 23; RGASPI f. 88, op. 1, d. 953, l. 8.

30. L. Belousov and A. Vatlin, *Propusk v rai: Sverkhoruzhie poslednei mirovoi: Duel' propagandistov na Vostochnom fronte* (Moscow: Vagrius, 2007), 53.

31. See, e.g., Tooze, *Wages of Destruction*, 467, 469, 476–80, 538–49; Timothy Snyder, *Bloodlands: Europe between Hitler and Stalin* (New York: Basic Books, 2010), 155–86.

32. Gotz Aly, *Hitler's Beneficiaries: Plunder, Racial War, and the Nazi Welfare State*, trans. Jefferson Chase (New York: Metropolitan Books, 2006), 104–5, 172–78.

33. Grossman, *Gody voiny*, 344.

34. RGASPI f. M-33, op. 1, d. 853, l. 259.

35. Nikulin, *Vospominaniia o voine*, 103. Nikulin also expressed "great satisfaction" at the sight of a field filled with enemy dead.

36. NA IRI RAN f. 2, r. XIII-z7, op. 9, d. 19.

37. See, e.g., Bair Irincheev and Denis Zhukov, *Palaces Destroyed: Nazi Occupation of Leningrad Region, 1941–1944* (St. Petersburg: Avrora-Design, 2011), 72, 92.

38. RGASPI f. 644, op. 1, d. 26, ll. 16–17.

39. Boris Suris, *Frontovoi dnevnik*, 72. *Dokhlyi* is generally used to describe animals, not humans.

40. NA IRI RAN f. 2, r. III, op. 5, d. 2a, l. 63.

41. NA IRI RAN f. 2, r. I, op. 174, d. 2, l. 5.

42. *Prokonvoirovanie voennoplennykh nemtsev cherez Moskvu* (Mosfilm, 1944), www.youtube.com/watch?v=6pcMdCkgVAo.

43. Erëmenko, *Dnevniki*, 239.

44. For example, portions of Kirovskii raion in St. Petersburg, Sotsgorod in Kazan', and the city center of Oktiabr'skii in Bashkortostan.

45. Niall Ferguson, "Prisoner Taking and Prisoner Killing in the Age of Total War: Towards a Political Economy of Military Defeat," *War in History* 11 (2004): 148–92.

46. NA IRI RAN f. 2, r. I, op. 16, d. 36, ll. 154–55.

47. Schivelbush, *Culture of Defeat*, 8.

48. RGASPI f. 74, op. 2, d. 95, ll. 150–51.

49. See, e.g., Samoilov, *Pamiatnye zapiski*, 267.

50. NA IRI RAN f. 2, r. I, op. 223, d. 4, l. 3. Pomerants recalled that for the first twenty minutes after taking a new position, no one took prisoners, something that soldiers in other armies also describe. See Pomerants, *Zapiski gadkogo utënka*, 116; and Bourke, *Intimate History of Killing*, 135.

51. NA IRI RAN f. 2, r. I, op. 227, d. 15, l. 2ob.

52. Tokarev, *Vesti dnevnik na fronte zapreshchalos'*, 169.

53. Ironically, Jews, who were among the most educated nationalities in the USSR and many of whom spoke Yiddish, often sat across the table interrogating German POWs. See, e.g., Suris, *Frontovoi dnevnik*; Dunaevskaia, *Ot Leningrada do Kënigsberga*; and Kopelev, *Khranit' vechno*.

54. TsAMO RF f. 2, op. 795437, d. 12, l. 199, in *Tyl Krasnoi Armii*, 583.

55. NA IRI RAN f. 2, r. X, op. 7, d. 13-b, l. 88.

56. Komskii, "Dnevnik 1943–1945," 28, 30, 57. The "bags" are likely map cases.

57. NA IRI RAN f. 2, r. I, op. 230, d.3 ll. 2, 12; Erëmenko, *Dnevniki*, 130, 297.

58. Erëmenko, *Dnevniki*, 239.

59. AP RF f. 3, op. 50, d. 561, ll. 10–20, in Kudriashov, *Voina*, 250–56.

60. *Trofei velikikh bitv (Vystavka obraztsov trofeinogo vooruzheniia)* (Moscow: Mosfilm, 1943), www.youtube.com/watch?v=w95csTqtOtI.

61. AP RF f. 3, op. 50, d. 561, ll. 49–50, in Kudriashov, *Voina*, 290.

62. Baklanov, *Zhizn'*, 53.

63. "Ispol'zovat' trofeinoe oruzhie," *Gvardiia*, April 18, 1944.

64. Chekalov, *Voennyi dnevnik*, 220.

65. Abdulin, *160 stranits*, 36–37.

66. Grossman, *Gody voiny*, 261–62.

67. Kopelev, *Khranit' vechno*, 1:14.

68. NA IRI RAN f. 2, r. III, op. 5, d. 14, l. 81.

69. NA IRI RAN f. 2, r. X, op. 7, d. 7-a, l. 6ob.

70. Slëzkin, *Do voiny i na voine*, 484.

71. Kiselëv, "Voina i zhizn' v predstavlenii 20-letnikh frontovikov," 1023.

72. *Inzh-43*, 154–68; Reutov, *Gvardeets*, 85.

73. See, e.g., the mixed media work by Kharis Iakupov, "V nemetskom blindazhe" (1944–1945); Iakupov, *Hater*, 159; and Inozemtsev, *Frontovoi dnevnik*, 144.

74. Abdulin, *Iz vospominanii soldata*, 21.

75. Brandenberger, *National Bolshevism*, 162.

76. GDA SB Ukraïni f. 9, spr. 18-sp, ark. 32–32 zv., in "Viddili viis'kovoï tsenzuri," 227–28.

77. Anonymous, *A Woman in Berlin* (New York: Metropolitan Books, 2005), 155.

78. Janina Struk, *Photographing the Holocaust: Interpretations of the Evidence* (New York: I. B. Taurus, 2004), 55, 63, 70–73.

79. "Eë zamuchili zlodei-nemtsy," *Bloknot agitatora Krasnoi Armii*, no. 5 (1945): 27–28. The rape and murder of female comrades was frequently recorded by soldiers: see, e.g., Suris, *Frontovoi dnevnik*, 223.

80. See, e.g., Hellbeck, "Diaries of Fritzes and Letters of Gretchens"; Berkhoff, *Motherland in Danger*, 123–25; and Samoilov, *Pamiatnye zapiski*, 244.

81. Tvardovskii, *Vasilii Tërkin*, 165.

82. This continued throughout the war. See, e.g., "Izuchat' taktiku vraga, sovershenstvovat' nashu taktiku," *Krasnaia zvezda*, March 1, 1944; and Pomerants, *Zapiski gadkogo utënka*, 126.

83. Slutskii, *O drugikh i o sebe*, 120.

84. Kovalevskii, "Nynche u nas peredyshka," 79.

85. S. V. Stepashin and V. P. Iampol'skii, *Vpered na zapad (1 ianvaria–30 iiunia 1944 g.): Organy gosudarstvennoi bezopasnosti SSSR v Velikoi Otechestvennoi voine* (Moscow: Kuchkovo pole, 2007), 5, bk. 1:426–31. Western Ukraine is treated like a foreign country in this document.

86. David-Fox, *Showcasing the Great Experiment*, 98–141, 206, 314.

87. NA IRI RAN f. 2, r. 3, op. 14, d. 2b, l. 124.

88. TsAMO RF f. 32, op. 795436, d. 11, ll. 36–39, in *Glavnye politicheskie organy vooruzhennykh sil SSSR v Velikoi Otechestvennoi voine 1941–1945 gg.*, 294.

89. Komskii, "Dnevnik 1943–1945 gg.," 58.

90. Michael David-Fox points out that Red Army soldiers at the war's end were practitioners of "cultural diplomacy from below" (*Showcasing the Great Experiment*, 317).

91. V. Ul'rikh, "Byt' bditel'nym vsegda i vsiudu," *Krasnoarmeets*, no. 5 (1945): 1–3; "Vyshe partiinuiu bditel'nost'," *Agitator i propagandist Krasnoi Armii*, no. 13 (1944):

4–5; TsAMO RF f. 32, op. 795436, d. 12, ll. 58–62, in *Glavnye politicheskie organy vooruzhennykh sil SSSR v Velikoi Otechestvennoi voine 1941–1945 gg.*, 283–86; TsAMO RF f. 32, op. 795436, d. 11, ll. 36–39, in *Glavnye politicheskie organy vooruzhennykh sil SSSR v Velikoi Otechestvennoi voine 1941–1945 gg.*, 294.

92. "Izuchai iazyk protivnika," *Gvardiia*, July 10, 1944; "Frontoviki izuchaiut inostrannye iazyki," *Krasnaia zvezda*, October 13, 1944, 2; "Vostochnaia Prussiia (spravka)," *Krasnaia zvezda*, October 26, 1944. See also Stepashin and Iampol'skii, *Vpered na zapad*, 426–31.

93. Slutskii, *O drugikh i o sebe*, 37.

94. RGASPI f. M-7, op. 2, d. 1261, l. 9. I. V. Turkenich had commanded the famous Young Guards of Krasnodar. He was killed in action in 1944.

95. RGASPI f. M-7, op. 2, d. 1261, l. 17.

96. P. G. Pustovoit, *Pëstrye liki voiny: Dnevnik voennogo perevodchika-radiorazvedchika* (Moscow: Dialog-MGU, 1998), 90.

97. M. G. Shumelishskii, *Dnevnik soldata* (Moscow: Kolos, 2000), 133–34.

98. Atabek, *K biografii voennogo pokoleniia*, 239.

99. Samoilov, *Podënnye zapiski*, 217–18.

100. Austin Jersild, "The Soviet State as Imperial Scavenger: 'Catch Up and Surpass' in the Transnational Socialist Bloc, 1950–1960," *American Historical Review* 116 (2011): 109–32.

101. RGVA f. 4, op. 11, d. 78, ll. 491–98, in *Prikazy narodnogo komissara oborony SSSR 1943–1945gg.*, 344–47.

102. Suris, *Frontovoi dnevnik*, 216.

103. TsAMO RF f. 241, op. 2618, d. 65, ll. 4–5, in *Tyl Krasnoi Armii*, 639–40.

104. Erëmenko, *Dnevniki*, 261; Tartakovskii, *Iz dnevnikov voennykh let*, 254.

105. Suris, *Frontovoi dnevnik*, 227–28. See also Kopelev, *Khranit' vechno*, 1:138; and Temkin, *My Just War*, 200.

106. Kiselëv, "Voina i zhizn' v predstavlenii 20-letnikh frontovikov," 1032.

107. Suris, *Frontovoi dnevnik*, 215.

108. Slutskii, *O drugikh i o sebe*, 100; Anonymous, *A Woman in Berlin*, 143–44. The author of this diary realized all her assailants were privileged members of rear-area units only when she saw frontline soldiers for the first time, men characterized by filthiness and utter indifference.

109. Suris, *Frontovoi dnevnik*, 233–35.

110. Tartakovskii, *Iz dnevnikov voennykh let*, 254–55.

111. NA IRI RAN f. 2, r. I, op. 30, d. 23, ll. 2ob.–3.

112. L. Nikulin, "Samovar," *Krasnoarmeets*, no. 9 (1945): 6.

113. Temkin, *My Just War*, 199. Temkin interpreted this resentment of wealth as a generally human, rather than specifically Soviet, phenomenon.

114. Kopelev, *Khranit' vechno*, 1:139.

115. Letter to V. Gel'fand, January 9, 1945, "Vladimir Gel'fand: Pis'ma 1941–1946. Skaner," http://www.gelfand.de/a/3/a.html.

116. Shaw, "Soldiers' Letters to Inobatxon and O'g'ulxon," 539; Ushakin and Golubev, *XX vek*, 157–58.

117. Erëmenko, *Dnevniki*, 281–82.

118. Inozemtsev, *Frontovoi dnevnik*, 211.

119. Frolov, *Vse oni khoteli zhit'*, 83.

120. Tvardovskii, *Ia v svoiu khodil ataku*, 344. The German term *kultur-traeger* is rendered into Cyrillic with Russian pluralization in the original text.

121. Tongur, *Frontovoi dnevnik*, 132.

122. Tvardovskii, *Ia v svoiu khodil ataku*, 344; Suris, *Frontovoi dnevnik*, 236.

123. Kopelev, *Khranit' vechno*, 1:56, 136–39; Suris, *Frontovoi dnevnik*, 236.

124. Tvardovskii, *Ia v svoiu khodil ataku*, 347.

125. Nikulin, *Vospominaniia o voine*, 189; Temkin, *My Just War*, 221. See Merridale, *Ivan's War*, 322–23; and Moine, "Of Loss and Loot," for more descriptions of loot.

126. Shumelishskii, *Dnevnik soldata*, 137.

127. Atabek, *K biografii voennogo pokoleniia*, 239.

128. Anonymous, *Woman in Berlin*, 224.

129. Shaw, "Making Ivan Uzbek," 195.

130. Inozemtsev, *Frontovoi dnevnik*, 208; Slutskii, *O drugikh i o sebe*, 30.

131. Kovalevskii, "Nynche u nas peredyshka," 99.

132. Tvardovskii, *Ia v svoiu khodil ataku*, 345.

133. Kopelev, *Khranit' vechno*, 1:140.

134. NA IRI RAN f. 2, r. III, op. 14, d. 2b, l. 122.

135. Dunaevskaia, *Ot Leningrada do Kënigsberga*, 369, 374.

136. Vladimir Bushin, *Ia posetil sei mir: Iz dnevnikov frontovika* (Moscow: Algoritm, 2012), 89.

137. RGASPI f. M-1, op. 32, d. 331, ll. 90–92, in *Zhenshchiny Velikoi Otechestvennoi voiny*, ed. N. K. Petrova (Moscow: Veche, 2014), 290.

138. Dunaevskaia, *Ot Leningrada do Kënigsberga*, 376.

139. Suris, *Frontovoi dnevnik*, 235; Dunaevskaia, *Ot Leningrada do Kënigsberga*, 15, photo insert 224–25.

140. "Gde na fronte sfotografirovat'sia," *Krasnaia zvezda*, June 30, 1943; Ushakin and Golubev, *XX vek*, 144; Baklanov, *Zhizn'*, 56.

141. Iakupov, *Frontovye zarisovki*, 44.

142. Baklanov, *Zhizn'*, 56.

143. NA IRI RAN f. 2, r. III, op. 14, d. 2b, l. 135.

144. NA IRI RAN f. 2, r. III, op. 14, d. 2b, ll. 171–72.

145. TsAMO RF f. 2, op. 795437, d. 12, l. 305, in *Tyl Krasnoi Armii*, 613–14.

146. Balkanov, *Zhizn'*, 76.

147. Lesin, *Byla voina*, 350.

148. Stezhenskii, *Soldatskii dnevnik*, 211.

149. Nikulin, *Vospominaniia o voine*, 190–91; Kiselëv, "Voina i zhizn' v predstavlenii 20-letnikh frontovikov," 1032.

150. Orme, *Comes the Comrade*, 74. This, of course, was not always the case: see, e.g., Merridale, *Ivan's War*, 309.

151. Pomerants, *Zapiski gadkogo utënka*, 120, 166.

152. Anonymous, *Woman in Berlin*, 79.

153. Slutskii, *O drugikh i o sebe*, 100–101; Naimark, *Russians in Germany*, 69–70.

154. Mark Edele, "Soviet Liberations and Occupations, 1939–1949," in *The Cambridge History of the Second World War*, vol. 2: *Politics and Ideology* (New York: Cambridge University Press, 2015), 493.

155. Naimark, *Russians in Germany*, 69–71, 114; Merridale, *Ivan's War*, 303–4, 309–20; Jeffrey Burds, "Sexual Violence in Europe in World War II, 1939–1945," *Politics &*

Society 37 (2009): 35–73, here 53–57; Mary Louis Roberts, *What Soldiers Do: Sex and the American GI in World War II France* (Chicago: University of Chicago Press, 2013), 7, 256.

156. Naimark, *Russians in Germany*, 113.

157. Slutskii, *O drugikh i o sebe*, 97.

158. Nikulin, *Vospominaniia o voine*, 187.

159. Temkin, *My Just War*, 221.

160. Inozemtsev, *Frontovoi dnevnik*, 208–9.

161. NA IRI RAN f. 2, r. III, op. 14, d. 2b, l. 143; Tartakovskii, *Iz dnevnikov voennykh let*, 243.

162. Pomerants, *Zapiski gadkogo utënka*, 171.

163. Komskii, "Dnevnik 1943–1945 gg.," 66–67.

164. Lesin, *Byla voina*, 317.

165. Orme, *Comes the Comrade*, 44–45, 48. Although as Komskii's statement attests, rage could be as important as keeping warm.

166. Orme, *Comes the Comrade*, 256.

167. Kopelev, *Khranit' vechno*, 1:102–3.

168. Nikulin, *Vospominaniia o voine*, 190–91.

169. Kopelev, *Khranit' vechno*, 1:102.

170. Inozemtsev, *Frontovoi dnevnik*, 210.

171. Naimark, *Russians in Germany*, 36. Discipline became a major topic in *Krasnaia zvezda* in the immediate aftermath of the war. See, e.g., "Vysoko derzhat' chest' i dostoinstvo sovetskogo voina," *Krasnaia zvezda*, May 18, 1945; and "Vnutrennii rasporiadok," *Krasnaia zvezda*, June 23, 1945.

172. P. Belov, "O preimushchestvakh sotsialisticheskoi sistemy khoziaistva," *Agitator i propagandist Krasnoi Armii*, no. 19–20 (1945): 21–29.

173. M. Nechkina, "Velikii russkii narod," *Agitator i propagandist Krasnoi Armii*, no. 13 (1945): 27.

174. I thank Anne Lounsbery for pointing this out. Natalie Moine has shown that Soviet officials accused of excessive looting after the war blamed the women in their lives, particularly their wives, for their actions, echoing discourse that attributed "survivals" of the past to women more than men ("Of Loss and Loot," 241).

175. RGVA f. 4, op. 11, d. 78, ll. 59–60, in *Prikazy narodnogo komissara oborony SSSR 1943–1945 gg.*, 304. It is possible that this was also a response to sex within the ranks of the Red Army, on which see Z. Zlatopol'skii, "Umei razoblachat' vraga," *Bloknot agitatora Krasnoi Armii*, no. 7 (1945): 19; Naimark, *Russians in Germany*, 97–100; and Dunaevskaia, *Ot Leningrada do Kënigsberga*, 380.

176. Slutskii, *O drugikh i o sebe*, 35–37; Samoilov, *Podënnye zapisi*, 226.

177. Slutskii, *O drugikh i o sebe*, 121.

178. Genatulin, "Vot konchitsia voina," 203; Sigmund Heinz Landau, *Goodbye, Transylvania: A Romanian Waffen-SS soldier in WWII* (Mechanicsburg: Stackpole Books, 2015), 118.

179. Pustovoit, *Pëstrye liki voiny*, 83.

180. Lesin, *Byla voina*, 351.

181. Nechkina, "Velikii russkii narod," 21.

182. AP RF f. 3, op. 50, d. 274, ll. 110–21, in Kudriashov, *Voina*, 393–99.

183. "Skhema postroeniia voisk na Parade Pobedy 24 iiunia 1945 g.," in Kudriashov, *Voina*, 403.

184. Richard Taylor, *Film Propaganda: Soviet Russia and Nazi Germany*, 2nd rev. ed. (New York: I. B. Tauris, 1998), 48–49, 100–101.

185. Joseph Brodsky, "Spoils of War," in his *On Grief and Reason: Essays* (New York: Farrar Straus Giroux, 1995), 8–11. See also Juliane Fürst, *Stalin's Last Generation: Soviet Post-war Youth and the Emergence of Mature Socialism* (New York: Oxford University Press, 2010), 205–07.

186. Stephen Lovell, *In the Shadow of War: Russia and the USSR, 1941–Present* (Malden: Blackwell, 2010), 288–89. Postwar *stiliagi* ("stylish ones"—a term for youth who adopted Western styles) were initially referred to as meshchane (Fürst, *Stalin's Last Generation*, 221).

187. Aleksandr Vasil'ev, *Russkaia moda: 150 let v fotografiiakh* (Moscow: Slovo, 2009), 337–38. 352–53; Moine, "Of Loss and Loot," 236.

188. Litvin, *800 Days on the Eastern Front*, 142.

189. Baklanov, *Zhizn'*, 77.

190. Brodsky, "Spoils of War," 5–7; Tartakovskii, *Iz dnevnikov voennykh let*, 255.

191. RGASPI f. 17, op. 3, d. 2198, ll. 28–29, http://www.alexanderyakovlev.org/fond/issues-doc/1002762.

192. V. N. Khaustov, V. P. Naumov, and N. S. Plotnikova, *Lubianka: Stalin i MGB SSSR, mart 1946–mart 1953* (Moscow: Mezhdunarodnyi fond "Demokratiia," 2007), 137–39.

193. Moine, "Of Loss and Loot," 258.

194. David M. Kennedy, *Freedom from Fear: The American People in Depression and War, 1929–1945* (New York: Oxford University Press, 1999), 742–45; Robert A. Pape, *Bombing to Win: Air Power and Coercion in War* (Ithaca: Cornell University Press, 1996), 258–83; Michael Bess, *Choices under Fire: Moral Dimensions of World War II* (New York: Knopf, 2006), 88–110.

Conclusion

1. Hellbeck, "Galaxy of Black Stars," 616; Slezkine, *House of Government*, 715–52.

2. Lomidze, "Druzhba, skreplënnaia krov'iu," 42–44.

3. Once passports were instituted, the most likely means of travel for peasants would have been military service, illegal travel to work at a factory, as part of a special delegation of shock workers or party members, or imprisonment in the Gulag.

4. These are all real people: A. B. Priadekhin, Fatikh Karimov, Makar Prianishnikov (from Samoilov, *Pamiatnye zapiski*, 206), Grigorii Baklanov, David Samoilov, Lev Slëzkin, Ivan Drushliak (Viola, *Stalinist Perpetrators on Trial*, 71), Bulat Okudzhava, Iuliia Zhukova and Alexandra Shliakhova, Rakhimzhan Koshkarbaev, and Ernst Neizvestnyi. (Those without references next to them are cited in footnotes or famous.)

5. Samoilov, *Pamiatnye zapiski*, 205.

6. Samoilov, *Pamiatnye zapiski*, 234; Mark Edele, "Veterans and the Village: The Impact of Red Army Demobilization on Soviet Urbanization, 1945–1955," *Russian History* 36 (2009): 159–82.

7. Ilya Gerasimov argues that a prerevolutionary "plebeian society" of urbanizing peasants was the Soviet blueprint for modernization and that skills acquired by these peasants were the only reason that the regime could survive its own horrendous mismanagement of resources (*Plebian Modernity: Social Practices, Illegality,*

and the Urban Poor in Russia, 1906–1916 [Rochester, NY: University of Rochester Press, 2018], 186–91).

8. Genatulin, "Tri protsenta," 10–11.

9. See, e.g., Siegelbaum, *Cars for Comrades*, 6–8; and Brown, *Plutopia*, 256.

10. Yekelchyk, *Stalin's Citizens*, 31–33.

11. Weiner, "Robust Revolution to Retiring Revolution," 228–31.

12. One could interpret the collapse of the Soviet project as, in part, a surrender to meshchanstvo—particularly the failure to dramatically reorganize society so that meshchanstvo lost its appeal. As Yuri Slezkine has argued, "In the Soviet Union, the decision to embrace private property left the emperor with no clothes at all." Slezkine concludes that the failure to offer an alternative to the family as the basic unit of society and the embrace of Russian culture by the generation after the Revolution proved fatal for the USSR. These two processes in many ways crystalized during the war. See Slezkine, *House of Government*, 951–57, esp. 957.

13. The years 1945–1953 were certainly the apex of what Michael David-Fox has dubbed the "Stalinist superiority complex"—the swinging pendulum of feeling superior or inferior to "the West" that characterized Soviet self-understanding and projection abroad and the institutions and practices that these sentiments drove. Interestingly, the moment of the zenith of this complex coincided with the most stringent xenophobia and isolationism in Russian and Soviet history, further cementing an immediate prewar situation in which "Superiority over the rest of the world and isolation from it were now virtually synonymous" (David-Fox, *Showcasing the Great Experiment*, 1–28, 285–311, 317, quotation 311).

14. Dmitrii Bykov, "Lichnoe delo kazhdogo: Luchshii iz tekh, kogo ia znal," *Novaia gazeta*, https://war.novayagazeta.ru/materials/32.

Index